# Advances in Photogrammetry, Remote Sensing and Spatial Analysis

# Advances in Photogrammetry, Remote Sensing and Spatial Analysis

Editor: Matt Weilberg

R CALLISTO REFERENCE

www.callistoreference.com

**Callisto Reference,**
118-35 Queens Blvd., Suite 400,
Forest Hills, NY 11375, USA

Visit us on the World Wide Web at:
www.callistoreference.com

ISBN: 978-1-64116-042-1 (Hardback)

**Cataloging-in-Publication Data**

Advances in photogrammetry, remote sensing and spatial analysis / edited by Matt Weilberg.
    p. cm.
Includes bibliographical references and index.
ISBN 978-1-64116-042-1
1. Photogrammetry. 2. Remote sensing. 3. Aerial photogrammetry. 4. Spatial analysis (Statistics).
I. Weilberg, Matt.

TA593.2 .A38 2019
526.982--dc23

# Table of Contents

# Preface

This book was inspired by the evolution of our times; to answer the curiosity of inquisitive minds. Many developments have occurred across the globe in the recent past which has transformed the progress in the field.

Advances in photogrammetry, remote sensing and spatial analysis have occurred due to the rapid technological developments in the last decade. While remote sensing refers to acquiring information from remote places instead of on-site accession, photogrammetry is used to reconstruct measurements of instruments like maps through photographs, particularly regaining exact surface points for precise design. All three fields are integral to each other and have relevance across many different fields of study such as architecture, engineering, space engineering, etc. This book is a valuable compilation of topics, ranging from the basic to the most complex advancements in these fields. It elucidates new techniques and their applications in a multidisciplinary manner. As the fields are emerging at a rapid pace, the contents of this book will help the readers understand the modern concepts and applications of the subject.

This book was developed from a mere concept to drafts to chapters and finally compiled together as a complete text to benefit the readers across all nations. To ensure the quality of the content we instilled two significant steps in our procedure. The first was to appoint an editorial team that would verify the data and statistics provided in the book and also select the most appropriate and valuable contributions from the plentiful contributions we received from authors worldwide. The next step was to appoint an expert of the topic as the Editor-in-Chief, who would head the project and finally make the necessary amendments and modifications to make the text reader-friendly. I was then commissioned to examine all the material to present the topics in the most comprehensible and productive format.

I would like to take this opportunity to thank all the contributing authors who were supportive enough to contribute their time and knowledge to this project. I also wish to convey my regards to my family who have been extremely supportive during the entire project.

Editor

# A COMPARISON OF CLOSE-RANGE PHOTOGRAMMETRY USING A NON-PROFESSIONAL CAMERA WITH FIELD SURVEYING FOR VOLUME ESTIMATION

S. Abbaszadeh [a], H. Rastiveis[a]*

[a] School of Surveying and Geospatial Engineering, University of Tehran, College of Engineering,, Tehran, Iran – (saeedabbaszadeh, hrasti)@ut.ac.ir

**KEY WORDS:** DEM, CRP Method, negative slops, Volume Estimation

**ABSTRACT:**

Rapid and accurate volume calculation is one of the most important requirements in many applications such as construction and mining industries. The accuracy of a calculated volume depends on the number of collected points on the object. Increasing the number of measured points undoubtedly requires higher cost and time. On the other hand, collecting surveying points might sometimes be difficult, dangerous or impossible. The aim of this study is to evaluate the close range photogrammetry (CRP) using a non-professional camera for DEM generation in comparison to the traditional field surveying technique (TST). For this purpose, a test area in Deralok hydropower planet site was considered and the process of DEM extraction in both methods was compared. The obtained results showed that although the CRP method in contrast with TST method was more time consuming, however, this method was able to successfully measure negative slops and berms and, consequently, calculated more accurate volume. Moreover, the relative error of 0.2% was reported.

## 1. INTRODUCTION

The volume calculations are an important requirement of the construction and mining industry (Fawzy 2015). The accurate volume estimation is important in many applications, for example, dam project, road project, and building applications (P. L. Raeva 2016). The traditional methods such as the trapezoidal method, traditional cross sectioning, and improved methods (Simpson-based, cubic spline, and cubic Hermite formula) have been used in volume computing. The main elements of these methods are to collect the points that appropriate distribution and density (M. Uysal 2015). These methods need more mathematical processes and take more time. Although, the high accuracy dependent on the number of collected points on the object(Karami 2014). More points require higher cost and time. In addition, collecting surveying points might be difficult, dangerous or impossible sometimes. (C. Arango 2015) Therefore, the classic surveying method is not always applicable to calculate excavation volume.

There are high accuracy requirements as far as heights are concerned due to volume calculations and therefore high resolution images. However, very precise terrestrial measurements could be extremely time-consuming process. On the other hand, by photogrammetric techniques, large areas can be covered in high details in less than an hour. (Patikova 2004)

In comparison to classical geodetic methods, close range photogrammetry is an efficient and fast method. It can significantly reduce the time required for collecting terrestrial data. The accuracy of the volume calculation is proportional to the presentation of the land surface. The presentation of the surface, on the other hand, is dependent on the number of coordinated points, their distribution, and its interpolation. (Yilmaz 2010)

The estimated accuracy of the comparison is up to 3% - 4%. In some countries, the legislation states that the volume should be calculated with a precision of ±3% of the whole material (P. L. Raeva 2016). However, this value depends on many factors such as the type of material being excavated in the quarry, the atmospheric conditions, etc. (M.Mazhdrakov 2007)

## 2. STUDY AREA

The study site is situated in the power house excavation of Deralok Hydropower Plant project in Iraqi Kurdistan and is located 80 kilometres northeast of Dohuk and the area is more than 1 hectare (lon 43.6562 °, lat 37.0689 °).  (Figure 1).

Figure 1. Deralok hydropower planet site, red boundary is power house

---

* Corresponding author

## 3. METHODOLOGY

Volume calculation is usually performed in two main steps: data acquisition and volume estimation. Time, cost and accuracy are very important parameters in volume estimation techniques. In this paper, the close range photogrammetry (CRP) using a non-professional camera and the traditional field surveying technique (TST) have been executed for estimation the volume in a sample project. In the following section, the details of these steps for both methods are described.

### 3.1 Data Acquisition

Data acquisition is an important step for volume estimation. The more accurate the collected data results in more accurate the estimated volume.

For the data acquisition in the field surveying method, a Leica TS-09 total station, with the specifications 1" angular accuracy and Enhanced measurement accuracy to prism 1.5 mm + 2 ppm was used and all the necessary points to estimate the volume were observed. In this case, around the excavation area with a prism and surveying characteristics break lines many points have been observed and, surveying trench wall with the total station non-prism tool was done.

In the CRP method a non-metric camera was used. We have been trying to use the camera at low prices with a high efficiency. The Table 1 shows the camera specifications.

| Camera Specification | |
|---|---|
| camera model | DSLR-A700 |
| Dimensions | 4272×2400 |
| Horizontal & vertical resolution | 72 dpi |
| Focal length | 35 mm |
| F-stop | f/11 |

Table 1. SONY- DSLR-A700 camera specifications.

For the study area, we walk around the power house excavation top of the trench and acquired 3-6 images in each station in differed view for increase overlap (Figure 2).

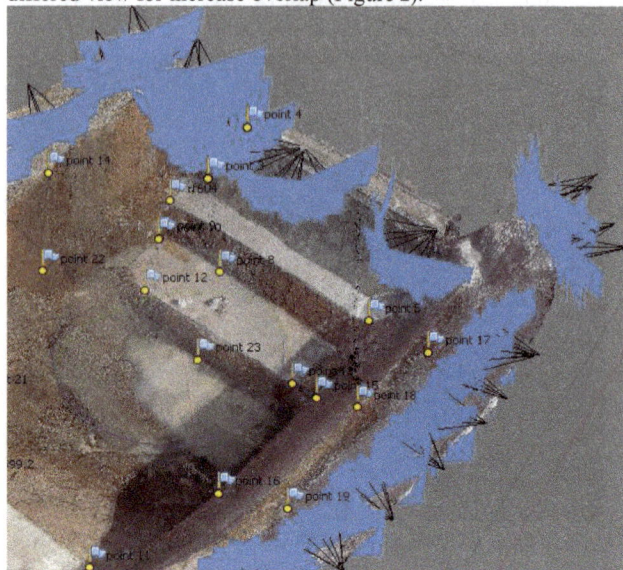

Figure 2. Camera locations and image network

### 3.2 Volume Estimation

To estimate the volume excavation many methods can be used. I.e. trapezoidal method for rectangular or triangular prism, Simpson's and improve methods using Simpson-based, cubic spline and cubic Hermite formula used for conventional method calculation (M.Yakar 2008).

### 3.2.2 Volume Estimation in TST Method

After obtaining the field data with the TST, the data was export to the TST with flash memory and saved in .idx format, after editing point and create a .txt file, this point is ready to import into Civil 3D software.

After import data, we fit a surface to points. In this step, editing the surface is an important step in volume computing because as long as we do not have a correct surface with all break lines, and etc. we cannot compute a correct volume. Figure 3 shows the flowchart of the volume calculation using Civil 3D software.

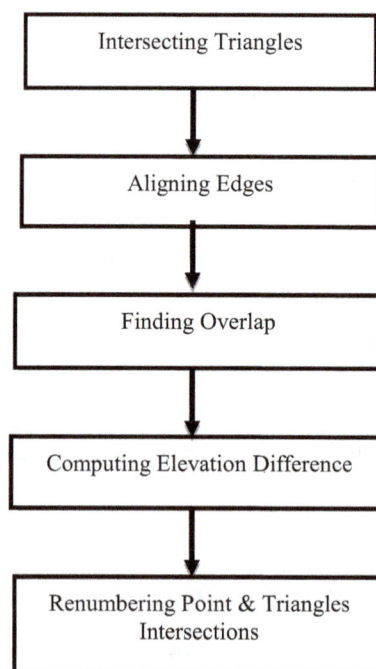

Figure 3. Volume computation flowchart

### 3.2.2 Volume Estimation with CRP Method

To obtain the values of the excavation volume, we adjust the image first with the GCP, these points was the same geo-referenced points around and into the excavation that we survey by TST, after that, we use AgiSoft software to process image, to generation an exact DSM we following this steps:

- Align photos
- Create dense cloud
- Build mesh
- Build tiled model
- Build DEM

## 4. RESULTS & DISCUSSION

Figure 4 show the number of images that cover the study area, as we see all of the excavation bottom, trench wall, berms exist in more than 9 Photos.

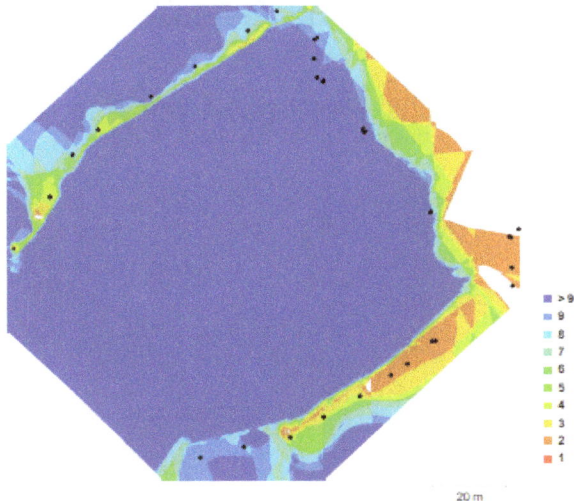

Figure 4.Camera locations and image overlap

In order to images georeferencing and assess the geometric accuracy (plane and height), a local network of control and cheek points have been considered according to Figure 5, and the coordinates of these points have been reading.

Figure 5. GCP locations

Basically, the principles of calculating the volume of an object in any kind of photogrammetric software slightly differ from the conventional methods. Unless a Point Cloud Densification is created. Agisoft output DSM include this negative slope, and TST method cannot modeling this negative slops as shown in Figure 6 DSM from Agisoft and Figure 7 shows Civil 3D DEM. After converting the agisoft DSM to DEM the volume was computed.

In both digital elevation models use contour line with 2.5 meters interval

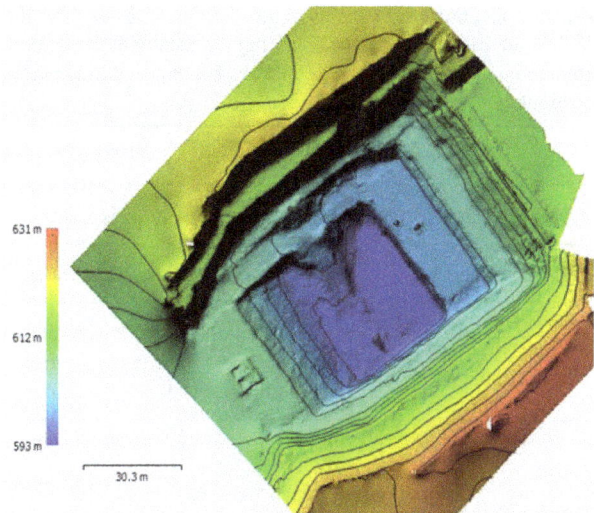

Figure 6.CRP Reconstructed digital elevation model with 2.5 contour line AgiSoft output

Figure 7. Land survey Reconstructed digital elevation model with 2.5 contour line Civil 3D output

### 4.1 Comparison of excavation Volumes

In this study, the traditional method with TST to estimate volumes of excavation were compared with CRP, data from the same site were taken and the post processing was done in civil 3D with a DEM that generated by AgiSoft with TST.

In this comparison, because we do not have the actual value of the volume excavated, our goal was to compare these two methods together only values compared and examine how close the numbers obtained.

**4.2 Time spent & cost for data acquisition**

The Table 1 shows a clear difference between the two methods of data collection, the CRP is about 3.5 times faster than the TST, in addition, the risks of obtaining the data with the camera are much lower, this because people are not exposed to unstable locations.

| Time taken for | Land Survey by TST | CRP by Camera |
|---|---|---|
| Preparation Equipment | 10 minutes | 5 minutes |
| Data collection | 4 hours | 35 minutes |
| Data processing to obtain DTM | 30 minutes | 1 hours |
| Create sections-calculate volume | 25 minutes | 25 minutes |

Table 2. Time comparison between both methods

Following CRP method is much less expensive than TST because the method requires fewer people and the device used to rent or buy cheaper.

## 4. CONCLUSION

In this study, we evaluated the CRP method using non-professional camera to generate excavation DEM in comparison to traditional field surveying method. We get 121 images around the study area and applied 23 control and check points.

The volume that obtained by the traditional method and CRP was 198424.48 $m^3$ and 198047.29 m3, respectively. The relative error of 0.2% can conclude that the estimated volume using CRP data was more accurate. The ability of the CRP method in considering negative slope due to loss right in the trenches, which cannot be measured by TST method, was promising.

## REFERENCES

C. Arango    , C. A. M. (2015). "Comparison Between Multicopter Uav And Total Station For Estimating Stockpile Volumes." The International Archives of the Photogrammetry, Remote Sensing and Spatial Information Sciences.

H. E.-D. Fawzy, "The Accuracy of Determining the Volumes Using Close Range Photogrammetry." IOSR Journal of Mechanical and Civil Engineering

Karami, A., sousouni, B. , Hosseininaveh, A. (2014). "A novel contactless approach for calculating volumes of objects." The 1st National Conference on Geospatial Information Technology.

M. Uysal , A. S. T., N. Polat a (2015). "DEM generation with UAV Photogrammetry and accuracy analysis in Sahitler hill."
M.Mazhdrakov (2007). "Mine engineering in open-pit quarries."

M.Yakar , H. M. Y. (2008). "Using In Volume Computing Of Digital Close Range Photogrammetry." The International Archives of the Photogrammetry, Remote Sensing and Spatial Information Sciences.

P. L. Raeva, S. L. F., D. G. Filipov (2016). "Volume Computation Of A Stockpile – A Study Case Comparing Gps And Uav Measurements In An Open Pit Quarry." The International Archives of the Photogrammetry, Remote Sensing and Spatial Information Sciences.

Patikova, A. (2004). Digital photogrammetry in the practice of open pit mining. Int. Arch. Photogramm. Remote Sens. Spat. Inf. Sci, 34, 1-4.

Yilmaz, H. M. (2010). Close range photogrammetry in volume computing. Experimental Techniques, 34(1), 48-54.

.

# THE EFFECT OF SHADOW AREA ON SGM ALGORITHM AND DISPARITY MAP REFINEMENT FROM HIGH RESOLUTION SATELLITE STEREO IMAGES

Nurollah Tatar [a*], Mohammad Saadatseresht[a], Hossein Arefi[a]

[a]School of Surveying and Geospatial Engineering, College of Engineering, University of Tehran
(n.tatar, msaadat, hossein.arefi @ut.ac.ir)

**KEY WORDS:** Shadow, Semi Global Matching, High Resolution Satellite Stereo Images, Disparity Map Refinement.

**ABSTRACT:**

Semi Global Matching (SGM) algorithm is known as a high performance and reliable stereo matching algorithm in photogrammetry community. However, there are some challenges using this algorithm especially for high resolution satellite stereo images over urban areas and images with shadow areas. As it can be seen, unfortunately the SGM algorithm computes highly noisy disparity values for shadow areas around the tall neighborhood buildings due to mismatching in these lower entropy areas. In this paper, a new method is developed to refine the disparity map in shadow areas. The method is based on the integration of potential of panchromatic and multispectral image data to detect shadow areas in object level. In addition, a RANSAC plane fitting and morphological filtering are employed to refine the disparity map. The results on a stereo pair of GeoEye-1 captured over Qom city in Iran, shows a significant increase in the rate of matched pixels compared to standard SGM algorithm.

## 1. INTRODUCTION

Dense stereo matching is the primary step in generating digital surface model from satellite images. The methods are categorized as Local (area and feature based) algorithms, Global and Semi-Global matching algorithms (Hirschmüller, 2008; Scharstein and Szeliski, 2002). The weakness of Local matching algorithms is in occlusion boundaries due to the depth discontinuity. Global matching algorithms are formulated as minimizing energy function, which includes discontinuity, smoothness, and occlusion (Brown et al., 2003). These algorithms usually have high computation cost. The SGM algorithm is faster than global matching algorithms (Hirschmüller and Scharstein, 2009). The SGM algorithm is popular due to the higher efficiency compared with other methods (Alobeid, 2011; Zhu et al., 2014).

The SGM algorithm comprised of four main steps including matching cost computation, cost aggregation, disparity computation and disparity map refinement (Hirschmüller, 2008). The capability of computed costs to match conjugate points has an ambiguity due to equal cost value of neighbor pixels and low entropy in objects. so, the aggregation of matching costs from 8 or 16 paths has been proposed (Hirschmüller, 2008). In this step, matching costs are aggregated based on matching cost and disparity. Equation 1 is used to aggregate the matching costs in each path. Then using equation 2 the costs are aggregated in all paths.

$$L_r(p,d) = C(p,d) + \min \begin{cases} L_r(p-r,d) \\ L_r(p-r,d \pm 1) + P_1 - \min_k L_r(p-r,k) \\ \min_i L_r(p-r,i) + P_2 \end{cases} \quad (1)$$

Where;
$p$: image location of interest pixel
$d$: disparity value
$Lr(p,d)$: cost path toward the actual pixel of path
$C(p,d)$: pixel-wise matching cost
$P1$: a small value penalizing disparity changes between neighbouring pixels of one pixel

$P2$: a large value penalizing disparity changes between neighbouring pixels of one pixel
$r$: actual path
$k$: pixels in each path

$$S(p,d) = \sum_{r=1}^{8 or 16} L_r(p,d) \quad (2)$$

Now, in equation 3 the disparity for each pixel is calculated by minimizing the aggregated cost values. Also, to estimate the disparity in sub-pixel level, a quadratic curve is fitted to the neighboring costs disparity, and the position of the minimum is obtained.

$$D = \min_d S(p,d) \quad (3)$$

After cost aggregation and generating disparity map, mismatching occurs in shadowing areas. The reason is the effect of pixels with lower entropy (such as shadowing areas) to those with higher entropy (such as building roof) to have the same disparity in a local neighbourhood. As shown in Figure 1, this is the reason to extend the roof of the buildings in the disparity map.

The SGM algorithm originally uses the Mean Shift algorithm to refine disparity map in large textureless areas (Hirschmüller, 2008). But this solution fails to solve the ambiguity of shadow effects. Also, some methods were proposed based on segmentation and statistical filtering (such as median of background) to enhance the disparity map in void areas (Bafghi et al., 2016; Krauß and Reinartz, 2010).

In this paper, after shadow detection and computing disparity map from high resolution satellite stereo images, a new method based on morphological filtering and RANSAC plane fitting is proposed to enhance the disparity map in shadow areas. In the following, the proposed method is described and then the result of experiments is presented. The paper will end with the conclusions.

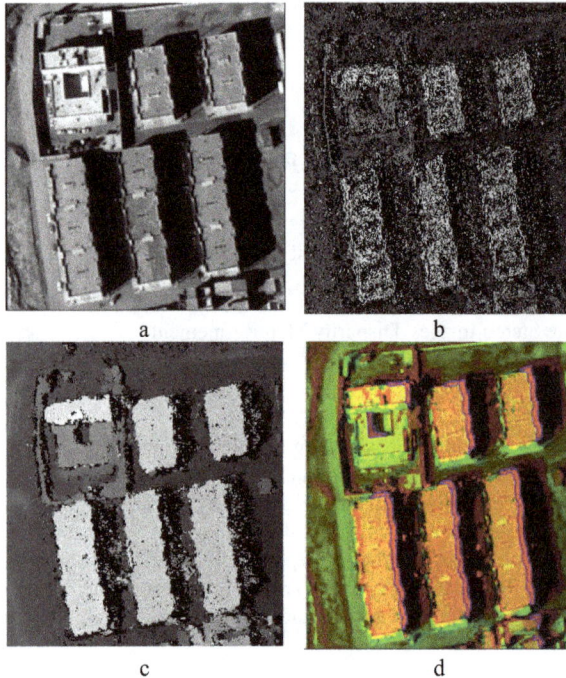

c                            d

Figure 1. The shadow effect on cost aggregation step (SGM algorithm). a) Panchromatic image; b) computed disparity map without cost aggregation; c) computed disparity map after cost aggregation; d) overlaid disparity map and panchromatic image (the extended roof).

## 2.  PROPOSED METHOD

The proposed method has three main steps, pre-processing, shadow detection, and disparity map refinement. Figure 2 shows the workflow. The details of each part will be discussed in the following.

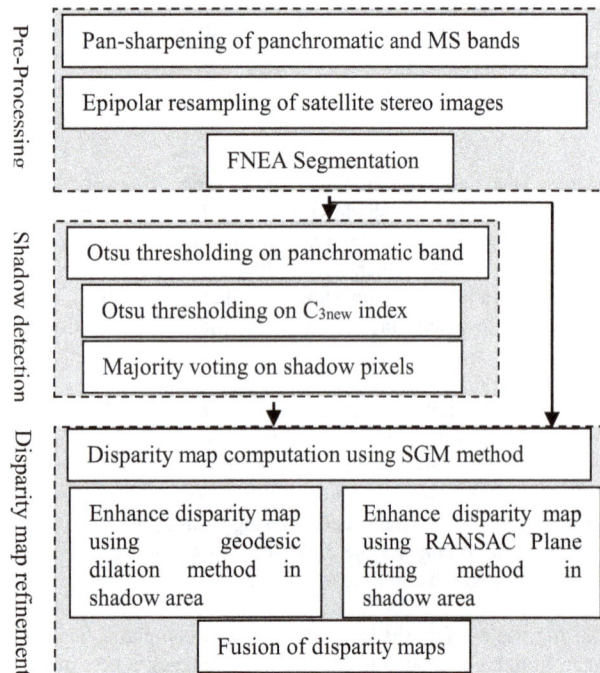

Figure 2. Flowchart of proposed method to enhance the disparity map in SGM algorithm

### 2.1  Pre-processing

In the pre-processing step, spectral bands are pan-sharpened to enhance the spatial resolution. In the proposed method, the HIS transformation algorithm is used due to its acceptable spatial quality (Strait et al., 2008; Tu et al., 2001).

Epipolar images are rectified images so that each row of the left image corresponds to the same row at the right image (i.e. parallax y is zero). The epipolar images are the input of SGM algorithm (Hirschmüller, 2008). Unlike the images with perspective geometry (Frame Camera), the epipolar geometry of linear array images could not be considered as a straight line. Recently, for linear consideration of epipolar geometry, a new strategy based on image tiling was proposed to produce epipolar images from high resolution satellite stereo images (Tatar et al., 2015b). The panchromatic and pan-sharpened multispectral images are rectified along the epipolar lines using this method.

Then the rectified panchromatic and spectral bands are introduced to Fractal Net Evolution Approach (FNEA) segmentation algorithm (Baatz and Schäpe, 2000; Benz et al., 2004), implemented in eCognition software. This paradigm known as object-based image analysis is used to detect the correct border of Shadow areas in urban area.

### 2.2  Shadow detection

Detecting shadow areas from high resolution satellite images is a critical role in disparity map refinement. Up to now, a lot of shadow detection methods were proposed, see complete review in (Shahtahmassebi et al., 2013). In this paper, an object-based shadow detection method is used based on our previous work (Tatar et al., 2015a). The method is based on thresholding on the panchromatic band and $C_{3new}$ index to find a binary shadow mask. Then a majority voting (Lam and Suen, 1997) analysis is used to detect shadow areas in object level.

The threshold value is calculated based on Otsu thresholding algorithm (Otsu, 1979). The shadow areas are considered as background and foreground in panchromatic and $C_{3new}$ index respectively. Also, the $C_{3new}$ index was defined by equation 4 (Tatar et al., 2015a).

$$C_{3new} = \tan^{-1}\left(\frac{B}{PAN}\right) \tag{4}$$

Where;
B: Blue band
PAN: Panchromatic band

After detecting shadow objects, connected component analysis (Gonzalez and Woods, 2002) is used to identify shadow regions, and unify them to use as the input for further processing.

### 2.3  Disparity map refinement

The aggregation step couldn't find the correct matching in shadow and large textureless areas. Therefore, some post processing on disparity map is performed to solve the problem. In this paper, only the effect of shadowing on disparity map will be improved. Two methods of RANSAC (Fischler and Bolles, 1981) plane fitting and geodesic dilation filtering (Arefi and Hahn, 2005) are employed and then the results are integrated to increases the confidence level.

The step by step RANSAC plane fitting is executed as follows.

1. Remove disparities of shadow areas
2. The dilation operator with a 3×3 structure element is applied to the disparity map
3. Using RANSAC algorithm, a plane is fitted to disparities of each shadow region.
4. Using an interpolation process, the shadow pixels getting interpolated disparities

At the same time with RANSAC plane fitting, the geodesic dilation filtering is employed as described below:

1. Detecting non-ground pixels from disparity map using geodesic dilation algorithm
2. Fitting a surface to disparities of ground pixels (estimate background of disparity map)
3. Replace disparities of non-ground pixels by original disparity map

The refined disparity map, the result of RANSAC plane fitting, as well as disparity of geodesic dilation algorithm, permits the determination of false disparities by performing a consistency check. Each disparity of RANSAC plane fitting is compared to its corresponding disparity in the result of geodesic dilation algorithm. The disparity is set to invalid if the difference more than one pixel. Otherwise the disparity is modified by averaging of two disparities.

## 3. EXPERIMENT AND ANALYSIS

### 3.1 Dataset

High resolution satellite stereo images from GeoEye-1 are used in the experiments. GeoEye-1 images with panchromatic and multispectral bands are prepared over urban area in Qom city in Iran. Table 1 gives detail information about images, including the acquisition date, Location, and spatial resolutions.

| Looking angle | time | Acquired date | Location Lat/Lon | GSD (m) | |
|---|---|---|---|---|---|
| | | | | Pan | MS |
| Afterward | 7:20 | 01/16/2014 | N 34.62 ° | 0.5 | 2 |
| Forward | 7:21 | | E 50.92° | 0.5 | 2 |

Table 1: Detail of dataset

### 3.2 Experimental results

In the pre-processing step, IHS pan-sharpening algorithm used to enhance the spatial resolution of spectral bands which helps to detect shadow areas and create a color point cloud with higher details. After the pan-sharpening, the panchromatic and pan-sharpened spectral bands are resampled along epipolar line. Due to the small size of corresponding stereo image tiles, the epipolar images can be generated by fundamental matrix (Loop and Zhang, 1999; Tatar et al., 2015b). The stereo anaglyph of epipolar images is shown in Figure 3.

In the next step, the epipolar panchromatic and pan-sharpened images are served as input to FNEA segmentation algorithm to produce the image objects. FNEA segmentation algorithm requires tuning the scale parameter, shape and compactness weight coefficients. The value of these parameters has a direct effect on the final segmentation and shadow detection result. The scale parameter, shape and compactness weight coefficients are selected as 80, 0.1 and 0.9 respectively and considered constant in the experiments based on our previous experiences on the same dataset (Tatar et al., 2015b; Tatar et al., 2016).

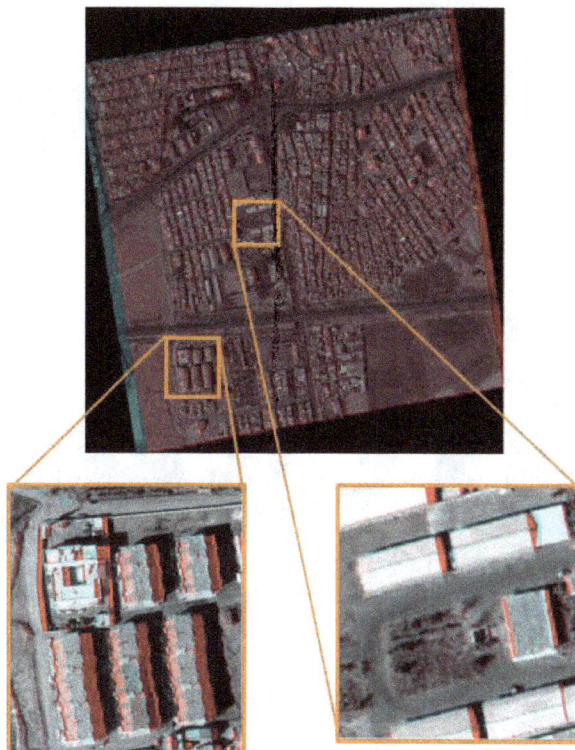

Figure 3. Stereo anaglyph of the generated epipolar images

Recognition of shadow objects in the proposed method depends on the detection of suspected shadow pixels. Thresholding on the panchromatic band and the $C_{3new}$ index are employed to this end. Thresholding on the panchromatic band detects a mixture of dark pixels including shadow pixels, asphalt road, and dark roofs. In the next step, thresholding on $C_{3new}$ index improves the accuracy of suspect shadow pixels by omitting the false shadow pixels among the suspicious shadow pixels.

Then the results of pixel-level shadow detection and FNEA image segments are integrated through the majority voting to recognize shadow areas in object level. This will improve the accuracy with omitting salt and pepper noise in suspect shadow detection. The threshold percentage of the shadow pixels in each image segment is needed to make decisions here. From previous studies, the best majority threshold is selected as 20% (Tatar et al., 2015a; Tatar et al., 2016). Figure 4 shows the comparison of pixel and object level shadow detection results.

After generating epipolar images for corresponding stereo image tiles, the SGM algorithm is applied to generate disparity map. As mentioned earlier, matching cost function, and penalty of $P_1$ and $P_2$ are critical inputs for SGM algorithm. The Census transformation with 7×9 kernel size is selected as the cost function. Also, the value of $P_1$ and $P_2$ are obtained 12 and 30 respectively in the experiment. The noises and outliers are removed using a median filter.

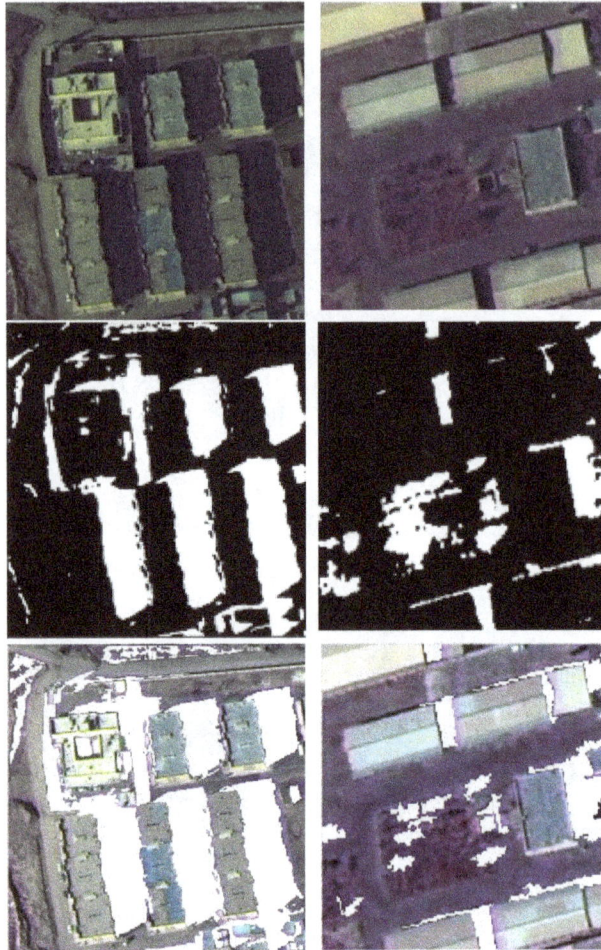

Figure 4. Shadow detection results. Top: pan-sharpened original images; middle: suspected shadow areas; bottom: overlaid object-based shadow areas on original images

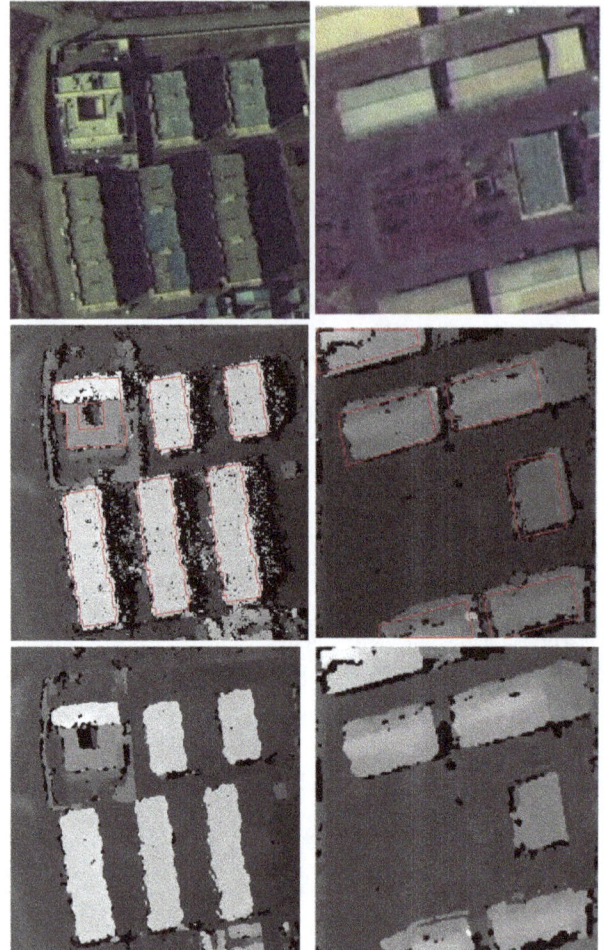

Figure 5. Disparity map refinement. Top: original epipolared left image; middle: Computed disparity map by SGM algorithm (the border of man-made objects drawn in red line); bottom: enhanced disparity map (proposed method)

Cost aggregation in SGM algorithm is sensitive to entropy values and it caused a shift in disparity value of shadow areas to get closer to those of neighbouring buildings. To solve this problem, as described in section 2.3, the disparity map refinement is applied to the disparity map. The generated disparity map of SGM algorithm and enhanced disparity map are shown in Figure 5.

Also to visual inspection of shadow effect on SGM algorithm, the corresponding points and images were intersected and color point cloud is generated (Figure 6). Due to the existence of rational polynomial coefficients (RPCs) for satellite stereo images, the corresponding points are intersected by RPCs in the object coordinate system.

Figure 6. 3D reconstruction of high resolution satellite stereo images. Left: 3D reconstruction with shadow effect (standard SGM algorithm); right: 3D reconstruction after disparity map refinement (proposed method).

### 3.3 Evaluation and Discussion

In this section, the result of epipolar resampling, shadow detection, and disparity map refinement are evaluated. Y-parallax values for corresponding points in epipolar images considered as evaluation measure to the results of epipolar resampling. The Y-parallax value of measured corresponding points is shown in Figure 7. The result of experiments in Figure 3 and 7 demonstrates that the epipolar images are generated with sub-pixel accuracy.

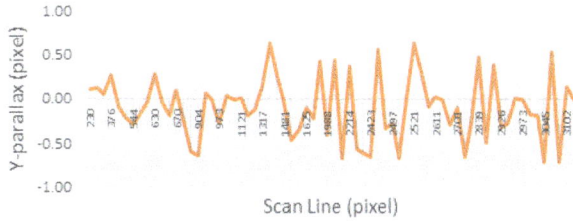

Figure 7. Y-parallax of corresponding points on the generated epipolar images

The confusion matrix is computed to evaluate the result of shadow detection and disparity map refinement. The confusion matrix for generation of two classes is shown in Table 2.

|  |  | Detected /generated | |
|---|---|---|---|
|  |  | Class 1 | Class 2 |
| Reference | Class 1 | TP | FP |
|  | Class 2 | FN | TN |

Table 2: Confusion matrix for evaluation of results

The confusion matrix in shadow detection includes: true positive (TP) number of correctly classified shadow pixels, false positive (FP) number of wrongly classified non-shadow pixels as shadow, false negative (FN) number of shadow pixels which detected as non-shadow, and true negative (TN) number of non-shadow pixels classified correctly.

Also, the confusion matrix is created for evaluation of disparity map refinement and computed disparity map by SGM algorithm. In this evaluation, the confusion matrix includes:
TP: number of correctly matched pixels
FP: number of mismatched pixels
FN: number of conjugate pixels which considered as occlusion
TN: number of occlusions truly found.

Completeness, correctness and F-measure criterions obtained from confusion matrix as below are used to evaluate the results.

$$Completeness = 100 \times \frac{TP}{TP+FN} \qquad (5)$$

$$Correctness = 100 \times \frac{TP}{TP+FP} \qquad (6)$$

$$F - measure = 100 \times \frac{2 \times TP}{2 \times TP+FN+FP} \qquad (7)$$

To evaluate the shadow detection results, 98 image objects in shadow class and 118 image objects in non-shadow class are selected manually. Compactness, correctness, and F-measure were calculated based on generated confusion matrix. Table 3 contains the result of object-based shadow detection. In this table, the F-measure is used as quantitative criteria due to the big gap between correctness and completeness.

| Completeness | Correctness | F-measure |
|---|---|---|
| 100% | 89% | 94% |

Table 3: Statistical analysis of object-based shadow detection

To evaluate the performance of proposed disparity map refinement method, 260 corresponding points are measured manually. The disparity value of measured points was compared to the disparity computed by SGM algorithm and the proposed method. Evaluation results for both results are presented in Table 4.

| Method | Completeness | Correctness | F-measure |
|---|---|---|---|
| SGM algorithm | 48% | 71% | 57% |
| Proposed method | 89% | 94% | 92% |

Table 4: Statistical analysis of computed disparity map by SGM algorithm and enhanced by proposed method

To evaluate the computed disparity map, completeness, correctness, and F-measure are used. It should be noted that the higher value of correctness, shows the non-occlusion pixels matched correctly. In the other word, the building roofs and land cover are matched correctly. According to the results presented in Table 4, lower correctness value in the case of SGM algorithm indicates the number of shadow pixels (wrongly matched) which added to all matched pixels. Due to the enhanced disparity map in shadow areas, the wrongly matched pixels are removed from all matched pixels. This will cause an increase in the correctness value in the case of proposed method.

Also, if completeness has a higher value, indicates the occlusion and shadow pixels are not matched as non-occlusion and roof pixels respectively. Quantitative evaluation in Table 4, shows the lower completeness in the case of original SGM algorithm. The reason is that the shadow pixels matched as roof pixels. However, in the case of proposed method, the completeness is increased due to disparity refinement in shadow areas.

## 4. CONCLUSION

In this paper, we investigate the effect of shadow area on SGM algorithm in the case of high resolution satellite stereo processes. The SGM algorithm causes shadow pixels have a disparity equal to the disparity of buildings roof. To solve this problem, FNEA segmentation was employed to detect shadow areas accurately in object level. Then a new method based on RANSAC plane fitting and morphological filtering used to enhance the disparity map in shadow areas. The result of experiments demonstrates that the proposed object-based disparity map refinement significantly increased the result of original SGM algorithm in shadow areas.

The proposed method in this paper focuses on disparity map refinement in shadow areas. For other areas such as mirrored surfaces and non-lamebrain surfaces, an adaptive disparity map refinement method seems to work. Also, the numerical assessment of the results was done by some corresponding points which measured manually. Creating a dataset with ground truth and evaluate the proposed method with ground truth will follow by the authors.

# REFERENCES

Alobeid, A., 2011. Assessment of matching algorithms for urban DSM generation from very high resolution satellite stereo images. Fachrichtung Geodäsie und Geoinformatik der Leibniz-Univ., Honnover, Germany.

Arefi, H., Hahn, M., 2005. A morphological reconstruction algorithm for separating off-terrain points from terrain points in laser scanning data. International Archives of Photogrammetry, Remote Sensing and Spatial Information Sciences 36.

Baatz, M., Schäpe, A., 2000. Multiresolution Segmentation: an optimization approach for high quality multi-scale image segmentation, in: Strobl, J. (Ed.), Angewandte Geographische Informationsverarbeitung XII. Beiträge zum AGIT-Symposium Salzburg 2000, Karlsruhe, Herbert Wichmann Verlag, pp. 12-23.

Bafghi, Z.G., Tian, J., d'Angelo, P., Reinartz, P., 2016. A New Algorithm For Void Filling in A DSM From Stereo Satellite Images in Urban Areas. ISPRS Annals of Photogrammetry, Remote Sensing & Spatial Information Sciences 3.

Benz, U.C., Hofmann, P., Willhauck, G., Lingenfelder, I., Heynen, M., 2004. Multi-resolution, object-oriented fuzzy analysis of remote sensing data for GIS-ready information. ISPRS Journal of photogrammetry and remote sensing 58, 239-258.

Brown, M.Z., Burschka, D., Hager, G.D., 2003. Advances in computational stereo. Pattern Analysis and Machine Intelligence, IEEE Transactions on 25, 993-1008.

Fischler, M.A., Bolles, R.C., 1981. Random sample consensus: a paradigm for model fitting with applications to image analysis and automated cartography. Communications of the ACM 24, 381-395.

Gonzalez, R.C., Woods, R.E., 2002. Digital image processing.

Hirschmüller, H., 2008. Stereo processing by semiglobal matching and mutual information. Pattern Analysis and Machine Intelligence, IEEE Transactions on 30, 328-341.

Hirschmüller, H., Scharstein, D., 2009. Evaluation of stereo matching costs on images with radiometric differences. Pattern Analysis and Machine Intelligence, IEEE Transactions on 31, 1582-1599.

Krauß, T., Reinartz, P., 2010. Enhancement of dense urban digital surface models from VHR optical satellite stereo data by pre-segmentation and object detection, Int Arch Photogramm Remote Sens Spat Inf Sci XXXVIII (Part 1), on CDROM.

Lam, L., Suen, S., 1997. Application of majority voting to pattern recognition: an analysis of its behavior and performance. IEEE Transactions on Systems, Man, and Cybernetics-Part A: Systems and Humans 27, 553-568.

Loop, C., Zhang, Z., 1999. Computing rectifying homographies for stereo vision, Computer Vision and Pattern Recognition, IEEE Computer Society Conference on. IEEE, Fort Collins, CO, USA.

Otsu, N., 1979. A threshold selection method from gray-level histograms. IEEE Transactions on System, Man, and Cybernetics SMC-9, 5.

Scharstein, D., Szeliski, R., 2002. A taxonomy and evaluation of dense two-frame stereo correspondence algorithms. International journal of computer vision 47, 7-42.

Shahtahmassebi, A., Yang, N., Wang, K., Moore, N., Shen, Z., 2013. Review of shadow detection and de-shadowing methods in remote sensing. Chinese Geographical Science 23, 403-420.

Strait, M., Rahmani, S., Merkurev, D., 2008. Evaluation of pan-sharpening methods. UCLA Department of Mathematics, Research Experiences for Undergraduates (REU2008).

Tatar, N., Saadatseresht, M., Arefi, H., Hadavand, A., 2015a. A New Object-Based Framework to Detect Shadows in High-Resolution Satellite Imagery Over Urban Areas. The International Archives of Photogrammetry, Remote Sensing and Spatial Information Sciences 40, 713.

Tatar, N., Saadatseresht, M., Arefi, H., Hadavand, A., 2015b. Quasi-Epipolar Resampling of High Resolution Satellite Stereo Imagery for Semi Global Matching. The International Archives of Photogrammetry, Remote Sensing and Spatial Information Sciences 40, 707.

Tatar, N., Saadatseresht, M., Arefi, H., Hadavand, A., 2016. Developing New Index to Object-Based Shadow Detection from High Resolution Satellite Images over Urban Area. Journal of Geomatics Science and Technology 5, 11-21.

Tu , T.-M., Su, S.-C., Shyu, H.-C., Huang, P.S., 2001. Anew look at IHS-like image fusion methods. Information Fusion 2, 10.

Zhu, K., der von der Ingenieurfakultät, V.A., Geo, B., Vermessungswesen, U.B.-u., 2014. Dense Stereo Matching with Robust Cost Funtions and Confidence-based Surface Prior. Universitätsbibliothek der TU München.

# 3

# HYBRID OPTIMIZATION OF OBJECT-BASED CLASSIFICATION IN HIGH-RESOLUTION IMAGES USING CONTINUOUS ANT COLONY ALGORITHM WITH EMPHASIS ON BUILDING DETECTION

E. Tamimi [a*], H. Ebadi [b], A. Kiani [c]

[a] MS.c. Student, Faculty of Geodesy & Geomatics Engineering, K.N.Toosi University of Technology, Tehran, Iran-elahe.tamimi@gmail.com

[b] Prof., Faculty of Geodesy & Geomatics Engineering, K.N.Toosi University of Technology, Tehran, Iran-ebadi@kntu.ac.ir

[c] PhD student, Faculty of Geodesy & Geomatics Engineering, K.N.Toosi University of Technology, Tehran, Iran-abbasekiani@yahoo.com

**KEY WORDS:** Hybrid Optimization, Object-based Classification, Ant Colony Optimization Algorithm, High Spatial Resolution Image.

**ABSTRACT:**

Automatic building detection from High Spatial Resolution (HSR) images is one of the most important issues in Remote Sensing (RS). Due to the limited number of spectral bands in HSR images, using other features will lead to improve accuracy. By adding these features, the presence probability of dependent features will be increased, which leads to accuracy reduction. In addition, some parameters should be determined in Support Vector Machine (SVM) classification. Therefore, it is necessary to simultaneously determine classification parameters and select independent features according to image type. Optimization algorithm is an efficient method to solve this problem. On the other hand, pixel-based classification faces several challenges such as producing salt-paper results and high computational time in high dimensional data. Hence, in this paper, a novel method is proposed to optimize object-based SVM classification by applying continuous Ant Colony Optimization (ACO) algorithm. The advantages of the proposed method are relatively high automation level, independency of image scene and type, post processing reduction for building edge reconstruction and accuracy improvement. The proposed method was evaluated by pixel-based SVM and Random Forest (RF) classification in terms of accuracy. In comparison with optimized pixel-based SVM classification, the results showed that the proposed method improved quality factor and overall accuracy by 17% and 10%, respectively. Also, in the proposed method, Kappa coefficient was improved by 6% rather than RF classification. Time processing of the proposed method was relatively low because of unit of image analysis (image object). These showed the superiority of the proposed method in terms of time and accuracy.

## 1. INTRODUCTION

Building detection by classifying Remote Sensing (RS) images has been attracted many researches' attention due to its extensive applications (e.g. automation of information extraction (Lari and Ebadi 2007, Hermosilla, Ruiz et al. 2011, Li, Wu et al. 2013), updating Geographic Information System (GIS) database (Gharibi, Arefi et al. 2016) ,change detection and urban management (Bouziani, Goïta et al. 2010, Huang, Zhang et al. 2014)). In recent years, the development of RS sensors results in obtaining high spatial resolution (HSR) images. For this reason, it is possible to detect many features from these images, so the extracted information from these images are useful in urban management. Despite of the mentioned advantage, pixel-based classification in HSR images has faced many limitations (e.g. producing salt- paper results due to high spectral diversity, high time processing and inability to interpret image due to weakness of pixel information) (Chen, Hay et al. 2012). Therefore, with the development of object-based techniques, the unit of image analysis changed from pixel to object (Blaschke 2010) and the limitations of pixel-based techniques have been solved (Hall and Hay 2003, Laliberte, Rango et al. 2004, Blaschke 2005, Desclée, Bogaert et al. 2006, Bontemps, Bogaert et al. 2008, Gamanya, De Maeyer et al. 2009). In object-based techniques, homogeneous image objects have more signal to noise ratio rather than image pixels. Therefore, the results obtained from object-based techniques are more accurate (Benz, Hofmann et al. 2004, Wang, Sousa et al.

2004, Ronczyk 2011, Petropoulos, Kalaitzidis et al. 2012). Since the input image is segmented into homogenous areas in object-based techniques, the earth's features are extracted based on reality. In other words, the main purpose of image processing is to extract features that match with reality (Castilla and Hay 2008).

There are two kinds of classification methods in terms of training samples availability (supervised and unsupervised) (Drăguţ, Csillik et al. 2014, Chutia, Bhattacharyya et al. 2015, Egorov, Hansen et al. 2015) and image analysis unit (pixel-based and object-based) (Lillesand, Kiefer et al. 2014). Thorough different classification methods, many researchers suggested Support Vector Machine (SVM) as the superior method (Xuegong 2000, Camps-Valls, Gómez-Chova et al. 2004, Wang 2005, Khatami, Mountrakis et al. 2016). The advantages of SVM are simple training, good generalization ability (Li and Yin 2013, Bin, Jian et al. 2014, Cheng and Bao 2014, Ghamisi and Benediktsson 2015, Chen and Tian 2016) and efficiency in high dimensional and nonlinear problems with small samples (Gao, Mandal et al. 2010). SVM classification, as well as other methods, has some parameters (e.g. penalty term and kernel parameters). These parameters involve in training process of SVM and affect the classification performance, so these parameters values are essential to be determined before training process. The process of finding the appropriate parameter value is known as model selection (MS) or parameter optimization (Cheng and Bao 2014). There are many MS techniques such as grid search, gradient descent algorithm and trial and error approach. The limitations of these approaches are high time processing and low-level automation. On the other hand, building detection from HSR images has faced various limitations due to low number of spectral bands. For this reason, using other features according to the unit of image analysis (e.g. spectral, textural and geometrical features) (Ji 2000, Shackelford and Davis 2003, Coburn and Roberts 2004) or use of other data sources (e.g. digital elevation data) have been suggested to improve accuracy (Geerling, Labrador-Garcia et al. 2007). However, adding the additional features will lead to increase the presence probability of dependent features, which may reduce the accuracy. Therefore, it is essential to use some techniques to detect and remove dependent features, which leads to improve accuracy. This task is known as Feature Selection (FS) (Wan, Wang et al. 2016). Selecting appropriate features by trial and error approach is based on expert knowledge and is dependent on image type and scene. MS of SVM and also FS cause to apply continuous Ant Colony Optimization (ACO$_R$) algorithm in order to solve the existence limitations. The advantages of ACO$_R$ are easy to realize and implement, parallel processing, suitable for continuous domain (Socha and Dorigo 2008, Zhang, Chen et al. 2010).

Many studies have been done in urban features detection based on image objects. For example, a comprehensive review of object-based features detection was presented by (Cheng and Han 2016). The limitations of the mentioned studies are low-automation level, SVM parameter determination by trial and error approach (Petropoulos, Kalaitzidis et al. 2012, Liu and Bo 2015), using independent features according to expert knowledge (Liu and Bo 2015). For example, Bouziani in 2009 used only features which is selected by expert knowledge for building change detection. The limitations of the mentioned study was low automation level in FS procedure and the limited expert knowledge in all cases (Bouziani, Goïta et al. 2010). On the other hand, Samadzadegan et al. have optimized pixel-based SVM classification by using binary ACO with aim of FS and MS in hyperspectral imagery (Samadzadegan, Hasani et al. 2012, Samadzadegan, Hasani et al. 2012). MS of SVM is a continuous problem and FS is a discrete problem, so simultaneously optimization of SVM is a continuous problem. In the mentioned studies, they transformed continuous problem into discrete problem by discretizing continuous domain. On the other words, the continuous range of variable, which should be optimized, are converted into finite sets. This is not always appropriate, especially if the initial possible range of variable is wide (Socha and Dorigo 2008). In addition, the mentioned studies were based on pixel-based processing. As mentioned before, pixel-based processing result in higher time consuming in optimizing classification because of unit of image analysis (image pixel) and producing salt and paper result.

To overcome the above limitations and according to importance of selecting independent features in object-based classification, in this paper, a novel method is proposed with the aim of building detection based on hybrid optimization of object-based SVM classification in HSR image and elevation data. ACO$_R$ was applied because it is appropriate for continuous optimization problems. The advantages of the proposed method include independency of image scene, relatively high automation level, relatively high speed processing because of unit of image analysis (image object), decreasing the probability of post processing with the purpose of building edge reconstruction. The results of the proposed method were evaluated by pixel-based SVM classification in terms of time, Ground Truth (GT) in terms of accuracy and precision and Random Forest (RF) classification in terms of the ability of

choosing independent features. The rest of this paper is organized as follows: the data are described in section 2. The methods and the proposed method are illustrated in Section 3 and 4, respectively. The results are given in Section 5, together with the discussion. Conclusions are drawn in Section 6.

## 2. DATA DESCRIPTION

The proposed method has been applied on subset of the digital aerial image and Digital Surface Model (DSM) of Bandar Anzali, Gilan Province (North of Iran). The digital aerial image was acquired by Ultracam D with a spatial resolution of 0.08 m and 3-band multispectral imagery (Red, Green, and Blue) (Figure 1). A look at Figure 1 reveals that the studied area is related to a complex urban area.

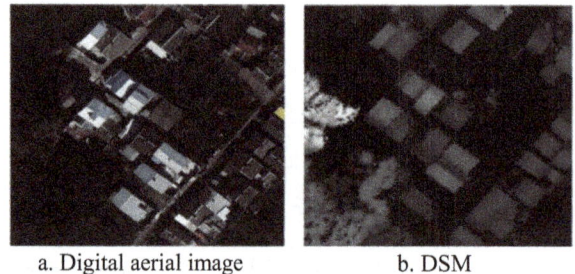

a. Digital aerial image        b. DSM

Figure 1. The study area: a. Digital Aerial image, b. DSM

## 3. METHODS

### 3.1 Classification

**3.1.1** Support Vector Machine Classification

SVM, a supervised machine learning algorithm, was proposed by Vapnik (Vapnik and Vapnik 1998). SVM finds an optimal separating hyper plane that maximizes the margin between two classes. In nonlinear problems, data are not separable in the original feature space. In these cases, SVM uses kernel functions (e.g. linear, sigmoid, polynomial and Radial Basis Function (RBF)) in order to map data into higher dimensional space, so data can be separable in the new space (Figure 2). In this paper, RBF kernel function is used due to achieving good results in terms of accuracy and time processing (Luo, Zhou et al. 2002, Pal 2002, Zhang, Chen et al. 2010) (Equation 4).

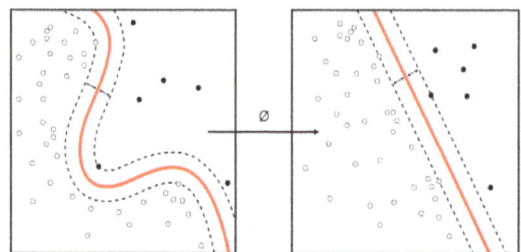

Figure 2. Map data from nonlinear space into higher dimensional space in SVM

Let consider a dataset with n samples $\{(x_i, y_i) \mid i = 1, ..., n\}$, where $x_i \in \mathbf{R}^k$ is a feature vector and $y_i \in \{-1, 1\}$ defines the label of $x_i$. The SVM finds a hyper plane in a high dimensional space that is able to separate the data with a maximum margin (Equation 1):

$$w^t \phi(x) + b = 0 \qquad (1)$$

In the above equation, $w$ is a weight vector, orthogonal to the hyper plan, $b$ is an offset term and $\phi$ is a mapping function. Maximizing the margin is equivalent to minimizing the norm of $w$. Thus, SVM is trained to solve the following minimization problem:

$$\min \left( \frac{\|w^2\|}{2} + c \sum_{i=1}^{n} \xi_i \right)$$
$$subject\ to: y_i(w^t \phi(x_i) + b) \geq 1 - \xi_i \qquad (2)$$
$$\xi_i \geq 0; \forall i$$

Where C is a regularization parameter that imposes a trade-off between the number of misclassification in the training data and the maximization of the margin and $\xi_i$ are stack variables. The decision function obtained through the solution of the minimization problem equation (3) is given by

$$f(x) = \sum_{i=1}^{nsv} \alpha_i y_i \phi(x_i) \phi(x) + b \qquad (3)$$

Where the constants are called Lagrange multipliers determined in the minimization process; SV corresponds to the set of support vectors, training samples for which the associated Lagrange multipliers are larger than zero. The kernel function compute dot products between any pair of samples in the feature space.

$$K_{RBF}(x_i, x_j) = \exp \left( -\frac{\|x_i - x_j\|}{2\sigma^2} \right) \qquad (4)$$

### 3.1.2 Random Forest Classification

RF classification was proposed by Breiman in 2001. It is an ensemble classifier that apply a bootstrap aggregated sampling technique ("bagging") to build many individual decision trees, from which a final class assignment is determined (Breiman 2001). Observations in the original training data which do not occur in the bootstrap sample are named Out-Of-Bag (OOB) observations. A random subset of predictor variables split apart the training data into homogenous subsets (Mellor, Haywood et al. 2013). The node-splitting variable that allows for the greatest variance is selected. This allows the overall model to increase its generalization capacity before and after the split (Walton 2008). The OOB sample data evaluates the performance by computing the accuracy and error rates averaged over all of the predictions (Breiman 2001) and estimates the importance of each variable in the classification. In RF classification, it is possible to select dependent features by variable importance.

### 3.1.3 ACOR Optimization Algorithm

ACO was introduced as a novel nature-inspired method by (Dorigo, Maniezzo et al. 1996, Dorigo and Sttzle 2004). It is based on the foraging behaviour of real ant (finding the shortest path between their nest and food sources (optimization problem)). The real ants initially explore the area surrounding their nest in a random manner when searching for food. As soon as an ant finds a food source, it carries some food back to the nest and deposits a pheromone trail on the ground during the return trip. The amount of deposited pheromone shows the quantity and quality of the path between the food source and their nest and guides others ants to find the food source. After a

while, the path that has the most pheromone is the shortest path between their nest and the food sources. Based on this, ant colonies choose this path as the shortest one. Indirect communication among ants via pheromone trails enables them to find the shortest path between their nest and food sources. This capability of real ant colonies has inspired the definition of artificial ant colonies that can find approximate solutions to optimization problems (Socha and Dorigo 2008).

Continuous optimization is hardly a new research field. ACO is suitable for discrete domain. For continuous domain, it was extended to continuous domains without making any major conceptual change to its structure. ACOR extended the population-based ACO with Gaussian Probability Density Function (PDF) as pheromone update and adopted the rank-based selection mechanism (Xiao and Li 2011). The advantages of ACOR is possible to solve problems where some variables are discrete and others are continuous (Socha and Dorigo 2008). ACOR uses initial population and the Gaussian PDFs in order to generate the best individuals of this population. After sampling from these Gaussian PDFs, the new individuals merge with the initial population. Finally, the predefined number of individual are selected from the new population sorted based on an objective function. The best solution is an individual that has the best objective function value in the stopping criterion. This procedure will continue until the stopping criterion is met. The general flowchart of ACOR is shown in Figure 3. Further details about ACOR can be found in (Socha and Dorigo 2008).

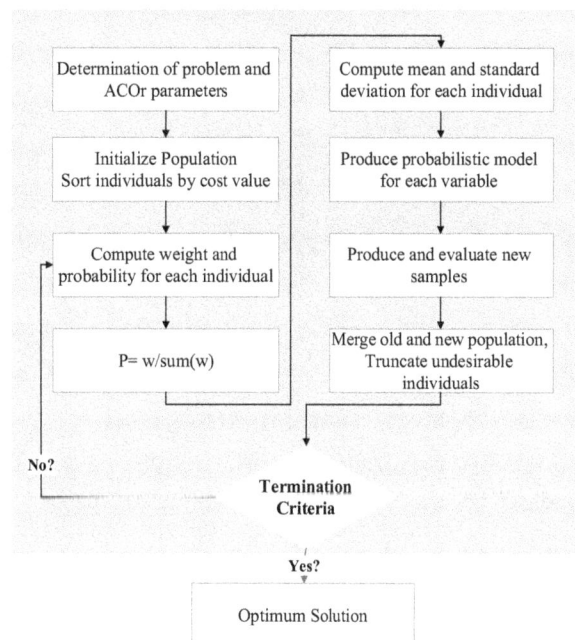

Figure 3. General Flowchart of ACOR

## 4. THE PROPOSED METHOD

The main aim of this paper is to propose a new method to optimize object-based SVM classification in FS and MS. The proposed method includes six steps: image pre-processing, normalized Digital Surface Model (nDSM) generation, segmentation, feature extraction and preparation, hybrid optimization, object-based SVM classification by optimum results and accuracy assessment. The flowchart of the proposed method is shown in Figure 4, which is discussed in the following sections.

## 4.1 Image Pre-processing

Image pre-processing (e.g. radiometric and geometric corrections) is the first step of most of the RS processing. The studied data sources were georefrenced. Histogram Equalization (HE) is a simple and effective radiometric correction method with the purpose of increasing contrast in images, which was applied on the image. It should be noted that the difference between the minimum and maximum value of brightness intensity (luminance) is low in image with low contrast. HE makes it possible to increase the contrast of input image. More details can be found in (Wang, Chen et al. 1999).

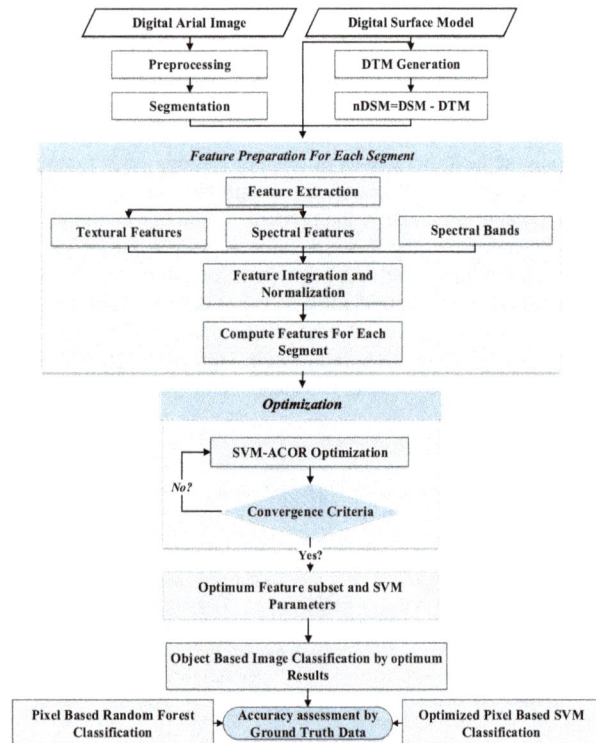

Figure 4. The flowchart of the proposed method

## 4.2 Normalized Digital Surface Model Generation

DSM and Digital Terrain Model (DTM) are essential to produce normalized DSM (nDSM). In order to identify the bare earth's surface and DTM generation, there are many filters which filter the ground and non-ground points. In this paper, Adaptive Triangulated Irregular Network Modelling (ATINM) was applied on DSM, which is capable of identifying the bare earth's points in complex urban areas. More detail about ATINM can be found in (Axelsson 2000). After that, DTM was produced by gridding the output results of applying ATINM on DSM using nearest neighbor method. Finally, nDSM was computed from Equation (5).

$$nDSM = DSM - DTM \qquad (5)$$

## 4.3 Image Segmentation

Image segmentation is an important step in object-based analysis because its quality directly affects the success or failure of those processing based on the object-based analysis (Witharana and Civco 2014). It is the process of labelling pixels in an image such that all of the pixels located in a homogenous area have a similar label (Haris, Efstratiadis et al. 1998). Among the proposed segmentation algorithms, multi resolution segmentation was used in this paper. The advantages of this algorithm includes easy realizing and implementation and producing results that match reality (Baatz and Schäpe 2000, Lefebvre, Corpetti et al. 2008). This method consecutively merges pixels or existing image objects. It is a bottom-up segmentation based on a pairwise region merging technique. Multi resolution segmentation is an optimization procedure which, for a given number of image objects, minimizes the average heterogeneity and maximizes their respective homogeneity. More detail about this method can be found in (Baatz and Schäpe 2000, Lefebvre, Corpetti et al. 2008).

By applying multi resolution segmentation, the input image was segmented into homogenous areas. Since the proposed method was applied on the over segments, the segmentation parameters (i.e. scale, shape and compactness) were determined by trial and error approach. For this reason, it was not necessary to determine accurate segmentation parameters in order to create segments that math reality.

## 4.4 Feature Extraction and Preparation

Generally, there are three kinds of features: spectral, textural and structural. In this paper, spectral and textural features were extracted to appropriately separate different classes because the proposed method was applied on the over segments due to determination of the segmentation parameters by trial and error approach. According to (Barzegar, Ebadi et al. 2015), the spectral features used in this paper includes Excessive Green Index (EGI), Difference Green Ratio (DGR), Normalized Difference Index (NDI), Brightness Index (BI), Saturation Index (SaI), Hue Index (HI), Correlation Index (CI), Redness Index (RI), Shadow Index (SI) and LAB space (Table 1).

| Formula | Spectral Index |
|---|---|
| EGI (Woebbecke, Meyer et al. 1995) | $3 * \left( (G) / (R + G + B) \right) - 1$ |
| DGR (Woebbecke, Meyer et al. 1995) | $(G - R) / (R + G + B)$ |
| NDI (Woebbecke, Meyer et al. 1995) | $(G - R) / (G + R)$ |
| BI (Mathieu, Pouget et al. 1998) | $\left( (R^2 + G^2 + B^2) / 3 \right)^{0.5}$ |
| SaI (Mathieu, Pouget et al. 1998) | $(R - B) / (R + B)$ |
| HI (Mathieu, Pouget et al. 1998) | $(2 * R - G - B) / (G - B)$ |
| CI (Mathieu, Pouget et al. 1998) | $(R - G) / (R + G)$ |
| RI (Mathieu, Pouget et al. 1998) | $R^2 / (B * G^3)$ |
| SI | $(R + B + G) / 3$ |

Table 1. The Spectral Features

The second type of features used in this paper were textural features which are extracted from Gray-Level Co-occurrence Matrix (GLCM). The idea of using GLCM was suggested by Haralick in the 1970s (Haralick, Shanmugam et al. 1973). Various textural features can be extracted from GLCM (e.g. mean, variance, homogeneity, contrast, dissimilarity, entropy, second moment, correlation). These features were extracted in five different kernel sizes (e.g. 3×3, 5×5, 7×7, 9×9 and 11×11) and four different directions (e.g. 0°, 45°, 90° and 135°). Due to removing the impact of different directions on the textural features, the average of textural features was computed for each segment. Then all of the extracted features were integrated with spectral bands and the input image was normalized so that all image pixel values would be located in the range of [0, 1]. After that, the extracted features were computed for each segment by averaging the values of all pixels located in each segment. This image was an input for the next step.

### 4.5 Optimized Object-Based Classification

#### 4.5.1 Hybrid Optimization
Due to using various features and determination of SVM parameters, hybrid optimization was applied to select independent features and to determine the parameters of RBF kernel function ($\gamma$) and penalty term (C) with relatively high automation level. Based on this, it was essential to collect some training and test samples segments in order to train SVM and hybrid optimization. These samples were selected by an expert in two classes (building and non-building) with respect to GT data. In this paper, $ACO_R$ algorithm was used due to reasons discussed in section 1 and 3.

In the initialize phase, the $ACO_R$ starts with an initial population. The initial population consisted of N individuals that were located in a D-dimensional search space. According to the optimization problem, each individual included three parts: mask of input features, C and $\gamma$. Thus, if the input image had (n) input features, the length of each possible solution will be (n+2). The (n) first position in each possible solution are related to selecting that feature or not. The $(n+1)^{th}$ and $(n+2)^{th}$ positions were related to C and $\gamma$. After initializing individuals in the search space, each of them was evaluated by an objective function. In this paper, classification error was used as an objective function to evaluate each individual. An individual of the population was the best member that had the lowest classification error, so the objective of this optimization problem was to minimize the cost function. Finally, after reaching to the stopping criteria (i.e. maximum iteration) in $ACO_R$, the best individual with the lowest cost value was the optimum solution, which includes optimum features, C and $\gamma$.

#### 4.5.2 Object-Based SVM Classification
After hybrid optimization, the object-based SVM classification was carried out by the optimum results. Then the classified image was evaluated by 5 different accuracy assessment criteria (e.g. Overall Accuracy (OA), Kappa Coefficient (KC), Quality Percentage (QP), Producer Accuracy (PA) and User Accuracy (UA)) with GT data. The results of the proposed method were also evaluated by pixel-based SVM and RF classification in terms of classification performance.

## 5. RESULTS AND DISCUSSION

The pre-processing procedures of the proposed method and the comparative methods were implemented using RS software (e.g. ALDPAT, Global Mapper, eCognition and ENVI) and MATLAB R2015a was used to implement the proposed method. To filter ground points from non-ground points, ATINM was applied on DSM in ALDPAT (Figure 5.a). The cell size of this filter was equal to the resolution of data sources (0.08 m). Then, DTM was produced by gridding the output results of applying ATINM on DSM using nearest neighbor method in Global Mapper. According to Equation 5, nDSM was computed (Figure 5.b).

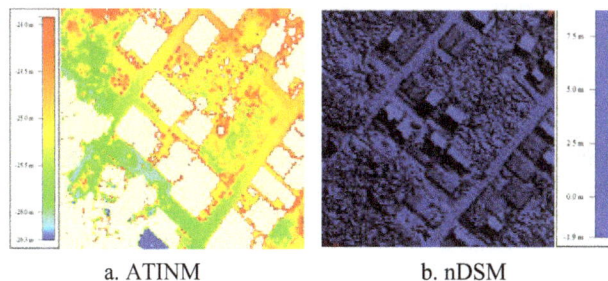

| a. ATINM | b. nDSM |

Figure 5. The result of a. ATINM, b. nDSM

As mentioned, it was essential to use HE method on the original image (Figure 6). After that, the spectral and textural features were extracted from HSR image and DSM in ENVI. Table 2 implies the characteristic these features.

| Feature | Number |
|---|---|
| Spectral Feature | 12 |
| Initial Textural Feature | 640 |
| Final Textural Feature | 160 |
| Spectral Bands and nDSM | 4 |
| Total Feature | 176 |

Table 2. The characteristic of input features

Next, multi resolution segmentation was applied on the image in order to create homogenous areas in eCognition. The parameters of this segmentation are scale, shape and compactness. It is essential to set these parameters accurately with the purpose of exact feature extraction. However, in over segmentation procedure, these parameters were determined by default values according to image size, area type (e.g. urban or non-urban) and interest feature class (e.g. line features such as roads or regular features such as buildings). Based on this, the values of scale, shape and compactness parameters were set 60, 0.8 and 0.3, respectively. The segmented image is shown in Figure 7. After that, all of the extracted features were computed for each over segment by averaging all of the pixels located in each segment so the segmented image with the features was prepared for the hybrid optimization step.

| Figure 6. The equalized image by HE method | Figure 7. The over segmented image |

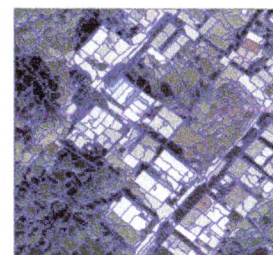

In hybrid optimization step, it was essential to determine some parameters of optimization algorithm (e.g. number of individuals, maximum iteration, etc.). The number of individuals and maximum iteration in $ACO_R$ was 100 and 50, respectively. As stated in section 4, in hybrid optimization step, each individual of population was evaluated based on the objective function (cost function). In each iteration, each individual of population that had the lowest cost function was selected as an optimum solution in that iteration. The cost function of object-based SVM-$ACO_R$ and pixel-based SVM-$ACO_R$ were 4.651 and 35.509 in the first iteration, respectively and reduced in the last iteration. It showed that ACO$_R$ tries to find the best solution with the lowest cost function or the highest fitness function based on swarm intelligence. Since the objective function of this paper was classification error for training and test samples, the optimum solution was the solution with the lowest cost function (the best individual in the last iteration).

The improvement of KC, OA and QP for different classification methods is shown in Figure 8. According to Figure 8, QP, OA and KC were improved by 9%, 2% and 30% in object-based SVM classification using 15 randomly selected features (Object-SVM-15R) rather than using all of the input features (Object-SVM-176), respectively. It should be noted that SVM parameters were determined in trial and error approach in this step. This showed the presence probability of dependent features among all of the input features and their effects on classification performance. The accuracy criteria were improved by using 15 randomly selected features, but the improvement was not entirely appropriate. For this reason, it is essential to apply FS and MS methods in high dimensional space. In comparison with Object-SVM-15R, the proposed method (Object-SVM-ACO$_R$) improved QP, OA and KC by 20%, 12% and 30%, respectively. These results showed the efficiency of the hybrid optimization in FS and MS procedure of SVM.

The optimized pixel-based SVM (Pixel-SVM-ACO$_R$) reduced QP, OA and KC by 17%, 10% and 24% in comparison with Object-SVM-ACO$_R$, respectively. These results showed the limitation of pixel-based classification in complex urban areas. Object-SVM-ACO$_R$ improved the KC by 6% rather than RF classification. According to good performance of RF in various RS studies, the relative improvement of the proposed method implied its superiority rather than RF in HSR images classification and clarified the efficiency of ACO$_R$ rather than other FS methods because ACO$_R$ is based on swarm intelligence that is more intelligent than other methods.

than pixel-based techniques in terms of accuracy and time processing.

The classified images obtained from different methods are shown in Figure 9. A look at Figure 9 reveals that the proposed method had better performance than pixel-based classification in building detection and didn't result in salt-paper results that produced in RF classification. According to figure 9, some over segments of shadow on building roof were assigned to building class in the proposed method while these over segments were not classified in building class in pixel-based classification. This showed the limitation and weakness of pixel-based techniques. As it can be seen, some of the segments and also some pixels were misclassified into building class while they must be classified into non-building class (vegetation class). This was due to unavailability of Near Infrared band (NIR) in order to separate building class from non-building class (vegetation class). To address this problem, NIR band or spectral features based on this band (e.g. normalized vegetation index, etc.) or structural features or features extracted from LiDAR data (e.g. laplacian, slope) can be used. Due to applying the proposed method on the over segments and relatively high automation level, it was not possible to use structural features for each segment. It is suggested to apply optimization method in order to determine the segmentation parameter, then it will be possible to extract structural features in the classification process. In Figure 9, the buildings edges were identified clearer in Object-SVM-ACO$_R$ than RF. This resulted in post processing reduction for building edge reconstruction.

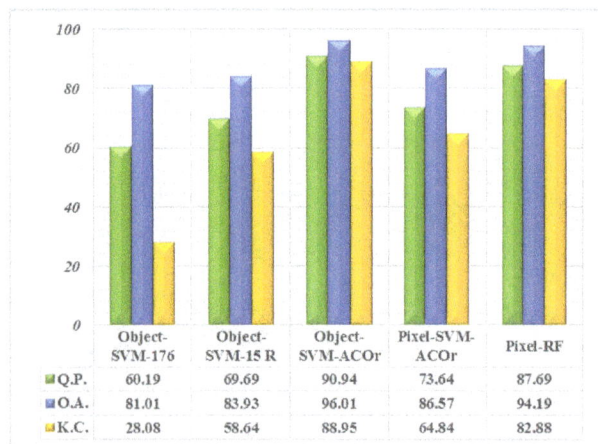

Figure 8. Chart of accuracy criteria in different methods

| | Object-SVM-176 | Object-SVM-15 R | Object-SVM-ACOr | Pixel-SVM-ACOr | Pixel-RF |
|---|---|---|---|---|---|
| Q.P. | 60.19 | 69.69 | 90.94 | 73.64 | 87.69 |
| O.A. | 81.01 | 83.93 | 96.01 | 86.57 | 94.19 |
| K.C. | 28.08 | 58.64 | 88.95 | 64.84 | 82.88 |

Different accuracy assessment criteria are shown in Table 3 for all of the methods. It should be noted that accuracy assessment was done by GT data. Time processing in the proposed method was relatively low in comparison with pixel-based methods. The reason of low time processing in the proposed method was the unit of image analysis (image object) because image objects reduced the computational complexity rather than pixels. While pixel was a unique unit in pixel-based techniques and this resulted in high time processing. The Object-SVM-ACO$_R$ provided more accurate results than pixel-based techniques. These showed the superiority of the proposed method rather

| Method | | Class | UA | PA | QP | Optimum Features |
|---|---|---|---|---|---|---|
| **Pixel Based Classification** | SVM-15R | Building | 19.22 | 99.61 | 19.21 | |
| | | Non-Building | 99.97 | 79.84 | 99.97 | 15 |
| | | Total | 59.60 | 89.72 | 59.59 | |
| | SVM-ACO$_R$ | Building | 79.46 | 68.90 | 58.49 | |
| | | Non-Building | 88.79 | 93.26 | 88.79 | 11 |
| | | Total | 84.13 | 81.08 | 73.64 | |
| | RF | Building | 78.46 | 96.47 | 76.27 | |
| | | Non-Building | 99.10 | 93.64 | 99.10 | 15 |
| | | Total | 88.78 | 95.06 | 87.69 | |
| **Object Based Classification** | SVM-176 | Building | 20.53 | 98.56 | 20.47 | |
| | | Non-Building | 99.91 | 80.09 | 99.91 | 176 |
| | | Total | 60.22 | 89.33 | 60.19 | |
| | SVM-15R | Building | 76.56 | 63.48 | 53.15 | |
| | | Non-Building | 86.23 | 92.17 | 86.24 | 15 |
| | | Total | 81.40 | 77.82 | 69.69 | |
| | SVM-ACO$_R$ | Building | 91.47 | 91.69 | 84.47 | |
| | | Non-Building | 97.41 | 97.34 | 97.41 | 12 |
| | | Total | 94.44 | 94.51 | 90.94 | |

Table 3. The accuracy assessment results obtained from different methods

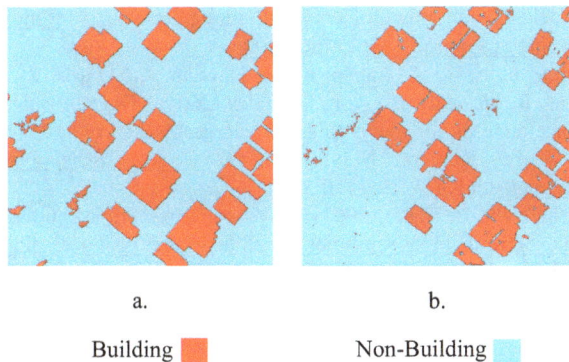

a.         b.

Building ■      Non-Building ■

Figure 9. The Classified images obtianed from a. Object -SVM-ACO$_R$ Clasification. b. RF Classification

## 6. CONCLUSION

Due to the existence limitation of pixel-based SVM classification in HSR images, object-based SVM classification was used in this paper. In nonlinear problems, it is essential to determine parameters of SVM because these parameters directly affect the results (MS). According to the limited number of spectral bands in HSR images, digital aerial images cannot provide information about classes with similar spectral behavior. Spectral and textural features or other data sources can be used to solve the limitations. However, adding some additional features will lead to increase the probability of dependent feature that may reduce the accuracy. Hybrid optimization of object-based SVM classification in MS of SVM and FS cause to apply ant colony optimization algorithm in order to solve the existence limitations. The advantages of the proposed method are relatively low time processing because of unit of image analysis, relatively high automation level, independency of image scene and type, post processing reduction for building edge reconstruction, hybrid optimization and accuracy improvement in comparison with pixel-based classification. The results showed that hybrid optimization of object-based SVM classification improved overall accuracy by 17% rather than object-based SVM by all of the input features. This showed the effect of hybrid optimization in classification performance and the presence probability of dependent features. Moreover, in comparison with hybrid optimization of pixel-based SVM classification, the overall accuracy and kappa coefficient were improved by 10% and 24% in the proposed method, respectively. This showed the superiority of optimized object-based SVM classification rather than optimized pixel-based SVM classification. Due to using over segments to satisfy automation level, there was no need to set fine values of the segmentation parameters that match reality. This resulted in lacking of structural features. For this reason, misclassification of some segments (misclassification of vegetation class into building class) were occurred. Therefore, it is recommended to propose a method to optimize the segmentation parameters, so structural features can be extracted for each segment in accordance with reality. Hence, the features can be used in classification that results in better separating between classes.

## REFERENCES

Axelsson, P. (2000). "DEM generation from laser scanner data using adaptive TIN models." International Archives of Photogrammetry and Remote Sensing **33**(B4/1; PART 4): 111-118.

Baatz, M. and A. Schäpe (2000). "Multiresolution segmentation: an optimization approach for high quality multi-scale image segmentation." Angewandte Geographische Informationsverarbeitung XII **58**: 12-23.

Barzegar, M., H. Ebadi and A. Kiani (2015). "Comparison of different vegetation indices for very high-resolution images, specific case UltraCam-D imagery." The International Archives of Photogrammetry, Remote Sensing and Spatial Information Sciences **40**(1): 97.

Benz, U. C., P. Hofmann, G. Willhauck, I. Lingenfelder and M. Heynen (2004). "Multi-resolution, object-oriented fuzzy analysis of remote sensing data for GIS-ready information." ISPRS Journal of photogrammetry and remote sensing **58**(3): 239-258.

Bin, W., Y. Jian, Z. Zhongming, M. Yu, Y. Anzhi, C. Jingbo, H. Dongxu, L. Xingchun and L. Shunxi (2014). "Parcel-Based Change Detection in Land-Use Maps by Adopting the Holistic Feature." Selected Topics in Applied Earth Observations and Remote Sensing, IEEE Journal of **7**(8): 3482-3490.

Blaschke, T. (2005). "Towards a framework for change detection based on image objects." Göttinger Geographische Abhandlungen **113**: 1-9.

Blaschke, T. (2010). "Object based image analysis for remote sensing." ISPRS journal of photogrammetry and remote sensing **65**(1): 2-16.

Bontemps, S., P. Bogaert, N. Titeux and P. Defourny (2008). "An object-based change detection method accounting for temporal dependences in time series with medium to coarse spatial resolution." Remote Sensing of Environment **112**(6): 3181-3191.

Bouziani, M., K. Goïta and D.-C. He (2010). "Automatic change detection of buildings in urban environment from very high spatial resolution images using existing geodatabase and prior knowledge." ISPRS Journal of Photogrammetry and Remote Sensing **65**(1): 143-153.

Breiman, L. (2001). "Random forests." Machine learning **45**(1): 5-32.

Camps-Valls, G., L. Gómez-Chova, J. Calpe-Maravilla, J. D. Martín-Guerrero, E. Soria-Olivas, L. Alonso-Chordá and J. Moreno (2004). "Robust support vector method for hyperspectral data classification and knowledge discovery." IEEE Transactions on Geoscience and Remote sensing **42**(7): 1530-1542.

Castilla, G. and G. Hay (2008). Image objects and geographic objects. Object-based image analysis, Springer: 91-110.

Chen, G., G. J. Hay, L. M. Carvalho and M. A. Wulder (2012). "Object-based change detection." International Journal of Remote Sensing **33**(14): 4434-4457.

Chen, W. and Y. Tian (2016). "Parameter Optimization of SVM Based on Improved ACO for Data Classification." International Journal of Multimedia and Ubiquitous Engineering **11**(1): 201-212.

Cheng, G. and J. Han (2016). "A survey on object detection in optical remote sensing images." ISPRS Journal of Photogrammetry and Remote Sensing **117**: 11-28.

Cheng, L. and W. Bao (2014). "Remote sensing image classification based on optimized support vector machine." Indonesian Journal of Electrical Engineering and Computer Science **12**(2): 1037-1045.

Chutia, D., D. Bhattacharyya, K. Sarma, R. Kalita and S. Sudhakar (2015). "Hyperspectral remote sensing classifications: a perspective survey." Transactions in GIS.

Coburn, C. and A. C. Roberts (2004). "A multiscale texture analysis procedure for improved forest stand classification." International journal of remote sensing 25(20): 4287-4308.

Desclée, B., P. Bogaert and P. Defourny (2006). "Forest change detection by statistical object-based method." Remote Sensing of Environment 102(1): 1-11.

Dorigo, M., V. Maniezzo and A. Colorni (1996). "Ant system: optimization by a colony of cooperating agents." IEEE Transactions on Systems, Man, and Cybernetics, Part B (Cybernetics) 26(1): 29-41.

Dorigo, M. and T. Sttzle (2004). "Ant Colony OptimizationMIT Press." Cambridge, MA.

Drăguţ, L., O. Csillik, C. Eisank and D. Tiede (2014). "Automated parameterisation for multi-scale image segmentation on multiple layers." ISPRS Journal of Photogrammetry and Remote Sensing 88: 119-127.

Egorov, A., M. Hansen, D. Roy, A. Kommareddy and P. Potapov (2015). "Image interpretation-guided supervised classification using nested segmentation." Remote Sensing of Environment 165: 135-147.

Gamanya, R., P. De Maeyer and M. De Dapper (2009). "Object-oriented change detection for the city of Harare, Zimbabwe." Expert Systems with Applications 36(1): 571-588.

Gao, H., M. K. Mandal and J. Wan (2010). Classification of hyperspectral image with feature selection and parameter estimation. Measuring Technology and Mechatronics Automation (ICMTMA), 2010 International Conference on, IEEE.

Geerling, G., M. Labrador-Garcia, J. Clevers, A. Ragas and A. Smits (2007). "Classification of floodplain vegetation by data fusion of spectral (CASI) and LiDAR data." International Journal of Remote Sensing 28(19): 4263-4284.

Ghamisi, P. and J. A. Benediktsson (2015). "Feature selection based on hybridization of genetic algorithm and particle swarm optimization." Geoscience and Remote Sensing Letters, IEEE 12(2): 309-313.

Gharibi, M., H. Arefi, H. Rastiveis and H. Hashemi (2016). "Building Map Updating Based on Active Contour Models." Journal of Geomatics Science and Technology 5(4): 211-225.

Hall, O. and G. J. Hay (2003). "A multiscale object-specific approach to digital change detection." International Journal of Applied Earth Observation and Geoinformation 4(4): 311-327.

Haralick, R. M., K. Shanmugam and I. H. Dinstein (1973). "Textural features for image classification." Systems, Man and Cybernetics, IEEE Transactions on(6): 610-621.

Haris, K., S. N. Efstratiadis, N. Maglaveras and A. K. Katsaggelos (1998). "Hybrid image segmentation using watersheds and fast region merging." IEEE Transactions on image processing 7(12): 1684-1699.

Hermosilla, T., L. A. Ruiz, J. A. Recio and J. Estornell (2011). "Evaluation of automatic building detection approaches combining high resolution images and LiDAR data." Remote Sensing 3(6): 1188-1210.

Huang, X., L. Zhang and T. Zhu (2014). "Building change detection from multitemporal high-resolution remotely sensed images based on a morphological building index." Selected Topics in Applied Earth Observations and Remote Sensing, IEEE Journal of 7(1): 105-115.

Ji, C. (2000). "Land-use classification of remotely sensed data using Kohonen self-organizing feature map neural networks." Photogrammetric Engineering and Remote Sensing 66(12): 1451-1460.

Khatami, R., G. Mountrakis and S. V. Stehman (2016). "A meta-analysis of remote sensing research on supervised pixel-based land-cover image classification processes: General guidelines for practitioners and future research." Remote Sensing of Environment 177: 89-100.

Laliberte, A. S., A. Rango, K. M. Havstad, J. F. Paris, R. F. Beck, R. McNeely and A. L. Gonzalez (2004). "Object-oriented image analysis for mapping shrub encroachment from 1937 to 2003 in southern New Mexico." Remote Sensing of Environment 93(1): 198-210.

Lari, Z. and H. Ebadi (2007). Automatic extraction of building features from high resolution satellite images using artificial neural networks. Proceedings of ISPRS Conference on Information Extraction from SAR and Optical Data, with Emphasis on Developing Countries, Istanbul, Turkey.

Lefebvre, A., T. Corpetti and L. Hubert-Moy (2008). Object-oriented approach and texture analysis for change detection in very high resolution images. Geoscience and Remote Sensing Symposium, 2008. IGARSS 2008. IEEE International, IEEE.

Li, C.-F. and J.-Y. Yin (2013). "Variational Bayesian independent component analysis-support vector machine for remote sensing classification." Computers & Electrical Engineering 39(3): 717-726.

Li, Y., H. Wu, R. An, H. Xu, Q. He and J. Xu (2013). "An improved building boundary extraction algorithm based on fusion of optical imagery and LiDAR data." Optik-International Journal for Light and Electron Optics 124(22): 5357-5362.

Lillesand, T., R. W. Kiefer and J. Chipman (2014). Remote sensing and image interpretation, John Wiley & Sons.

Liu, X. and Y. Bo (2015). "Object-based crop species classification based on the combination of airborne hyperspectral images and LiDAR data." Remote Sensing 7(1): 922-950.

Luo, J.-c., C.-h. Zhou, Y. Leung and J.-h. Ma (2002). "Support vector machine for spatial feature extraction and classification of remotely sensed imagery." JOURNAL OF REMOTE SENSING-BEIJING- 6(1): 55-61.

Mellor, A., A. Haywood, C. Stone and S. Jones (2013). "The performance of random forests in an operational setting for large area sclerophyll forest classification." Remote Sensing 5(6): 2838-2856.

Pal, M. (2002). Factors influencing the accuracy of remote sensing classifications: a comparative study, University of Nottingham.

Petropoulos, G. P., C. Kalaitzidis and K. P. Vadrevu (2012). "Support vector machines and object-based classification for obtaining land-use/cover cartography from Hyperion hyperspectral imagery." Computers & Geosciences 41: 99-107.

Ronczyk, M. (2011). "Object-based classification of urban land cover extraction using high spatial resolution imagery." URL: https://bismarck. nyme. hu.

Samadzadegan, F., H. Hasani and T. Schenk (2012). "Determination of optimum classifier and feature subset in hyperspectral images based on ant colony system." Photogrammetric Engineering & Remote Sensing 78(12): 1261-1273.

Samadzadegan, F., H. Hasani and T. Schenk (2012). "Simultaneous feature selection and SVM parameter determination in classification of hyperspectral imagery using Ant Colony Optimization." Canadian Journal of Remote Sensing 38(2): 139-156.

Shackelford, A. K. and C. H. Davis (2003). "A hierarchical fuzzy classification approach for high-resolution multispectral data over urban areas." IEEE transactions on geoscience and remote sensing 41(9): 1920-1932.

Socha, K. and M. Dorigo (2008). "Ant colony optimization for continuous domains." European journal of operational research 185(3): 1155-1173.

Vapnik, V. N. and V. Vapnik (1998). Statistical learning theory, Wiley New York.

Walton, J. T. (2008). "Subpixel urban land cover estimation." Photogrammetric Engineering & Remote Sensing **74**(10): 1213-1222.

Wan, Y., M. Wang, Z. Ye and X. Lai (2016). "A feature selection method based on modified binary coded ant colony optimization algorithm." Applied Soft Computing **49**: 248-258.

Wang, L. (2005). Support vector machines: theory and applications, Springer Science & Business Media.

Wang, L., W. Sousa and P. Gong (2004). "Integration of object-based and pixel-based classification for mapping mangroves with IKONOS imagery." International Journal of Remote Sensing **25**(24): 5655-5668.

Wang, Y., Q. Chen and B. Zhang (1999). "Image enhancement based on equal area dualistic sub-image histogram equalization method." IEEE Transactions on Consumer Electronics **45**(1): 68-75.

Witharana, C. and D. L. Civco (2014). "Optimizing multi-resolution segmentation scale using empirical methods: Exploring the sensitivity of the supervised discrepancy measure Euclidean distance 2 (ED2)." ISPRS Journal of Photogrammetry and Remote Sensing **87**: 108-121.

Woebbecke, D., G. Meyer, K. Von Bargen and D. Mortensen (1995). "Color indices for weed identification under various soil, residue, and lighting conditions." Transactions of the ASAE **38**(1): 259-269.

Xiao, J. and L. Li (2011). "A hybrid ant colony optimization for continuous domains." Expert Systems with Applications **38**(9): 11072-11077.

Xuegong, Z. (2000). "Introduction to statistical learning theory and support vector machines." Acta Automatica Sinica **26**(1): 32-42.

Zhang, X., X. Chen and Z. He (2010). "An ACO-based algorithm for parameter optimization of support vector machines." Expert Systems with Applications **37**(9): 6618-6628.

4

# A NOVEL 3D INTELLIGENT FUZZY ALGORITHM BASED ON MINKOWSKI-CLUSTERING

Shima Toori[1], Ali Esmaeily[2]

[1]M.Sc. in GIS Engineering, Graduate University of Advanced Technology, Kerman, Iran
Email: st69626@yahoo.com
[2]Assistant professor, Dept. of Remote Sensing Engineering, Graduate University of Advanced Technology, Kerman, Iran
Email: aliesmaeily@kgut.ac.ir

**KEY WORDS:** Fuzzy, 3D NDVI, Minkowski Clustering, Algorithm, Distance

**ABSTRACT:**

Assessing and monitoring the state of the earth surface is a key requirement for global change research. In this paper, we propose a new consensus fuzzy clustering algorithm that is based on the Minkowski distance. This research concentrates on Tehran's vegetation mass and its changes during 29 years using remote sensing technology. The main purpose of this research is to evaluate the changes in vegetation mass using a new process by combination of intelligent NDVI fuzzy clustering and Minkowski distance operation. The dataset includes the images of Landsat8 and Landsat TM, from 1989 to 2016. For each year three images of three continuous days were used to identify vegetation impact and recovery. The result was a 3D NDVI image, with one dimension for each day NDVI. The next step was the classification procedure which is a complicated process of categorizing pixels into a finite number of separate classes, based on their data values. If a pixel satisfies a certain set of standards, the pixel is allocated to the class that corresponds to those criteria. This method is less sensitive to noise and can integrate solutions from multiple samples of data or attributes for processing data in the processing industry. The result was a fuzzy one dimensional image. This image was also computed for the next 28 years. The classification was done in both specified urban and natural park areas of Tehran. Experiments showed that our method worked better in classifying image pixels in comparison with the standard classification methods.

## 1. Introduction

Using high resolution remote sensing datasets in order to detect vegetation change offers an access to monitoring vegetation change dynamics and to some degree plant diversity. Vegetation mapping also presents valuable facts for comprehending the natural and man-made environments through measuring vegetation cover from local to global scales at a certain time point or over a continuous period.

Based on recent researches, half of the world's population lives in cities, and in the recent past, the main raise of the urban population took place in underdeveloped and developing countries. Cities are the most obvious human environment, and now almost 50% of global populations live in cities (Willem et al., 2006). From an ecological point of view, urbanization has a large impact on the ecosystem as it always accompanied by land use changes, altering its combination and structure, and consequently influencing ecosystem processes and functioning (Alberti, 2005). Land use and land cover changes affects the original matter and energy cycles of the ecosystem (Wackernagel and Yount, 1998; Pielke et al., 1999; Imhoff et al., 2000). In the direction of urbanization, the transmutation of land to urban use mainly decreases photosynthesis of the ecosystem in regions with productive forests (Wear and Greis et al., 2001; Nizeyaimana et al., 2001).

The world is encountering urban extend and experiencing rapid population growth, which is transforming regional natural landscapes. For instance, large patches of fertile forests or cropland tend to be fragmentized due to quick urbanization (Imhoff et al., 1997; Folke et al., 1997). Loss of cropland in some areas has an important affect on climate conditions, food security, and the environment (Yu et al., 2006). Scientists have done many researches about the effect of urbanization on matter cycles, energy flow and ecosystem service (Schimel et al., 2000).

Nowadays remote sensing has become a useful alternative to collect spatial information of vegetation dispersion during a period of time, by means of studying definite spatial and temporal resolution. Regional land reclamation programmers are progressively combining remotely sensed imagery for monitoring current and historical vegetation dynamics. Many researchers used Spectral vegetation index data to monitor vegetation, government and minerals industry (Willem et al.,2006). AlBakri and Taylor (2003) used the time series data including NOAA AVHRR images in order to monitor vegetation conditions for 15 years and they find that NDVI is a proper index to monitor vegetation conditions.

Specifically, this research focuses on Tehran's vegetation mass and its changes within 29years with the use of Remote sensing technology.

## 2. Study Area

The study area is Tehran province, located in North of Iran, at 35 40 N Latitude and 51 26 E Longitude, which is the capital of Iran (Figure1).

Figure1: location of study area, Tehran

Tehran's population was about 800,000. It had increased to 3 million by 1966 and by 1986 migrants brought the population to 6million. Today, the metropolitan region has more than 10 million residents, more than the sum of the Iran's next five major metropolitan provinces combined. This explosive growth has public health and environmental consequences, including air pollution and water impurity and the loss of arable land. Figure2 and Figure3 illustrate Tehran transportation network and Tehran green areas respectively.

Figure2: Tehran Transportation network

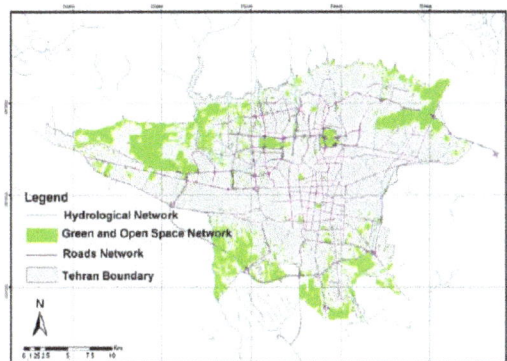

Figure3: Tehran green areas

The NASA's Landsat 5 satellite with Thematic Mapper sensor, acquired an images of Tehran on August 2, 1989 (figure4 top), and another one on July 19, 2009 (figure4 bottom). In both images, vegetation appears bright green, urban areas range in color from gray to black, and unproductive areas appear brown. Whereas non-urbanized areas border the earlier image, urbanization fills almost the entire frame of the later image. Major roadways crisscrossing the city in 1989 remain visible in 2009, but many additional roadways have been added, particularly in the north.

Figure4: False color images of Tehran acquired on August 2, 1989 (top), and on July 19, 2009 (bottom)

## 3. METHODS

The main purpose of this paper is to measure the changes in vegetation mass. By the use of different high resolution remote sensing Landsat8 and LandsatTM datasets from three continuous days, vegetation impact and recovery can be identified for years 1988 to 2016. For this purpose the first step is Classification. Compared to previous attempts, the suggested algorithm of this research differs in the similarity matrix construction. It does not use the resampling technique, but it uses the Minkowski distance to measure the input data. Each pixel belongs to the class with shortest distance. The new intelligent algorithm evaluates the accuracy using distribution. In low preciseness conditions, the intelligent algorithm changes the class center. Depending on the number of pixels in each class, if we do not reach the desired accuracy, the cycle is repeated 100 times. The final respond optimized by the algorithm will be considered.

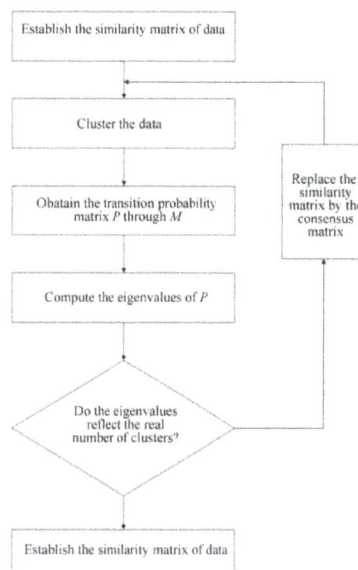

Figure5: Flow graph of iterative Minkowski clustering

The Landsat images received on 9 June 1988 to 11 June 1988 - three continuous days- with 30 meter resolution, processed and the NDVI factor computed using pixel values of bands 3 to 5. No significant deviation must be observed in NDVI values, because of the three day period. This was repeated for the years 1989 to 2016. The outcome of this step is a 3D NDVI image, with one dimension for each day NDVI. Classification is the next step. Classification is a complex process of sorting pixels into a finite number of individual categories of data, based on their data values. We used an intelligent Minkowski weighted distance algorithm in order to identify the four NDVI class centers. If a pixel satisfies a certain set of standards, the pixel belongs to the class that corresponds to those criteria. The Minkowski distance is the distance between the two measured NDVI dimensions (De-Gang X, 2017).

The similarity matrix does not adopt one specific Minkowski distance, but makes use of the different forms of Minkowski distances by adjusting the parameters in the equation. In this paper, we use the Minkowski distance to construct the similarity matrix:

$$M_P(x,y) = \left( \sum_{i=1}^{n} |x_i - y_i|^P \right)^{1/P} \quad (1)$$

$$S_{M_P}(x,y) = \mu\, exp\left(-\mu M_P(x,y)\right) \quad (2)$$

| | |
|---|---|
| ■ | Unclassified |
| ■ | Class1 |
| ■ | Class2 |
| ■ | Class3 |
| ■ | Class4 |

$\mu$ values can be selected from{0.1, 0.2, 0.5, 0.8, 0.9}. In case P=1, the first equation is the Manhattan distance, which reflects the sum of the absolute value of the difference between pixel $i$ and pixel $j$. If P=2, formula (1) is the shortest distance between pixel i and pixel j, namely, the diagonal distance. If P $\rightarrow \infty$, then equation (1) is the Chebyshev distance, which reflects the maximum deviation between $i$ and $j$ in a special dimension. Additionally, P could have a value that is less than 1, and the categorizing method, by taking different values of $P$, will provide different outcomes. $\mu$ is an adjustable element. Because there are two variables, the distance equation can reflect similar information for different aspects by adapting the parameters. Minkowski distances can define the similarity of the special data in advance. We can use this equation to establish different similarity matrices and acquire the real similarity information picked from different clustering effects. This method is proper for the idea of consensus clustering, namely, selecting the appropriate matrix from the different results.

In order to obtain the clusters:

1. Different similarity matrices based on the Minkowski distance with different parameters ($p$ and $\mu$) must be established.
2. One of the above matrices must be categorized with the clustering algorithms and the clustering numbers ($n$ numbers), and obtain the $3 \times n$ matrices.
3. The consensus similarity matrix $M$ has been obtained by analyzing the $3 \times n$ matrices (which reflect the information of the clustering).
4. The transition probability matrix have been obtained.
5. The transition probability matrices have been obtained corresponding to the other similarity matrices by using the above methods.
6. The number of clusters must be obtained by analyzing the Eigen values of the different transition probability matrices.

## 4. RESULTS AND DISCUSSION

The main goal of this study is to evaluate the changes in vegetation mass. The research was carried out in Tehran. A number of varied high resolution remote sensing Landsat TM and Landsat8 datasets from three continuous days were used to identify vegetation impact and retrieval from 1989 to 2016. The Landsat images received on 9 June 1988 to 11 June 1988 - three consecutive days- with 30-meter resolution, processed and the NDVI factor calculated using pixel values of bands 3to5.

Because of the 3day period duration of the proposed algorithm, no significant changes must be observed in NDVI values. We used an intelligent Minkowski distance algorithm in order to identify the four NDVI class centers. If a pixel satisfies a certain set of criteria, the pixel is allocated to the corresponding class to those criteria.

Three calculated NDVI values must be clustered for each pixel of the 3D NDVI image and as they were for three consecutive days, there should not be any significant alters in their classes. The intelligent algorithm decides the possible NDVI class for each pixel of 3D NDVI image. The four class names were determined by the class center NDVI value. Every pixel with significant deviation between their three measured NDVI values is categorized as unclassified.

The result was a fuzzy one dimensional image. This image was also computed for the next 26 years, the classification *was done* in both specified urban and natural park areas of Tehran that showed detailed measurable vegetation, re-growth and reduction.

Intelligent algorithm automatically accomplished classification in four classes, which Class 1 covers large and healthy plants and Class 4 determines poor vegetation areas.

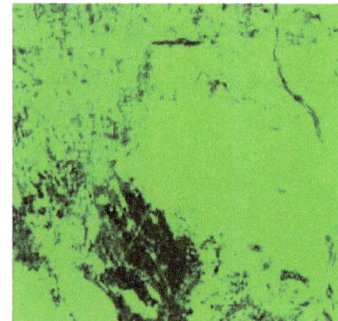

1988

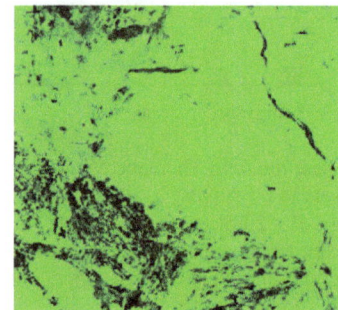

2000

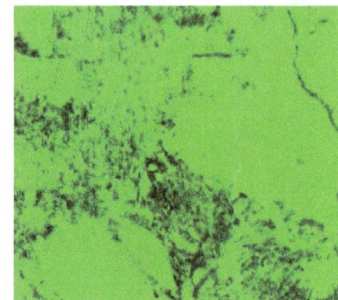

2006

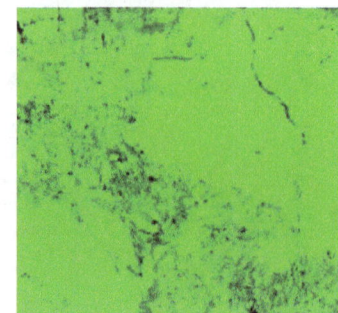

2016

Figure6: The final output of algorithms, fuzzy images

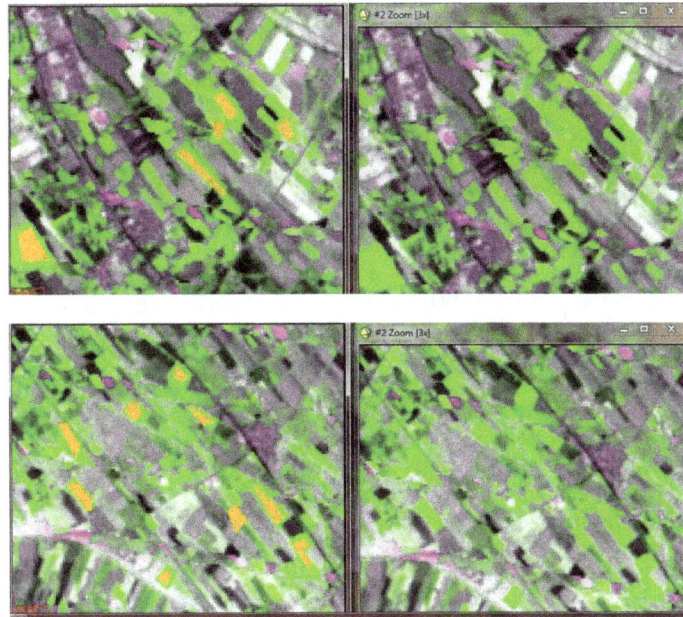

Figure7: fuzzy Minkowski classification test pixels

For evaluation of the accuracy of output, a number of test pixels were selected (Figure7). The pixels were categorized in two separate sets; one related to urban vegetation and the second related to the pixels of the natural vegetation, urban parks and plants. The table1&2 describe that the algorithm has performed very successful in detecting intervals.

| Test | Total Pix | ▮ | ▮ | ▮ | ▮ | ▮ |
|---|---|---|---|---|---|---|
| 1988-DAY1 | 948 | 841 | 76 | 12 | 5 | 14 |
| % | - | 88.71% | 8.01% | 1.2% | 0.52% | 1.47% |
| 1988-DAY3 | 659 | 614 | 13 | 9 | 16 | 7 |
| Urban | - | 93.17% | 1.97% | 1.36% | 2.42% | 1.06% |
| 2000- DAY3 | 1473 | 1341 | 94 | 4 | 13 | 21 |
| % | - | 91.03% | 6.38% | 0.27% | 0.88% | 1.42% |
| 2006-DAY2 | 1649 | 1498 | 104 | 17 | 20 | 11 |
| % | - | 90.84% | 6.3% | 1.03% | 1.21% | 0.66% |
| 2016-DAY1 | 993 | 941 | 24 | 6 | 15 | 7 |
| % | - | 94.76% | 2.41% | 0.6% | 1.51% | 0.7% |
| 2016-DAY3 | 878 | 814 | 29 | 11 | 8 | 16 |
| Urban | - | 92.71% | 3.3% | 1.25% | 0.91% | 1.82% |

**Table1: Test pixels**

| Area | 1988 | 2000 | 2006 | 2016 |
|---|---|---|---|---|
| ▮ | 4.97% | 5.64% | 5.66% | 1.53% |
| ▮ | 7.12% | 8.48% | 6.89% | 4.84% |
| ▮ | 9.34% | 8.51% | 9.14% | 8.48% |
| ▮ | 8.39% | 9.14% | 7.67% | 5.34% |
| ▮ | 68.82% | 68.23% | 70.64% | 79.81% |

**Table2:Test pixels result**

## 5. CONCLUSION

The aim of this study is to focus on Tehran's vegetation mass and its changes within 29 years using mathematics and remote sensing technology, a key that extends possible data archives from present time to over several decades back. According to the previous studies, the combination of a 3D fuzzy NDVI clustering and weighted Minkowski distance operators have never been used on images bands, in order to detect changed pixels. Our proposed algorithm, repeated for the same three days of years 1989 to 2016. The outcome of this step is a 3D NDVI image, with one dimension for each day NDVI. The next step was classification. The four class names were determined by the class center NDVI value. Also pixels with significant deviation between their three measured NDVI values were classified as unknown.

The outcome was a fuzzy one dimensional image, which was also computed for the remained 28years, the classification was done in both specified urban and natural park areas of Tehran that denoted detailed measurable vegetation, re-growth and reduction. The results shows that the method of this paper works better in classifying image pixels compared to some of the standard classification methods. The results of this study emphasize the need for more attention to changes and vegetation reduction in that areas. In order to improve the results obtained and continue working in this direction, it is suggested that future studies can be done by adding parameters such as the change of seasons.

## REFERENCES

Al-Bakri JT and Taylor JC (2003) Application of NOAA AVHRR for monitoring vegetation conditions and biomass in Jordan. *Journal of Arid Environments* 54: 579–593

Alberti, M., 2005. The effects of urban patterns on ecosystem function. *International Regional Sciences*. Rev. 28, 168–192.

De-Gang X., Pan-Lei Z., Chun-Hua Y., Wei-Hua G., Minkowski-distance-based Consensus Clustering Algorithm, *International Journal of Automation and Computing*, 14(1), February 2017, 33-44

Folke, C., Jannson, A., Larsson, J., Costanza, R., 1997. Ecosystem appropriation by cities. *AMBIO 26*, 167–172.

Imhoff, M.L., Tucker, C.J., Lawrence, W.T., Stutzer, D.C., 2000. The use of multisource satellite and geospatial data to study the effect of urbanization on primary productivity in the United States. *IEEE Trans. Geoscience Remote sensing* 38 (6), 2549–2556.

Imhoff, M.L., Lawrence, W.T., Stutzer, D.C., Elvidge, C.D., 1997. Using nighttime DMSP/OLS images of city lights to estimate the impact of urban land use on soil resources in the US. *Remote Sensing Environment*. 59, 105–117.

Imhoff, M.L., Bounoua, L., DeFries, R., Lawrence, W.T., Stutzer, D., Tucker, C.J., Ricketts, T., 2004. The consequences of urban land transformation on net primary productivity in the United States. *Remote Sensing Environment*. 89, 434–443.

Nizeyaimana, E., Petersen, G.W., Imhoff, M.L., Sinclair, H., Waltman, S., Reed- Margetan, D.S., Levine, E.R., Russo, P., 2001. Assessing the impact of land conversion to urban use on soils with different productivity levels in the USA. *Soil Science Soc*. 65, 391–402.

Pielke, S.R.A., Walko, R.L., Steyaert, L.T., Vidale, P.L., Liston, G.E., Lyons, W.A., Chase, T.N., 1999. The influence of anthropogenic landscape changes on weather in south Florida. Mon. *Weather Rev*. 127, 1663–1673.

Schimel, D., Enting, I.G., Heimann, M., 1995. CO2 and the carbon cycle. In: Climate Change 1994, *Intergovernmental Panel on Climate Change*, Cambridge UniversityPress, Cambridge.

Schimel, D., Melillo, J., Tian, H.Q., 2000. Contribution of increasing CO2 and climate to carbon storage by ecosystems in the United States. *Science*, 287 (5460), 2004– 2006.

Wackernagel, W., Yount, D., 1998. The ecological footprint: an indicator of progress toward ecological sustainability. *Environment Monitoring Assessment*. 51, 511–529.

Wear, D. N., Greis, J. G., 2001. The Southern Forest Resource Assessment Draft Summary Report. http://www.srs.fs.fed.us/sustain.

Willem JD, Barron J, Stuart E and Stefanie M (2006) Multi-sensor NDVI data continuity: Uncertainties and implications for vegetation monitoring applications. *Remote Sensing of Environment* 100: 67–81

Yu, D.Y., Pan, Y.Z., Liu, X., Wang, Y.Y., Zhu, W.Q., 2006. Ecological capital measurement by remotely sensed data for Huahou and its socio-economic application. *Plant Ecology*. 30 (3), 404–413 (in Chinese, with English abstract)

# 5

# A FUZZY AUTOMATIC CAR DETECTION METHOD BASED ON HIGH RESOLUTION SATELLITE IMAGERY AND GEODESIC MORPHOLOGY

N. Zarrinpanjeh [a, *], F. Dadrassjavan [b]

[a] Department of Geomatics Engineering, Qazvin Branch, Islamic Azad University, Qazvin, Iran - nzarrin@qiau.ac.ir
[b] School of Surveying and Geospatial Engineering University College of Engineering University of Tehran - fdadrasjavan@ut.ac.ir

KEY WORDS: Car Detection, Fuzzy Inference System, Geodesic Morphology, Satellite Imagery

ABSTRACT:

Automatic car detection and recognition from aerial and satellite images is mostly practiced for the purpose of easy and fast traffic monitoring in cities and rural areas where direct approaches are proved to be costly and inefficient. Towards the goal of automatic car detection and in parallel with many other published solutions, in this paper, morphological operators and specifically Geodesic dilation are studied and applied on GeoEye-1 images to extract car items in accordance with available vector maps. The results of Geodesic dilation are then segmented and labeled to generate primitive car items to be introduced to a fuzzy decision making system, to be verified. The verification is performed inspecting major and minor axes of each region and the orientations of the cars with respect to the road direction. The proposed method is implemented and tested using GeoEye-1 pansharpen imagery. Generating the results it is observed that the proposed method is successful according to overall accuracy of 83%. It is also concluded that the results are sensitive to the quality of available vector map and to overcome the shortcomings of this method, it is recommended to consider spectral information in the process of hypothesis verification.

## 1. INTRODUCTION

Since the advent of High Resolution Satellite Imagery (HRSI) and the spread of satellite image acquisition systems, researchers have been enthusiastic in finding approaches to use information from these sources to detect and recognize objects. The number of detectable objects in satellite images, have increased due to the improvements in spatial and spectral resolutions in recently developed satellite systems. Cars are of the most desired items for object recognition which are found noticeable inspecting urban areas in many aspects such as traffic monitoring and road extraction (Hinz, 2005). The problem of computing the number of cars passing a route is important to be solvable in all passages throughout the city while direct systems are only available on major roads. On the other hand, solving the problem of road extraction is fortified when violating objects such as cars at the road scene are detected and removed. Cars are detectable using both aerial and also satellite images. In aerial images cars are usually reflected into rectangular regions with 13-26 pixels in length and width(with respect to image scale), but in 0.5 meter Ground Sample Distance (GSD) satellite imageries, cars are reflected as 9×5 pixel objects (Eikvil, 2009).

Car detection from scanned aerial images was firstly studied and practiced before any attempt to extract cars from HRSI was made (Hinz, 2005). In such approaches each pixel usually has 0.15 meter GSD in average. These approaches either use implicit or explicit vehicle models. The appearance-based implicit model uses sample images of vehicles to derive grey-value or texture features and their statistics assembled in vectors. These vectors are used as reference to test computed feature vectors from image regions. In addition, training images

of cars are extracted and introduced. Then, through computing some distinguishing features, the classification process is performed and cars are extracted. Approaches using an explicit model, describe vehicles in 2D or 3D by filter or wire-frame representations. The model is then matched "top-down" to the image or extracted image features are grouped "bottom-up" to create structures similar to the model. A vehicle is detected, if there is sufficient support of the model in the image (Hinz, 2005).

Regarding to HRSI and their specific resolutions, spatial and grey value features are extracted for car detection and manual/automatic classification solutions are studied and overall performance of the approach is evaluated (Eikvil, 2009). In another study, spot detectors are used to extract car hypotheses and considering vector map of the region and objects' direction, true car hypotheses are verified (Stilla, 2004). An encouraging approach for single vehicle detection uses morphological filtering to distinguish between vehicle pixels and similar non-vehicle pixels (Michaelsen, 2001). This approach achieves high completeness and correctness but is not able to extract vehicles in queues or parking lots. Moreover, some thresholding approaches are also tested for car detection using HRSI (Jin, 2004).

In some recent researches vehicle detection is done using high-resolution aerial images through a fast sparse representation classification method and a multi order feature descriptor that contains information of texture, colour, and high-order context. To speed up computation of sparse representation, a set of small dictionaries, instead of a large dictionary, containing all training items, is used (Chen, 2016). Vehicle detection is also practiced using a probabilistic classification method followed by a

---

\* Corresponding author

refinement, based on object segments using registered aerial RGB images and airborne LiDAR data. Pixel-wise vehicle probability estimation is achieved using Gaussian process (GP) classification and object segments are obtained by applying a gradient based segmentation algorithm (GSEG)(Liu, 2016).

In this paper, morphological operators and more specifically, Geodesic dilation for the purpose of car detection are introduced and tested. In the next sections, introducing Geodesic dilation, the proposed car detection approach is explained. Then, using sample satellite images and vector maps the proposed method is developed and tested and finally, the results of the study are discussed.

## 2. GEODESIC MORPHOLOGY FOR CAR DETECTION

Mathematical morphology is a tool for extracting image components that are useful in the representation and description of region shape such as boundaries and convex hull (Gonzalez 2003). Geodesic dilation is one of the processing that seems to be useful for robust car detection. Morphological Geodesic dilation employs two input image known as marker and mask images. Both images must have the same size and the mask image must have intensity values greater or equal to the marker image. The marker image is dilated by an elementary isotropic structuring element and the resulting image is forced to remain below the mask image (Arefi 2005). In equation 1 the marker image is denoted by $J$ and the mask image by $I$. The Geodesic dilation of size $1$ of the marker image $J$ with respect to mask image $I$ is defined as

$$\delta_I^{(1)}(J) = (J \oplus B)^\wedge I \tag{1}$$

In this equation $\wedge$ stands for the point-wise minimum operation between the dilated marker image and the mask image and $J \oplus B$ is the dilation of $J$, using the elementary isotropic structuring element $B$. The Geodesic dilation of size $n$ of the marker image $J$ with respect to a mask image $I$ is obtained by performing n successive Geodesic dilations of size $1$ of $J$ with respect to $I$ (Arefi, 2005).

$$\delta_I^{(n)}(J) = \delta_I^{(1)}(J) {}^\circ \delta_I^{(1)}(J) {}^\circ \delta_I^{(1)}(J) {}^\circ \dots {}^\circ \delta_I^{(1)}(J) \tag{2}$$

The desired reconstruction is achieved by carrying out Geodesic dilations until stability is reached. In other words, this process can be thought of as repeated dilations of the marker image until the contour of the marker image fits under the mask image. When further dilations do not change the marker image any more, the iterative process is finished (Arefi, 2005).

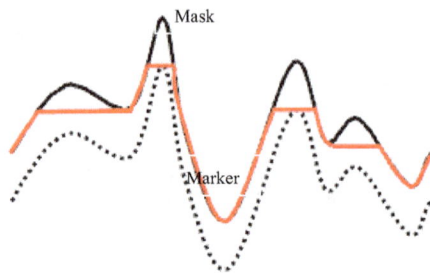

Figure 1. Mask and marker images in Geodesic dilation.

## 3. CAR DETECTION BASED ON GEODESIC DILATION

As previously mentioned this paper is dedicated to detection and recognition of cars based on HRSI and roads vector maps using morphological operators. The flowchart of the proposed method is presented in figure 2.

At the first layer of processing, roads are extracted overlaying vector map and HRSI. Therefore, nodes from vector map are projected to the image and pixels under each road segment is extracted. Although sometimes available maps are not updated or may suffer from some geometrical distortions to avoid distractions from the basic goals of this research road maps are assumed to be error free.

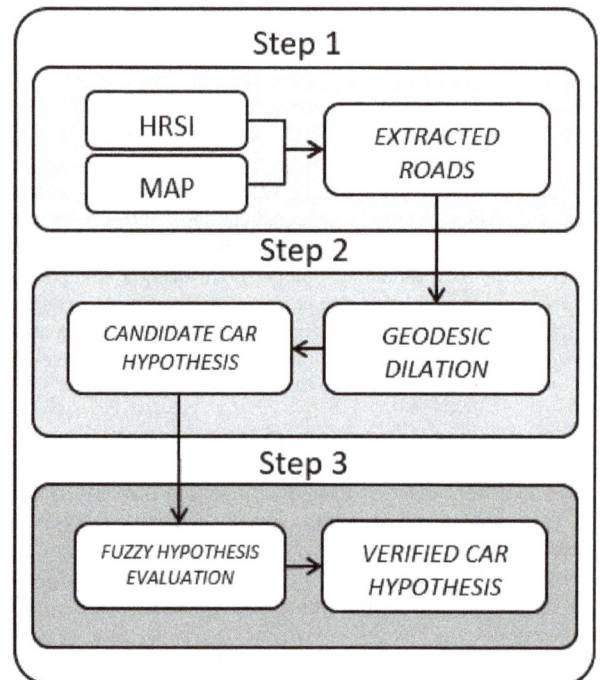

Figure 2. The flowchart of the proposed method.

At the next step, Geodesic dilation is applied on the image of each extracted road. Thresholding the results, candidate car pixels are extracted. Then, as cars are supposed to be recognized by the shape and size in this specific study, the results of the previous process should go through segmentation to label each connected candid pixels into candid regions to be evaluated and verified in the final step of processing.

At the final step, the labelled regions are introduced to a fuzzy decision making engine to evaluate and verify regions which are assumed to be car items. This can take place according to the shape features and intensity values of each region. This process is established through inspecting three main shape features of each region; major and minor axes and orientation. As it is predicted, a car has a specific size with respect to sensor GSD and is supposed to lie towards specific orientation considering centreline axis direction of the road. Using a fuzzy solution, car hypothesis verification is performed. The mentioned fuzzy solution takes shape features as input variables and produces the possibility of true detection as output. Input and output parameters, linguistic variables and labels in depicted in Table 1. Fuzzy membership functions of the input and output variables are depicted in figure 3. The defined rules for the fuzzy inference system are listed in Table 2.

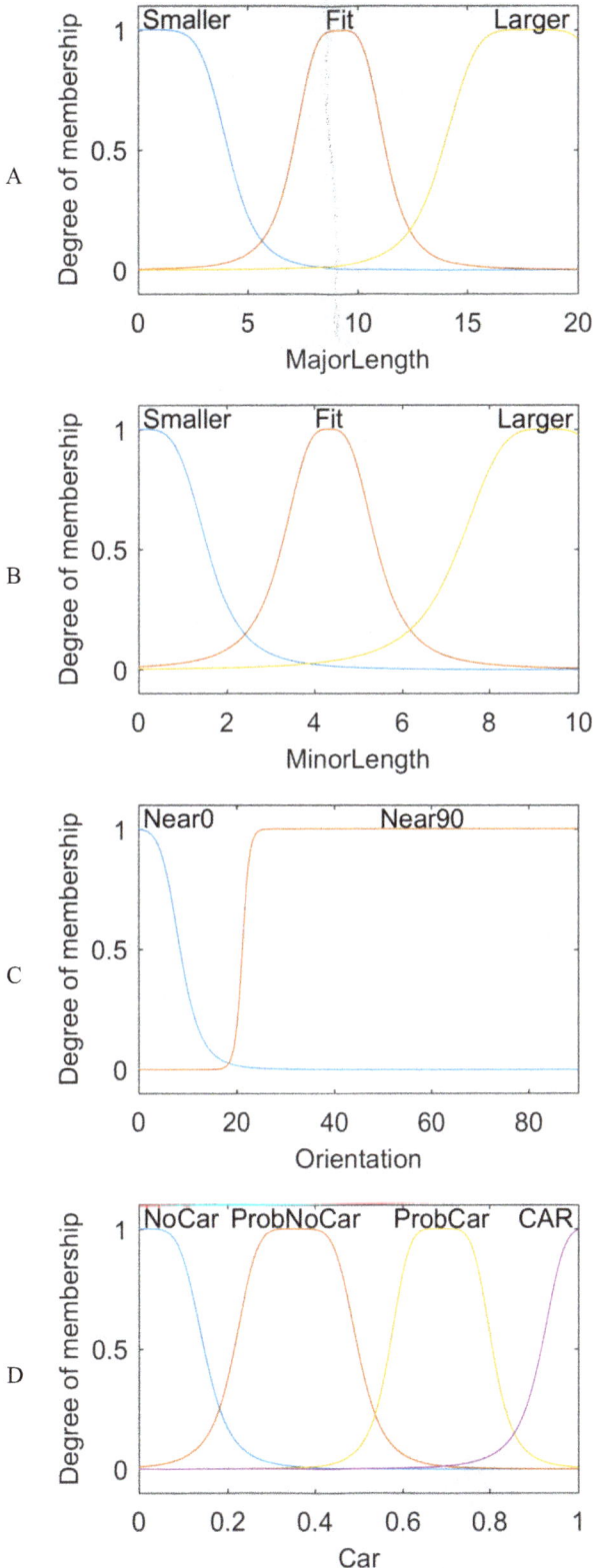

A

B

C

D

Figure 3. a) Membership function of Major axis length of the region. b) Membership function of Minor axis length of the region. c) Membership function of Orientation of the major axis. d) Membership function of output Car.

Table 1. Linguistic variables and labels for the designed Fuzzy Inference system.

| Linguistic variable | Linguistic Labels |
|---|---|
| Input #1: Major Length | Smaller- Fit- Larger |
| Input #2: Minor Length | Smaller- Fit- Larger |
| Input #3: Orientation | Near 0 – Near 90 |
| Output: CAR | NoCar-ProbNoCar-ProbCar-Car |

Table 2. Fuzzy rules for candidate car verification.

| Rules |
|---|
| 1. If (MajorLength is Fit) and (MinorLength is Fit) then (Car is CAR) (1) |
| 2. If (MajorLength is Smaller) and (MinorLength is Fit) and (Orientation is Near0) then (Car is ProbCar) (1) |
| 3. If (MajorLength is Larger) and (MinorLength is Fit) and (Orientation is Near0) then (Car is ProbCar) (1) |
| 4. If (MajorLength is Fit) and (MinorLength is Smaller) and (Orientation is Near0) then (Car is ProbCar) (1) |
| 5. If (MajorLength is Fit) and (MinorLength is Larger) and (Orientation is Near0) then (Car is ProbCar) (1) |
| 6. If (MajorLength is Smaller) and (MinorLength is Smaller) and (Orientation is Near90) then (Car is NoCar) (1) |
| 7. If (MajorLength is Larger) and (MinorLength is Larger) and (Orientation is Near90) then (Car is NoCar) (1) |
| 8. If (MajorLength is Smaller) and (MinorLength is Larger) and (Orientation is Near0) then (Car is ProbNoCar) (1) |
| 9. If (MajorLength is Larger) and (MinorLength is Smaller) and (Orientation is Near0) then (Car is ProbNoCar) (1) |
| 10. If (MajorLength is Fit) and (MinorLength is Fit) and (Orientation is Near90) then (Car is ProbCar) (1) |
| 11. If (MajorLength is not Fit) and (MinorLength is Fit) and (Orientation is Near0) then (Car is ProbCar) (1) |
| 12. If (MajorLength is Fit) and (MinorLength is not Fit) and (Orientation is Near0) then (Car is ProbCar) (1) |
| 13. If (MajorLength is Smaller) and (MinorLength is Smaller) then (Car is ProbNoCar) (1) |
| 14. If (MajorLength is Larger) and (MinorLength is Larger) then (Car is ProbNoCar) (1) |
| 15. If (MajorLength is Smaller) then (Car is ProbNoCar) (1) |
| 16. If (MajorLength is Larger) then (Car is ProbNoCar) (1) |
| 17. If (MinorLength is Smaller) then (Car is ProbNoCar) (1) |
| 18. If (MinorLength is Larger) then (Car is ProbNoCar) (1) |

## 4. IMPLEMENTATION AND RESULTS

To inspect the capabilities of the proposed method for car detection using HRSI, it is implemented and tested using GeoEye-1 pan-sharpen satellite image of urban area in the city of Tehran. The road map of roads of the corresponding region is also available at the scale of 1:2000. The overlaying road map and satellite image road regions with respect to each element in the map are extracted. The test data set and overlaid vector map are presented in figure 4.

As experimented, a number of 208 road segments are individually located and extracted in the test dataset. In figure 5, a sample extracted road segment from satellite imagery is illustrated. The main achievement of this process could be summarized in minimizing the search space for car detection as it is not expected to find cars in non-road regions. This directly reduces the number of candidate cars in non-road regions which have higher possibility of being rejected during the verification process. From another point of view it is important to detect cars in the region near roads and streets usually according to the cardinal goals of car detection such as traffic monitoring.

Figure 4. Satellite image and overlaid vector map.

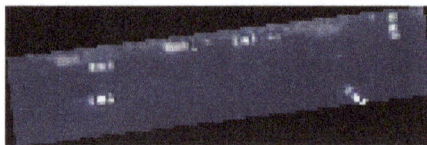

Figure 5. Extracted road segments.

Each extracted road segments are then inspected for car primitives. Geodesic dilation is applied on each segment of extracted roads. The extracted image is considered as mask and therefore reduced smoothed image is considered as marker. Then, Gaussian smoothing filter is applied on the image and digital number of each pixel is reduced by 50 values to generate marker image in 11 bit depth satellite image. Geodesic dilation is repeatedly applied until the results reach stability. The results of Geodesic dilation are presented in figure 6. Thresholding the results, with a threshold set to 30% of the maximum peak, pixels which have high probability of being a car item are marked. To reduce noise, the resulted thresholded image is morphologically closed by a 3x3 structure element.

The resulted candidate pixels are then checked with eight neighbour connectivity and connected elements are labelled, so car primitives are generated. Shape properties of each labelled region is computed and in accordance with data resolution optimum size and direction of objects are extracted and inserted into a fuzzy system via membership functions and rules to provide the capability of car hypothesis verification. The results of each stage of this process is depicted in figure 6.

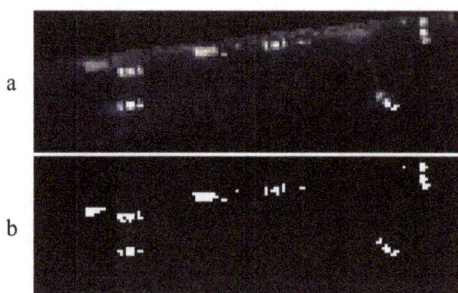

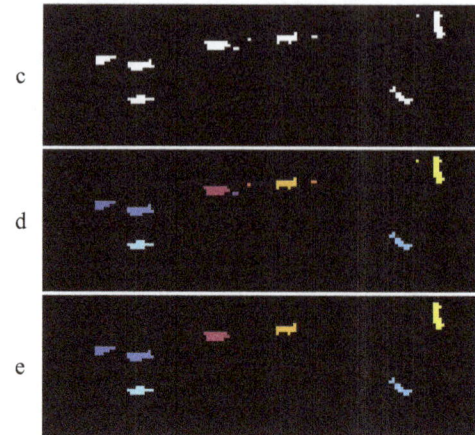

Figure 6. a) The results of Geodesic dilation on test image b) Thresholding of the Geodesic Dilation, c) Closed Geodesic dilation by a 3x3 isotropic structure element, d) Results of segmentation and labelling and candidate car items generation. e) Verified car items.

Table 3 demonstrates the final results of car detection through confusion matrix. Considering the mentioned confusion matrix the overall accuracy of 83.38% is achieved.

Table 3: Confusion matrix of the proposed car detection.

|  | Car | No car | Sum | Commission Error |
|---|---|---|---|---|
| Car | 270 | 7 | 277 | 2.52 |
| No Car | 51 | 21 | 72 | 70.83 |
| Sum | 321 | 28 | 349 |  |
| Omission Error(%) | 15.88 | 25 |  |  |

### 4.1   Discussions

The followings briefly address the benefits, problems, achievements and shortcomings of the proposed method:

Geodesic dilation presents less redundant results in comparison with classic morphological operations such as top-hat filtering in terms of extracting primitives. As depicted in figure 7, road surface is covered by noise in Top-hat filtering while on the other hand the surface of the road is entirely removed and only car or some similar other objects remained. In addition, the effect of moving cars in Geodesic dilation image is vividly visible while no trace of such object is detectable in classic Top-Hat filtering. This means that detecting moving cars is potentially applicable through Geodesic dilation.

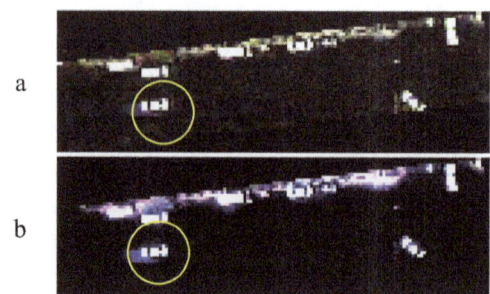

Figure 7. Comparison of Top Hat with Geodesic dilation.a) Top-Hat filtering results. b) Geodesic dilation results.

The proposed car detection method is successful in marking objects making high grey value contrast to road surface. In other words, in case of detecting dark cars an increase in the rate of errors is observed. This usually happens when car region is not acquired completely due to low luminance or shades as shown in figure 8.

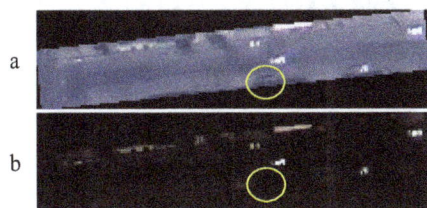

Figure 8. Dark cars are missed due to low illumination. a)Test image. b) Missed Car.

Car detection is weak when non-road areas are inspected. As a matter of fact, there are many similar objects to items when building or bare-land areas are inspected. This indicates that using cars for detecting roads from building regions is not recommendable. As depicted in figure 9, it is observed that efficient car detection is closely bounded to how updated and accurate the geometric information from vector map is. Car detection is not conducted appropriately if road regions are mistakenly chosen.

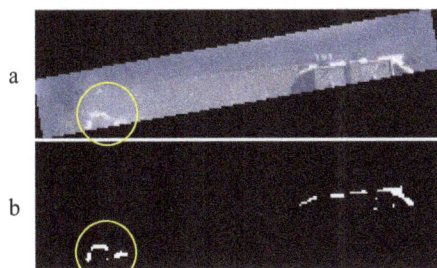

Figure 9. False car detection due to investigating non-road areas. a)Mistakenly extracted road. b) Mistakenly extracted car pixels.

As each road has its specific width, the extraction of search area should be accomplished with respect to each roads specific width range. As illustrated in figure 10, the absence of compatibility between search area and road width, non-road areas would be inspected and car detection would inevitably results in false hypothesis verification. Therefore the existence of a GIS ready vector road map with enriched attribute would facilitate true car detection.

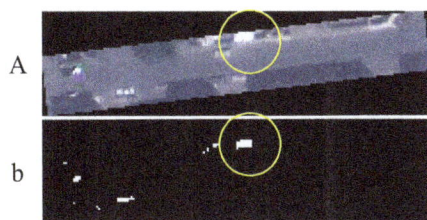

Figure 10. False hypothesis verification due to road width incompatibility.

## 5. CONCLUSIONS

In this paper, it is attempted to inspect and develop an automatic car detection technique based on HRSI using Geodesic morphology. The proposed method is successful in automatic car detection when cars are vividly noticeable and brightly detectable and also when vector map accurately locates roads. For a more robust detection of cars, it is recommended to use Spectral information to make more robust distinguishes between car and non-car objects. This can be inserted to the fuzzy engine when car primitives are being evaluated. An updated road map helps more robust car detection. Therefore implication of an updated vector map is recommended.

## REFERENCES

Arefi, H., Hahn, M., 2005. A morphological reconstruction algorithm for separating off-terrain points from terrain points in laser scanning data. In: Proceedings of the ISPRS Workshop Laser Scanning", 2005, pp. 120-125.

Chen, Z., Wang, C., Luo, H., Wang, H., Chen, Y., Wen, C., Yu, Y., Cao, L. and Li, J., 2016. Vehicle Detection in High-Resolution Aerial Images Based on Fast Sparse Representation Classification and Multiorder Feature. IEEE Transactions on Intelligent Transportation Systems, 17(8), pp.2296-2309.

Eikvil, L., Aurdal, L., Koren, H., 2009. Classification-based vehicle detection in high-resolution satellite images. ISPRS Journal of Photogrammetry and Remote Sensing, 64(1), pp. 65-72.

Gonzalez, R. C., Woods, R. E. , 2003. Digital Image Processing. Second. Edition.

Hinz, S., 2005. Detection of vehicles and vehicle queues for road monitoring using high resolution aerial images", In: The 9th World Multiconference on Systemics, Cybernetics and Informatics July 10 - 13, 2005

Jin, X.,  Davis, C.H., 2004. Vector-guided vehicle detection from high resolution satellite imagery. In: Geoscience and Remote Sensing, Symposium, IGARSS '04, Proceedings 2004 IEEE International , Vol. II, pp.1095 -1098.

Liu, Y., Monteiro, S.T. and Saber, E., 2016, July. Vehicle detection from aerial color imagery and airborne LiDAR data. In Geoscience and Remote Sensing Symposium (IGARSS), 2016 IEEE International, pp. 1384-1387

Michaelsen, E., Stills, U., 2001. Estimating Urban Activity on High-Resolution Thermal Image Sequences Aided by Large Scale Vector Maps. In: IEEE/ISPRS Joint Workshop on Remote Sensing and Data Fusion over Urban Areas, pp. 25 – 29.

Stilla, U., Michaelsen, E., Soergal, U., Hinz, S., Ender, HJ., 2004. Airborne Monitoring of vehicle activity in urban areas. In: Altan MO (ed) International Archives of Photogrammetry and Remote Sensing, pp. 973-979.

6

# QUANTITATIVE RISK MAPPING OF URBAN GAS PIPELINE NETWORKS USING GIS

P. Azari [a]*, M. Karimi [a]

[a] Faculty of Geodesy and Geomatics Engineering, K.N.- Toosi University of Technology, Tehran, Iran
(peymanazari1993@gmail.com; mkarimi@kntu.ac.ir)

**KEY WORDS:** Urban Gas Pipeline, Quantitative Risk, GIS, Risk Mapping

**ABSTRACT:**

Natural gas is considered an important source of energy in the world. By increasing growth of urbanization, urban gas pipelines which transmit natural gas from transmission pipelines to consumers, will become a dense network. The increase in the density of urban pipelines will influence probability of occurring bad accidents in urban areas. These accidents have a catastrophic effect on people and their property. Within the next few years, risk mapping will become an important component in urban planning and management of large cities in order to decrease the probability of accident and to control them. Therefore, it is important to assess risk values and determine their location on urban map using an appropriate method. In the history of risk analysis of urban natural gas pipeline networks, the pipelines has always been considered one by one and their density in urban area has not been considered. The aim of this study is to determine the effect of several pipelines on the risk value of a specific grid point. This paper outlines a quantitative risk assessment method for analysing the risk of urban natural gas pipeline networks. It consists of two main parts: failure rate calculation where the EGIG historical data are used and fatal length calculation that involves calculation of gas release and fatality rate of consequences. We consider jet fire, fireball and explosion for investigating the consequences of gas pipeline failure. The outcome of this method is an individual risk and is shown as a risk map.

## 1. INTRODUCTION

One of the important components of an urban infrastructure for transmission and distribution of gas is natural gas pipeline network. With growing modern cities, the usage of natural gas as a major source of energy has become more (Ma, Li et al. 2013). Natural gas network consists of three main parts:

- Gathering pipelines: The work that is done in this part is gathering and transmitting gas from production site to central collection point. It can be considered as an initial step in natural gas production.
- Transmission pipelines: This section of natural gas pipeline network undertake two tasks, firstly, it transmits gas in order to refine, process and store in the storage facilities and secondly, it transmits gas from storage facilities to consumers or distribution part.
- Distribution pipelines: Natural Gas Distribution Company is responsible for receiving gas from transmitting pipelines and distributing it to end users. The end users can be commercial consumers or residential ones.

Figure 1 shows the components of natural gas pipeline networks and their processes. The starting point of these pipelines is in production site and the ending point is in the house of consumers. Distribution pipelines pass in urban area and there are many people and building in the vicinity of them. Therefore occurring accident for distribution pipelines will cause damage to people and their property. On the other hand, the frequency of natural gas distribution networks accident is relatively higher than the other pipelines and it involves 80 percent of all the accidents that occur in natural gas pipelines (Amir-Heidari, Ebrahimzadih et al. 2014). In this paper we investigated the risk of distribution pipelines.

By increasing urbanization in large cities, the demand for natural gas as an important energy has increased. Growing demand has led to dense urban natural gas pipeline networks. So, it is important to design urban gas pipeline networks carefully (Ma, Cheng et al. 2013). It is obvious that any infrastructure development in urban regions needs protection and maintenance. Therefore it is better to design various methods to prevent the probability of accident in urban natural gas pipeline networks. In order to better management of accidents and to precisely predict them, it is important to create risk map for pipeline network. Security of life and property in city life is another need for people. Therefore, the issue of safety and security of urban natural gas pipeline networks must be adequately investigated (Vianello and Maschio 2014). So risk mapping of urban natural gas pipeline networks is an important issue for risk analysis and preventing dangers for better management.

In this paper, a quantitative risk assessment for urban natural gas pipeline networks is implemented. In the second section of this paper, the history of risk assessment in gas pipeline networks is presented. The methodology and the steps to calculate risk and mapping are presented in section 3. In section 4 implementation results on a sample pipeline network are presented and in the last section conclusions are discussed.

## 2. RELATED WORKS

Many studies have been performed about risk analysis of urban natural gas pipeline networks. Jo and Ahn (2002) studied area of hazard that created from rupture of high pressure pipelines. In that study hazard distance determined between 20 meters for pipelines with low pressure and small diameter, to 300 meters for high pressure pipeline with large diameter (Jo and Ahn 2002). Jo, Park and Ahn analysed risk quantitatively and introduced fatal length and cumulative fatal length parameters. Results of their work were individual and social risk (Jo, Park et al. 2004). Jo and Ahn proposed a new approach for quantitative risk analysis and they applied the approach to a pipeline with one meter diameter and 50 bar pressure in depth of 130 cm from ground. In their study they used geometry of pipelines and population density for risk analysis (Jo and Ahn 2005). A comparison between quantitative and qualitative risk analysis has been done and implemented on small and large urban regions. In the quantitative approach many consequences are considered so the outcome has a high precision and in the qualitative method many reasons of failure considered and it is more effective (Han and Weng 2011).

Ma et al. focused on a new method of quantitative risk assessment for urban natural gas pipeline networks based on grid difference of pipeline sections (GDPSs). They used graph concepts for creating relations between pipelines and stations. For any point of the region a number was determined as an individual risk and after that contour lines were created using ArcGIS. The results indicated that more pipelines produce more risk and it was shown in contour lines (Ma, Li et al. 2013). After that, they also presented a novel method for quantitative risk assessment for urban gas pipeline networks using geographical information systems. This method consists of three sections: calculation of failure rate, quantitative analysis model of accident consequences and determination of individual and societal risks. GIS has an important role in better management and controlling accidents (Ma, Cheng et al. 2013).

In 2014, risk assessment for distribution gas network of Iran was performed. They focused on the consequences such as jet fire and explosion similar to Jo and Ahn (2005) and by the presented method, they determined individual and societal risk (Amir-Heidari, Ebrahimzadih et al. 2014).

Vianello and Maschio assess risk of distribution network of Italian, quantitatively. Their method consists of three parts: description of the system, risk identification, estimation of failure frequency and estimation of consequences. For the failure, they considered three cases as full, medium and small rupture and because of not penetrating of gas to ground level they ignored small case. They used PHAST software for simulation of accidents (Vianello and Maschio 2014).

A key problem with much of the literature in relation to risk analysis of urban natural gas pipelines is the question that how several pipelines influence on the risk value of an area. This raises many questions about whether the risk around a single main pipeline is higher or several secondary pipelines. Within the framework of these criteria we tried to investigate the relation between risk value and the density of pipelines in a study area.

## 3. METHODOLOGY

As mentioned in the previous section, urban gas pipeline risk assessment is crucial to better manage the incidents and to prevent them, as well as for more efficient urban planning. There are three methods for risk assessment of gas pipelines: quantitative, qualitative and hybrid. Qualitative risk assessment consists of indexing system. In this method, there are three index levels: causation index, an inherent risk index, a consequence index and their corresponding weights. The outcome of the qualitative method is a qualitative risk value and relatively presented. The quantitative method consists of a probability assessment, a consequence analysis and a risk evaluation. The outcome of this method is individual and societal risk (Han and Weng 2011). Generally, risk is defined as a mathematical function that calculates probability of a pipeline rupture and the magnitude of death (Ma, Li et al. 2013). Risk has various types such as individual risk, societal risk, economic risk and average rate of death. Individual risk and societal risk are the most important types of risk and we implemented a method to calculate individual risk for the entire pipelines in an urban pipeline networks.

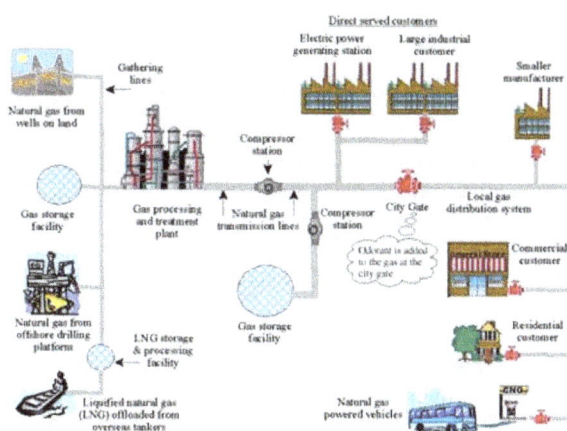

Figure 1. Natural Gas Pipeline Network Components

In this paper we focus on qualitative risk assessment of urban natural gas pipeline networks. In Figure 2, we represent the process of quantitative risk assessment for urban natural gas pipeline networks.

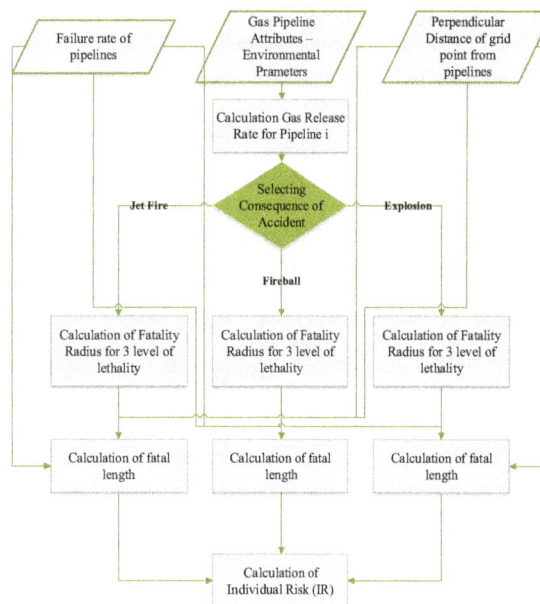

Figure 2. Process of quantitative risk assessment

### 3.1 Calculation of gas release rate

The process of leakage is an isentropic adiabatic expansion process and for any failure type, release rate from gas pipeline can be calculated by leakage models. Gas release rate from natural gas pipelines occur after rupture and it may cause dangerous accidents. In order for quantitative risk assessment of urban natural gas pipeline networks, gas release rate should be calculated. There are many methods for this calculation. In this paper we use the method presented by Ma et al. in 2013.

For rupture case, Small Hole model is a model that is widely used in engineering. The amount of gas release depends on whether gas flow is sonic or subsonic. In order to determine that a gas flow is sonic or subsonic we can use equation 1 (Dong, Gao et al. 2003):

$$\begin{cases} \dfrac{P_0}{P_1} \leq \left(\dfrac{2}{k+1}\right)^{\frac{k}{k-1}} & sonic \\[3mm] \dfrac{P_0}{P_1} > \left(\dfrac{2}{k+1}\right)^{\frac{k}{k-1}} & subsonic \end{cases} \qquad (1)$$

Where $P_0$ = environmental pressure in Pa

$P_1$ = pressure inside the gas pipeline in Pa

k = adiabatic index (1.28 for natural gas)

After choosing that gas has sonic or subsonic flow we can calculate gas release rate using equation 2:

$$\begin{cases} Q = C_0 A P_1 \sqrt{\dfrac{kM}{RT}\left(\dfrac{2}{k+1}\right)^{\frac{k+1}{k-1}}} & sonic \\[5mm] Q = C_0 A P_1 \sqrt{\dfrac{2M}{RT}\left(\dfrac{k}{k-1}\right)\left[\left(\dfrac{P_0}{P_1}\right)^{\frac{2}{k}} - \left(\dfrac{P_0}{P_1}\right)^{\frac{k+1}{k}}\right]} & subsonic \end{cases} \qquad (2)$$

Where Q = gas release rate in kg/s

A = area of the leakage opening in $m^2$

M = molecular weight of gas in kg/mol (0.016 kg/mol)

R = gas constant (8.314 J/mol K)

T = temperature of gas inside the pipeline in K

$P_0$ = environmental pressure in Pa

$P_1$ = pressure inside the gas pipeline in Pa

k = adiabatic index (1.28 for natural gas)

### 3.2 Failure rate

Failure rate of a pipeline is the number of times the pipeline failure occurs per unit length every year, assuming that all of conditions along the pipeline are equal (Jo and Ahn 2005). There are numerous methods for assessment of failure rate such as AHP, Fault Tree Model, Event Tree Analysis and empirical formula based on historical observations (Ma, Cheng et al. 2013). For any of failure cause there is a failure rate that can be estimated by equation 3 (Amir-Heidari, Ebrahimzadih et al. 2014):

$$\varphi_i = \sum_j \varphi_{i.j.0} K_j(a_1. \, a_2. \, a_3. \, \cdots) \qquad (3)$$

Where $\varphi_i$ = failure rate of pipeline network that is caused by failure type $i$

$i$ = assumed failure type (small hole, large hole and fracture)

$j$ = reasons for the failure (including external interference, construction defects, corrosion, ground movement, and others)

$\varphi_{i.j.0}$ = probability of different failure types resulting from specific failure causes

$K_j$ = modifying equation of corresponding failure causes

$a_1$ = parameters of modifying equation

It should be considered that failure rate changes with changing the conditions along the pipeline and is not equal. Therefore, the first step for calculating failure rate is dividing pipeline to sections that has equal conditions. Parameters that have influence on varying condition are soil, design, age of pipeline, depth of cover and others. But considering all of these parameters in our calculation is not possible due to the lack of statistical data. So, we use historical data. In the Table 1, reasons of failure and failure rates are shown:

| Failure causes | Failure rate (1/year km) | Percentage (%) | Rates of occurrence of different hole size (%) | | |
|---|---|---|---|---|---|
| | | | Small | Medium | Large |
| External interference | $1.8 \times 10^{-4}$ | 49.6 | 25 | 56 | 19 |
| Construction defects | $6.5 \times 10^{-5}$ | 16.5 | 69 | 25 | 6 |
| Corrosion | $6 \times 10^{-5}$ | 15.4 | 97 | 3 | < 1 |
| Ground movement | $2.5 \times 10^{-5}$ | 7.3 | 29 | 31 | 40 |
| Other factors | $4 \times 10^{-5}$ | 11.2 | 74 | 25 | < 1 |
| Total failure rate | $3.7 \times 10^{-4}$ | 100 | 48 | 39 | 13 |

Table 1. Examples of different causes of failure and the corresponding rates of failure types (EGIG 2008)

### 3.3 Consequences of pipeline ruptures

Distribution network in urban region have a dense pipelines with high and low pressure. Because of flammable nature of gas in pipelines, there are many consequences that may occur in an accident. The consequences of natural gas pipeline accidents are usually: thermal radiation of fire and confined explosion.

#### 3.3.1 Fire

One of the most important and most common adverse events that led to a serious danger to human life in the process industries is fire phenomenon. Fire contains a chemical reaction in which a combustible substance combines with oxygen and a huge amount of energy releases from this reaction. Fire phenomena include 4 types:

- Pool fire: as a result of leakage or rupture in gas pipeline, the substances in those pipelines drain outside. If the fluid is flammable, it begins to vaporize and form a gas cloud around leakage point. If this cloud fires, pool fire occurred.
- Jet fire: Drain fluid under pressure from a small opening on an equipment containing flammable materials, creates jet of fluid that when arrives to the ignition source, creates a continuous fire that is jet fire. Thermal radiation flux at a specific point is as equation 4 (Ma, Cheng et al. 2013):

$$I = \frac{\eta \tau_a Q_h H_c}{4\pi r'^2} \qquad (4)$$

Where I = thermal radiation flux in W/m2

$\eta$ = ratio of radiation heat to the heat released by the fire (0.2 for methane)

$\tau_a$ = atmospheric transmissivity (value = 1)

$H_c$ = combustion heat of natural gas in J/kg

r = distance between the target and the center of the flame zone

- Fire ball: Fire ball formation is possible when large amounts of flammable materials extend out in the presence of spark to the environment. Thermal radiation flux at a specific point is as equation 5 (Ma, Li et al. 2013):

$$I = \frac{F_r \Delta H_c m \tau}{4\pi R^2 t_f} \qquad (5)$$

Where I = thermal radiation flux in W/m2

$\tau$ = atmospheric transmissivity (value = 1)

$F_r$ = percentage of radiation for flammable gas (normally 90%)

$\Delta H_c$ = combustion heat of natural gas in J/kg
m = mass of the gas combustible gas cloud in kg
R = distance between the target and the center

- Flash fire: Short-term ignitions of flammable gases which are flammable limits are the sudden fire. The fire causes no formation of shock waves. The fire did not last long, more than a few tenths of a second.

### 3.3.2 Explosion

Pipeline rupture in an urban gas network has various consequences. Explosion can be considered as one of the important consequences. Pressure created by explosion can be extremely high and cause damage to people and buildings around the explosion point. Therefore its consequences should be considered in risk analysis of urban gas pipeline network. In the case of restricted and flammable vapor cloud that is mixed with air, the explosion is likely to occur. Using equation 6, we can calculate the volume of confinement that is created by explosion (Vianello 2011):

$$V = \pi \times L \times \left(\frac{d_{pipe}}{2} + d_{depth}\right)^2 \qquad (6)$$

Where L = length of pipeline
$d_{pipe}$ = internal diameter of pipeline
$d_{depth}$ = depth of pipeline under ground

### 3.4 Lethality rate calculation

For determining the danger for people surrounding the accident point that caused by pipeline rupture, we should calculate lethality rate for all of pipeline rupture consequences such as jet fire, fire ball and explosion. The effects of gas leakage (poisoning, thermal effect and explosion) on human beings can be shown by probability. The following probability unit function creates a relation between effects of pressure, heat and poisoning on people surrounding the accident point (Ma, Li et al. 2013):

$$P_r = a + b \ln D \qquad (7)$$

Where a, b = empirical constant
D = dose of the load for a given exposure time
The value of $P_r$ is between 2 to 9 (HSE 2010). This value is a probability value and we need to present it in a percentage of lethality. Equation 8 converts probability value to a percentage of morality:

$$P = \frac{1}{\sqrt{2\pi}} \int_{-\infty}^{P_r-5} e^{-\frac{s^2}{2}} ds \qquad (8)$$

Where $P_r$ = probability unit function
For classification of lethality we consider three levels for morality rate: 100% lethality, 50% lethality and 1% lethality. What is important to individual risk assessment is that all of consequences of pipeline accidents can be taken into account so that the result has a high degree of accuracy.
In 2013, Ma et al. received the following formula using the probability unit 7.33, 5 and 2.67 for calculation of radius of fatality 100%, 50% and 1%:

$$r_{jet,99} = 3.891\sqrt{Q}, r_{jet,50} = 5.498\sqrt{Q}, r_{jet,1} = 7.767\sqrt{Q} \quad (9)$$

Where Q = mass flow rate of leakage in kg/s.

$$\frac{(r_{Fireball,99})^{\frac{4}{3}}}{m^{1.106}} = 2.855, \frac{(r_{Fireball,50})^{\frac{4}{3}}}{m^{1.106}} = 4.518,$$

$$\frac{(r_{Fireball,1})^{\frac{4}{3}}}{m^{1.106}} = 7.149 \qquad (10)$$

Where m = mass of gas combustible gas cloud in kg.

$$\frac{r_{explosion,99}}{\sqrt[3]{m_{TNT}}} = 2.855, \frac{r_{explosion,50}}{\sqrt[3]{m_{TNT}}} = 2.861,$$

$$\frac{r_{explosion,1}}{\sqrt[3]{m_{TNT}}} = 3.017 \qquad (11)$$

Where $m_{TNT}$ is the TNT equivalent in kg that can be calculated by equation 12:

$$m_{TNT} = \frac{m_d \Delta H_d}{Q_{TNT}} \qquad (12)$$

Where $m_d$ = mass of the gas involved in the explosion in kg
$\Delta H_d$ = explosion heat of the gas in J/kg
$Q_{TNT}$ = the calorific value of the standard TNT explosion source (4.2 MJ/kg)

### 3.5 Fatal length calculation

Fatal length is a weighted length in a pipeline that if an accident occur for that pipeline, person at a specific location will die because of that effect (Jo and Ahn 2005). There is a simple way to calculate fatal length. Parameters that is necessary for calculating fatal length is radius of fatality for each consequences of pipeline rupture and the perpendicular distance of specific point to pipeline. We represented the relation between those parameters in Figure 3:

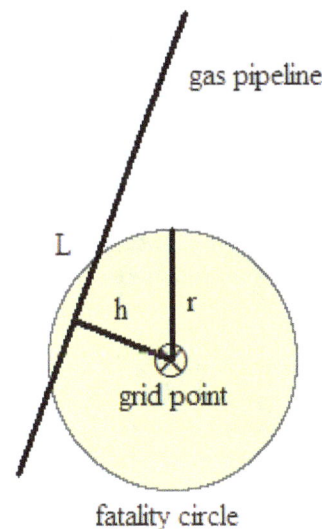

Figure 3. Relation Between Pipeline Parameters and Fatality Circle

Fatal length in each lethality level for any consequence can be estimated by equations 13 (Ma, Li et al. 2013):

$$l_{i,100-99} = 2\sqrt{(r_{i,99})^2 - h^2}, l_{i,99-50} = 2\sqrt{(r_{i,50})^2 - h^2},$$

$$l_{i,50-1} = 2\sqrt{(r_{i,1})^2 - h^2} \qquad (13)$$

Where i = consequence type of pipeline rupture

$r$ = fatality radius

$h$ = the perpendicular distance of specific point to pipeline

After calculation of fatal length for each lethality level, we should consider the average fatality of those zones. We use the following values and calculate final fatal length for every consequence similar to Ma et al. does in 2013:

$$L_{Fatal\ Length,jet\ fire}$$
$$\approx l_{jet\ fire,100-99} + 0.805 l_{jet\ fire,99-50}$$
$$+ 0.172 l_{jet\ fire,50-1}$$

$$L_{Fatal\ Length,fireball}$$
$$\approx l_{fireball,100-99} + 0.805 l_{fireball,99-50}$$
$$+ 0.172 l_{fireball,50-1}$$

$$L_{Fatal\ Length,explosion}$$
$$\approx l_{explosion,100-99} + 0.805 l_{explosion,99-50}$$
$$+ 0.172 l_{explosion,50-1}$$

## 3.6 Individual risk assessment

Risk assessment is a process that (https://www.ccohs.ca/oshanswers/hsprograms/risk_assessment.html):

- Identifies hazard and parameters of risk that create danger
- Analyses and evaluates the risk that is caused by the identified hazards
- Determines appropriate methods to get rid of those dangers or control them in case that they cannot be eliminated

In Figure 4 we show the main in risk assessment.

Figure 4. Risk Assessment Steps

There are various types of risk, such as individual risk, societal risk, economy risk and environmental risk. As stated in previous section we focus on individual risk assessment for natural urban gas pipeline network. Individual risk is defined as frequency of specific hazards that threat an unprotected person in a particular location (Jo and Ahn 2005). It is calculated by multiplying the probability of pipeline rupture by the magnitude of death at any particular position (x, y) using equation 15 (Ma, Li et al. 2013):

$$IR(x,y) = \sum_i \sum_{conequence} \int_{l_-}^{l+} \varphi_i P_i(x,y) dL \qquad (15)$$

Where $i$ = denote accident scenario

(x, y) = specific location

$\varphi_i$ = failure rate per unit length of the pipeline by accident scenario $i$

L = length of pipeline

$P_i$ = the lethality associated with accident scenario $i$

$l \pm$ = the ends of interacting section of pipeline

In equation 16 we use fatal length that makes calculation easier (Ma, Li et al. 2013):

$$IR(x,y) = \sum_i \varphi_i \sum_{consequence} L_{FL,i} \qquad (16)$$

Where $L_{FL,i}$ = fatal length associated with accident scenario $i$.

## 4. RESULTS

We proposed this approach in order to evaluate risk of urban gas pipeline on a small part of Tehran city. The pipeline networks consist of two types of pipeline: main pipeline with 12 inch diameter and secondary pipeline with 8 inch diameter. The gas pressure in these pipelines is different. The main pipeline has 250 psi pressure and the secondary pipeline 60 psi. In this paper we considered jet fire, fireball and explosion as the consequences of small failure of pipeline. The result of our processing is a risk map for pipelines that you can see in the Figure 5, Figure 6 and Figure 7. Blue pipelines are main ones and pipelines with black colour are secondary pipelines. In all of cases, we use regular grid points in order to determine risk of pipelines.

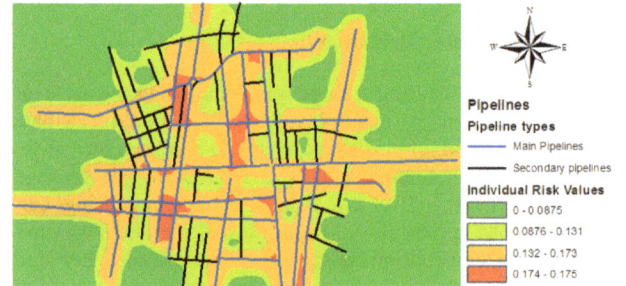

Figure 5 Individual Risk Map for Explosion

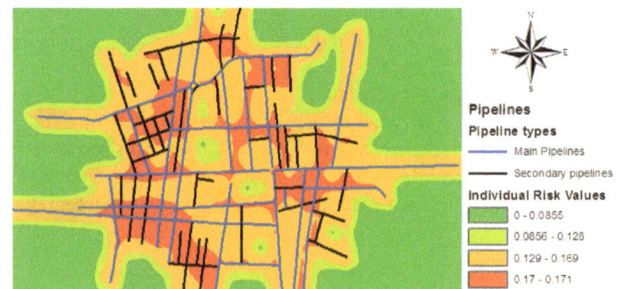

Figure 6 Individual Risk Map for Fireball

Figure 7 Individual Risk Map for Jet Fire

Figure 8 Individual Risk Map - All of Consequences

## 5. DISCUSSION AND CONCLUSION

Urban natural gas pipeline networks are vital infrastructure in large cities. The probability of occurring pipelines failure and accidents in these networks is high and therefor it is necessary to calculate risk of pipelines in order for urban planning and management. In this paper we implemented a quantitative method on a small part of the pipelines of Tehran.

All the analyses are investigated using regular grid points and the results showed that the risk value around pipelines is higher compared to the areas that are too far from pipelines. Our experiments are consistent with previous results (Jo and Ahn 2005, Ma, Li et al. 2013). In risk map for fireball we see that the values of risk are higher than the other two consequences and the values of explosion are higher than the jet fire. The probability of occurring accident when a rupture takes action is different for each consequence. This value for jet fire, fireball and explosion is 0.5, 0.3 and 0.2, respectively (Han and Weng 2011). Figure 8 shows individual risk for all of consequences by considering the probability of occurring accidents. It can be indicated that the risk around several secondary pipelines is less than around a single main pipeline. It showed that there is a relation between the pressure of pipelines and the value of risk around them. Main pipelines have larger pressure than secondary pipelines. So it is reasonable that the risk around main pipelines is larger than area of secondary pipelines.

Although our work is about prediction of risk in region and visualization of it, but for validation of our result we can select several random point in our region and calculate risk value of those points by using software such as ALOHA or Phast and comparing outputs with our results of proposed method in this paper.

The current study was not specifically designed to consider direction of wind and all of consequences of accidents. Future studies on the current topic are therefore required in order to investigate the effects of all the consequences and also wind direction on risk value of urban natural gas pipeline networks.

## REFERENCES

Amir-Heidari, P., M. Ebrahimzadih, H. Farahani and J. Khoubi (2014). "Quantitative Risk Assessment in Iran's Natural Gas Distribution Network." Safety Science and Technology **4**: 59-72.

Dong, Y. H., H. L. Gao, J. E. Zhou and Y. R. Feng (2003). "Mathematical modelling of gas release through holes in pipelines." Chemical Engineering **92**: 237-241.

EGIG, E. G. P. I. D. G. (2008). Gas pipeline incidents 7th report 1970-2007.

Han, Z. Y. and W. G. Weng (2011). "Comparison study on qualitative and quantitative risk assessment methods for urban natural gas pipeline network." Hazardous Materials **189**: 509-518.

HSE (2010). "Methods of Approximation and Determination of Human Vulnerability for Offshore Major Accident Hazard Assessment." Health and Safety Executive.

Jo, Y.-D. and B. J. Ahn (2002). "Analysis of hazard areas associated with high-pressure natural-gas pipelines." Loss Prevention in the Process Industries **15**: 179-188.

Jo, Y.-D. and B. J. Ahn (2005). "A method of quantitative risk assessment for transmission pipeline carrying natural gas." Hazardous Materials **123**: 1-12.

Jo, Y.-D., K.-S. Park and B. J. Ahn (2004). "Risk assessment for a high-pressure natural gas pipeline in an urban area."

Ma, L., L. Cheng and M. Li (2013). "quantitative risk analysis of urban natural gas pipeline networks using geographical information systems." Loss Prevention in the Process Industries **26**: 1183-1192.

Ma, L., Y. Li, L. Liang, M. Li and L. Cheng (2013). "A novel method of quantitative risk assessment based on grid difference of pipeline sections." Safety Science.

Vianello, C. (2011). Risk analysis of gas distribution network, Università degli Studi di Padova.

Vianello, C. and G. Maschio (2014). "Quantitative risk assessment of the Italian gas distribution network." Loss Prevention in the Process Industries **32**: 5-17.

# 7

# OPTIMIZING ENERGY CONSUMPTION IN VEHICULAR SENSOR NETWORKS BY CLUSTERING USING FUZZY C-MEANS AND FUZZY SUBTRACTIVE ALGORITHMS

Aref Ebrahimi [a], Parham Pahlavani [b*], Zohreh Masoumi [c]

[a] MSc. Student in GIS Division, School of Surveying and Geospatial Eng., College of Eng., University of Tehran, Tehran, Iran - aref.ebrahimi@ut.ac.ir
[b] Assistant Professor, Center of Excellence in Geomatics Eng. in Disaster Management, School of Surveying and Geospatial Eng., College of Eng., University of Tehran, Tehran, Iran - pahlavani@ut.ac.ir
[c] Assistant Professor, Department of Earth Sciences, Institute for Advanced Studies in Basic Sciences, Zanjan, Iran - z.masoumi@iasbs.ac.ir

**KEY WORDS:** Wireless Sensor Networks, Traffic Information Systems, Optimization, Clustering, Energy Consumption, Clustering, Fuzzy C-means Algorithm, Fuzzy Subtractive Algorithm

**ABSTRACT:**

Traffic monitoring and managing in urban intelligent transportation systems (ITS) can be carried out based on vehicular sensor networks. In a vehicular sensor network, vehicles equipped with sensors such as GPS, can act as mobile sensors for sensing the urban traffic and sending the reports to a traffic monitoring center (TMC) for traffic estimation. The energy consumption by the sensor nodes is a main problem in the wireless sensor networks (WSNs); moreover, it is the most important feature in designing these networks. Clustering the sensor nodes is considered as an effective solution to reduce the energy consumption of WSNs. Each cluster should have a Cluster Head (CH), and a number of nodes located within its supervision area. The cluster heads are responsible for gathering and aggregating the information of clusters. Then, it transmits the information to the data collection center. Hence, the use of clustering decreases the volume of transmitting information, and, consequently, reduces the energy consumption of network. In this paper, Fuzzy C-Means (FCM) and Fuzzy Subtractive algorithms are employed to cluster sensors and investigate their performance on the energy consumption of sensors. It can be seen that the FCM algorithm and Fuzzy Subtractive have been reduced energy consumption of vehicle sensors up to 90.68% and 92.18%, respectively. Comparing the performance of the algorithms implies the 1.5 percent improvement in Fuzzy Subtractive algorithm in comparison.

## 1. INTRODUCTION

Nowadays, transportation is a subject that people are associated with it. In addition, by the development of cities, the need for public services and facilities has been increased. This issue draws attentions to the subjects such as urban transportation (Seredynski and Bouvry, 2011).

Due to the increasing volume of urban traffic and its undesirable effects on economy, environment, and community health, effective management of urban traffic has become a remarkable subject. In recent years, the use of new technologies and techniques in management of urban traffic in most countries has been considered as the best solution to the metropolises' traffic volume. One of the newest and most effective techniques of traffic management, which takes the advantages of information technology, is to use intelligent transportation systems. Generally, the goal of design and implementation of a traffic management system is collecting, estimating, and propagating the traffic information. Particularly, in a cooperative traffic management system, this information is collected and transferred between vehicles (Bali et al., 2014).

Due to the use of wireless sensor networks to develop traffic information systems, paying attention to the general aspects of these networks is required. One of the most challenging issues of design criteria is resource constraints that as a consequence, leads to traffic volume balancing and energy consumption issues (Bali et al., 2014). Sometimes, some parts of the network have huge volumes of traffic and information. In order to resolve the bandwidth issue in high-volume parts of the network, clustering techniques are useful. Using clustering, cluster heads are responsible for gathering and aggregating the information of clusters. Then, it transmits the information to the

data collection center. The use of clustering decreases the volume of transmitting information, and consequently, it balances the traffic volume in busy parts of the network. Due to this, the amount of energy consumption of sensors to transmit the data is also reduced (Vodopivec et al., 2012; Lee et al., 2016; Abbasi and Younis, 2007; Boyinbode et al., 2011).

A lot of researches have been conducted on using wireless sensor networks in traffic management area, using different clustering methods.

Fundamental limitation and problem of wireless sensor networks is extending the life time of sensors provided from batteries. Fundamental solution to extend the life of these networks is optimizing the energy consumption of sensors. Karthikeyan et al. (2013) compared five hierarchical routing protocols for energy efficiency routing, developed by the classical protocol LEACH. The researchers focus was on how to increase lifespan and reduce the power consumption of wireless sensor networks. Wireless sensor networks consist of hundreds of thousands of sensor nodes that are responsible to collect important data including temperature, location, and etc. These networks are used in different fields such as health monitoring, military applications, and etc. Another problem in this area is recharging or replacing the sensor nodes that have limited battery capacity. Therefore, energy efficiency is a key issue in maintaining the network (Park et al., 2013).

Clustering as technique to group adjacent nodes, firms the network (Robust), and makes it scalable. Vodopivec et al. (2012) reviewed clustering techniques and proposed a clustering algorithm for Ad-hoc vehicle networks. They investigate various clustering algorithms, pointed out each one's goals, characteristics, advantages, and limitations.

Bali et al. (2014) considered clustering mechanism in vehicle sensor networks as a grouping method based on density, speed, and geographical position and due to the above factors, investigated the challenges and solutions in the clustering of vehicle sensor networks.

Khan and Seth (2014) have proposed a clustering technique on the network with less communication to sink nodes and operate network hierarchically. As a result, network costs and energy consumption will be reduced. The techniques used K-means and FCM.

Park et al. (2013) proposed an efficient method to select cluster-heads using the K-means algorithm in their research to maximize energy efficiency of wireless sensor networks. The proposed method is based on minimization of the sum of the distances between cluster-heads and member nodes. The researchers demonstrated that the proposed method outperformed the existing protocols such as LEACH and HEED.

Bandyopadhyay and Coyle (2013) suggested a distributed, randomized clustering algorithm to organize the sensors in a wireless sensor network. They extended this algorithm to generate a hierarchy of cluster heads and observed that the energy savings increase with the number of levels in the hierarchy.

Given the notes above, the goal of this study is to optimize the traffic data transfer in vehicular sensor networks using spatial clustering. This study aims, in particular, at optimizing the traffic data transfer from an energy consumption perspective. The clustering is applied using FCM and Fuzzy Subtractive algorithms and the results are compared with each other. Besides, this study has been conducted based on two assumptions: (a) the number of sinks is considered as a value, and (b) all of the vehicles have been embedded with GPS devices.

The rest of this study is as follows: Section 2 describes the clustering, energy consumption modelling, and FCM and Fuzzy Subtractive algorithms. In Section 3, the methodology of the research is discussed. Section 4 introduces the dataset and the study area. The results and discussions are provided in Sections 5 and 6, respectively. Finally, Section 6 also concludes the study and provides some suggestions for future works.

## 2. FUNDAMENTALS OF THE RESEARCH

In literature of sensor networks, clustering is the process of grouping close or similar sensors. Similarity criteria of clustering can be different depending on the research target. In this study, clustering is used as a way to reduce sensors energy consumption. Hence, in following of the paper, fundamental concepts of the employed clustering methods will be discussed.

### 2.1 Fuzzy C-MEANS Clustering Algorithm

Fuzzy C-MEANS (FCM) is a clustering algorithm for non-crisp data such that similar to hard mode of this algorithm, a cost function of dissimilarity criterion is to be minimized. In clustering using this algorithm (and in general, fuzzy clustering algorithms), each point can be a member of different clusters with different membership degree in the range of zero to one. It is based on minimization of the following (Equation 1) objective function (Chaung et al., 2006).

$$J_m = \sum_{i=1}^{N} \sum_{j=1}^{C} u_{ij}^m D\left(x_i, c_j\right), \qquad 1 \le m \le \infty \qquad (1)$$

where $m$ is any real number greater than 1, $u_{ij}$ is the degree of membership of $x_i$ in the cluster $j$, $x_i$ is the $i^{th}$ of $d$-dimensional measured data, and $c_j$ is the $d$-dimension center of the cluster. Also, $D$ is a cost function or distance function between $x_i$ and $c_j$ points (Usually the Euclidean distance function).

In fact, fuzzy partitioning is carried out through an iterative optimization of the objective function shown above, with the update of membership $u_{ij}$ and the cluster centers $c_j$ using Equation 2.

$$u_{ij} = \frac{1}{\sum_{k=1}^{c}\left(\frac{\|x_i - c_j\|}{\|x_i - c_k\|}\right)^{\frac{2}{m-1}}}, \qquad c_j = \frac{\sum_{i=1}^{N} u_{ij}^m x_i}{\sum_{i=1}^{N} u_{ij}^m} \qquad (2)$$

This process is repeated until a stopping criterion is met.

### 2.2 Fuzzy Subtractive Clustering Algorithm.

Fuzzy Subtractive clustering algorithm considers a set of points as candidates for center of cluster. Initially, it is assumed that all of the points are possible to be chosen as the center of cluster. Then, a competency criterion, termed density measure, is computed for all the points using Equation 3.

$$D_i = \sum_{i=1}^{n} e^{\left(\frac{-\|x_i - x_j\|^2}{\left(\frac{r_a}{2}\right)^2}\right)} \qquad (3)$$

where $r_a$ is a positive constant and refers to the neighbourhood radius, and $n$ is the total number of data points. The more the data points in the vicinity of a point, the more the value of the density measure for that point and consequently, the more the competency of that point to be chosen as the first center of cluster. After computing the density measure for all points, the point corresponding to the highest value of competency will be chosen as the center of cluster. When the first center of cluster is found, the algorithm searches to find the second and, similarly, the other centers of clusters. For this purpose, the density measure values of other points should be first updated using Equation 4, such that the points that are closer to the center of cluster have a lower chance to be chosen.

$$D_i = D_i - D_{c_1} \times e^{\left(\frac{-\|x_i - x_{c_1}\|}{\left(\frac{r_b}{2}\right)^2}\right)} \qquad (4)$$

where $r_b$ defines the neighbourhood of the previously chosen center in which the values of the density measure is decreased. The value of $r_b$ is usually set up to 1.5 times of $r_a$. Now, by taking into account the values of density measure for other points, the second center is chosen (Hammouda and Karray, 2000; Pal and Chakraborty, 2000).

## 3. METHODOLOGY

In this part the methodology of the research and how clustering methods have been structured will be discussed.

## 3.1 Cost function of the clustering algorithms

Energy consumption in sensor networks is a function of two factors, computation and data transmission. Sending data uses much more energy than computation. The energy consumption of each sensor during data transmission is a function of sensor's hardware features and depends on two factors: size of the data packets and transmission range (Zytoune et al., 2010).

In the energy modelling of this study, distance to destination of each sensor (cluster-head or sink node) is determined and the range of transmission is changing according to that. This would spend less energy to send data to closer destinations. In many papers (Zytoune et al., 2010; Le Borgne et al., 2007), the power consumption of sensors can be defined as Equation 5:

$$E_T\left(\ell,d\right) = \begin{cases} E_{TX} + \ell\varepsilon_{fs}d^2 & \text{if } d < d_0 \\ E_{TX} + \ell\varepsilon_{mp}d^4 & \text{if } d > d_0 \end{cases} \quad (5)$$

where $E_{TX}$ (Power Electronics) is a constant that depends on sensor characteristics such as Modulation, filtering, signal propagation, and so on, $\ell\varepsilon_{fs}d^2$ and $\ell\varepsilon_{mp}d^4$ are amplifier power depend on the distance between the transmitter and receiver and acceptable bit error value. Also, value of $d_0$ can be calculated using Equation 6.

$$d_0 = \sqrt{\frac{\varepsilon_{fs}}{\varepsilon_{mp}}} \quad (6)$$

where, $\varepsilon_{fs}$ and $\varepsilon_{mp}$ are the activating energies for power amplifiers in multi-path and open space cases, respectively (Fadaei et al., 2016).

Finally, energy consumption for receiving 1 data bits is calculated by Equation 7:

$$E_R\left(\ell\right) = E_{RX} + \ell E_{elec} \quad (7)$$

in which, $E_{RX}$ and $E_{elec}$ are constant values and depend on the hardware characteristics of the sensor. This energy consumption model exactly matches the model of methods used by Heinzelman et al. (2002) and Zytoune et al. (2010).

At the end, clustering algorithms has been run using the mentioned cost function.

## 3.2 Structure of transmitting packets

As previously mentioned, the power consumption during data transmission depends on the length of the packets. As a result, it is necessary to determine an overall structure for data packets so that the packets length can be achieved. The first hypothesis of this study is that common nodes only pack the sensed data from the environment into the package and send it to the cluster-head. Cluster-heads gather the collected data from the members of the cluster and send them to the sink node (information collection center). For this purpose, the easiest method is to consider equal and constant length for data packets. The major flaw of this method is that advantages and disadvantages of aggregated data in cluster-heads cannot be presented correctly. When it is assumed that cluster-heads are responsible for aggregating the information, it is expected that transmitting packets of a cluster-head are larger than the transmitting packet of a common node. Also a cluster with a greater number of members is expected to have a larger transmitting packet, too.

Thus, a simple model is used to support variable packet data length.

On the selected model of this paper, each packet has a header and a body. Packets sent from a common node have a header and a data. Cluster-head packets have also a header and include all the data from the cluster members. In summary, the length of each section is as follows:

For variable packet length, each packet header is 24 bytes, and any data sensed by a sensor is 24 bytes (Le Borgne et al., 2007). Also, integration of all the data in cluster-head is considered as all member sensor data plus the header.

## 3.3 Fuzzy C-means and Fuzzy subtractive algorithms coordination with problem

Considering the dataset and the type of problem, to compare the performances of the algorithms, 5 clusters were determined.

To implement the FCM algorithm, the Distance between each sensor (vehicle) and cluster head, is Obtained based on the Euclidean distance function. According to these distances, the amount of energy consumption of data transfer will be calculated. Also, the degree of fuzziness (m) was chosen to be of 2.

The algorithm was applied for 100 times, each time with different initial values and in 100 iterations, on the sensors.

Similarly, to implement the fuzzy subtractive algorithm, the Distance between each sensor (vehicle) and cluster head, is Obtained based on the Euclidean distance function. And as mentioned earlier, $r_a$ is considered equal to 75 m and the value of $r_b$ is set up to 1.5 times of $r_a$.

## 4. IMPLEMENTATION

The algorithms and mechanisms have been implemented using MATLAB in a computer with a 4 GB RAM and an Intel Core 2 Duo T8100 / 2.1 GHz.

## 4.1 The study area and dataset

The dataset used in this study involves the passing vehicles' trajectories toward the west via the interstate highway 80 of the Emeryville in the California State. The data have been collected on 13 April 2005 during a 15 minutes period. In this study, the collected data from the hour 16 to 16:15 have been used. This dataset has been gathered via 7 cameras installed on top of a 30-storey building in the vicinity of the study area. The location of the sink is also considered on the top of this building. The trajectory data have been extracted from the video data using a dedicated program. This program automatically recognizes almost all the vehicles and tracks their path. In this way, the vehicles' trajectories in the specified range can be obtained. Those vehicles that are not recognized can be manually tracked, so that their trajectories are obtained. The dataset includes the (x, y) positions of vehicles with the temporal resolution of 0.1 of a second. For each vehicle in any recorded position, information such as the length and the width of the vehicle, instantaneous velocity and acceleration of the vehicle, the ID of the band that the vehicle is passing, the number of leading and pursuing vehicles in the same band, the distance (and time) to the vehicle in front are also recorded. In this study, the positions of 181 vehicles in a frame have been used. The locations of vehicles (sensors) and the sink are shown in Figure 1.

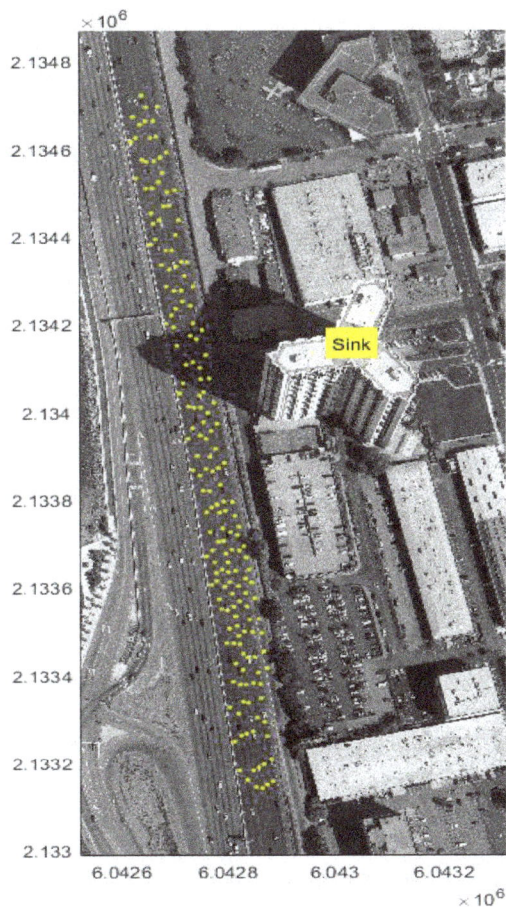

Figure 1. The location of vehicular sensors and the sink

## 4.2 Definition of the simulation parameters

The parameters used in the simulation are presented in Table 1.

| Description | Parameter | Value |
|---|---|---|
| Energy consumed by the amplifier to send to close destinations | $\varepsilon_{fs}$ | 10 $pJ/bit/m^2$ |
| Energy consumed by the amplifier to send to farther destinations | $\varepsilon_{mp}$ | 0.0013 $pJ/bit/m^4$ |
| Energy consumption in electronic circuits to send or receive data | $E_{elec} = E_{TX} - E_{RX}$ | $50nJ$ |
| Energy consumption for data integration in the cluster head | $E_{DA}$ | $5nJ/bit$ |

Table 1. Energy Consumption Parameters in the Simulation

## 5. RESULTS

In this study, the FCM and Fuzzy Subtractive algorithms were used to cluster the vehicular sensors and their impact on decreasing the energy consumption of sensors was explored.

For this dataset, if no clustering had been made to the sensors and each sensor directly transmits its own information to the sink, the amount of energy consumed by the whole network would be equal to 20.6 units. The best solution is presented in Table 2. The performance of the FCM algorithm has been enhanced, up to iteration 50, by about 90.68 percent in energy consumption, after that the algorithm has been converged. The clustering obtained by the FCM algorithm applied on the dataset is shown in Figure 2. In addition, the decreasing trend of the cost function (the total amount of energy consumption of sensors) in iterations 1 through 100 has been shown in Figure 3.

| Algorithm | FCM | | | |
|---|---|---|---|---|
| The number of iteration | 10 | 20 | 50 | 100 |
| Total energy consumption of Network | 2.23 | 2.00 | 1.92 | 1.92 |
| The amount of reduction of energy consumption | 89.17% | 90.29% | 90.68% | 90.68% |

Table 2. The performance of the FCM algorithm in decreasing the amount of energy consumption of sensors

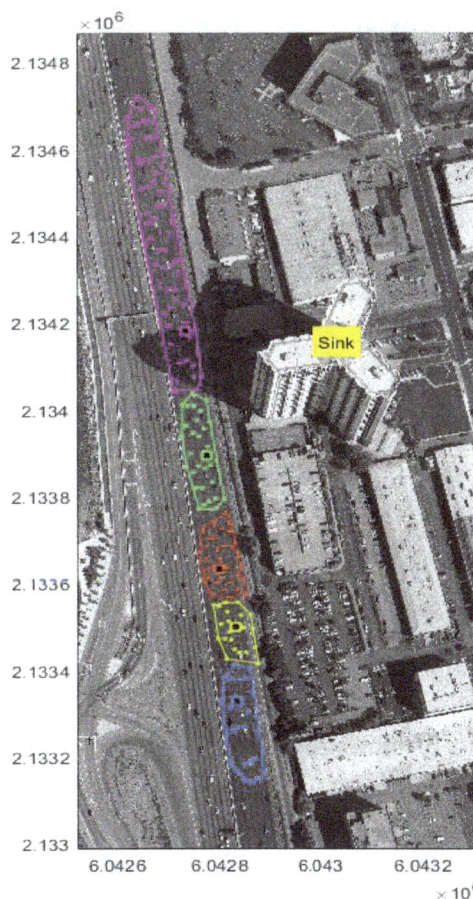

Figure 2. The best clustering result obtained by applying the FCM algorithm over 100 iterations

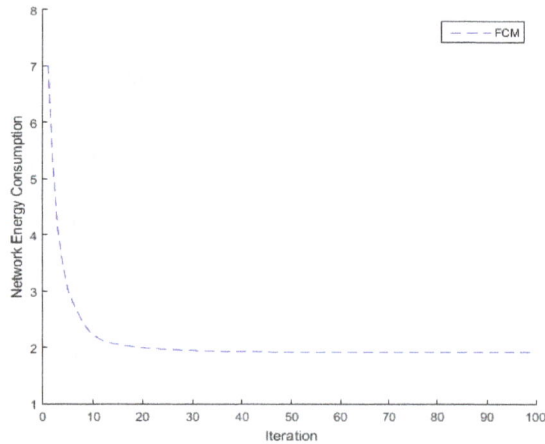

Figure 3. The decrease in the total amount of energy consumed by the network over 100 iterations using the FCM algorithm

The result of the Fuzzy Subtractive algorithm is shown in Figure 4. Moreover, the effect of the neighbourhood radius $r_a$ on the Fuzzy Subtractive algorithm has been explored. Figure 5 shows the effect of $r_a$ on the total energy consumed by sensors.

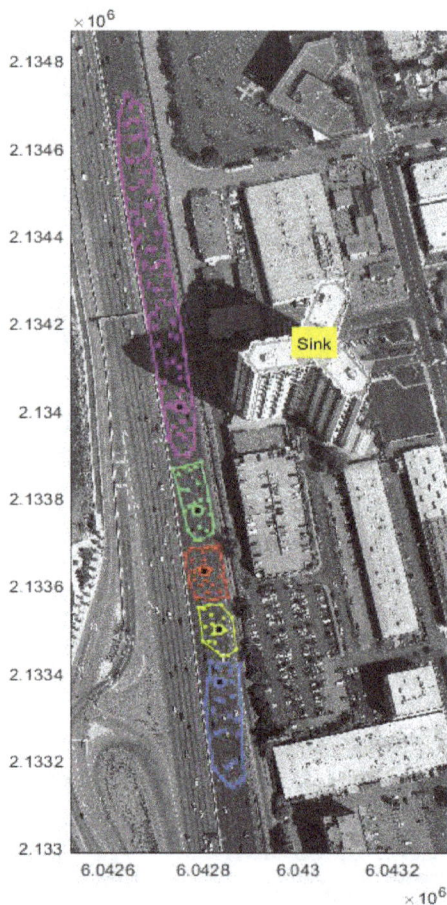

Figure 4. The clustering obtained by applying the Fuzzy Subtractive algorithm (with $r_a$=75 m )

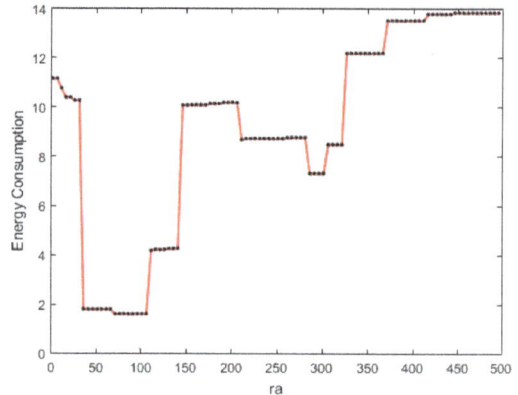

Figure 5. Relationship between Neighbourhood radius ($r_a$) in Fuzzy Subtractive clustering and cost function (energy consumption)

It is clear from the results of this algorithm that choosing a very small or very large $r_a$ will result in more energy consumption because if $r_a$ is chosen very small the density function will not take into account the effect of neighbouring data points; while if it is chosen very large, the density function will be affected all the data points in the data space. So, a value between 70 and 100 should be adequate for the radius of neighbourhood.

The best clustering, which corresponds to the least amount of energy consumption, has obtained using $r_a$ equal to 75 m and is shown in Figure 4. The least amount of energy consumption in this clustering is equal to 1.5 units and results in a decrease by about 92.18 percent in the energy consumption.

As it was shown, both algorithms have significant effects on the total amount of energy consumption and have reduced the consumption about 90.68% and 92.18%, respectively. Generally, for the dataset used in this study, the Fuzzy Subtractive algorithm provided better results than those achieved by FCM algorithm by about 1.50% progressive. Table 3 and Figure 6 compare the results of two algorithms in a numerical and visual manner. As shown in Table 3, there is no significant different between two algorithms results but any how the results of Fuzzy Subtractive algorithm is better. Also, elapsed time for implementation of the two algorithms, FCM and Fuzzy Subtractive, is 3.36 and 2.22 seconds, respectively. Hence, from run-time point of view, the fuzzy Subtractive algorithm is executed faster.

| The amount of reduction of energy consumption | | | Fuzzy Subtractive algorithm has better performance than FCM |
|---|---|---|---|
| Algorithm | FCM | Fuzzy Subtractive | |
| Total energy consumption of network | 1.92 | 1.61 | -0.31 |
| The amount of reduction of energy consumption | 90.68% | 92.18% | +1.50% |

Table 3. Comparison of the best performance of the FCM and Fuzzy Subtractive algorithms in decreasing the amount of energy consumption of sensors

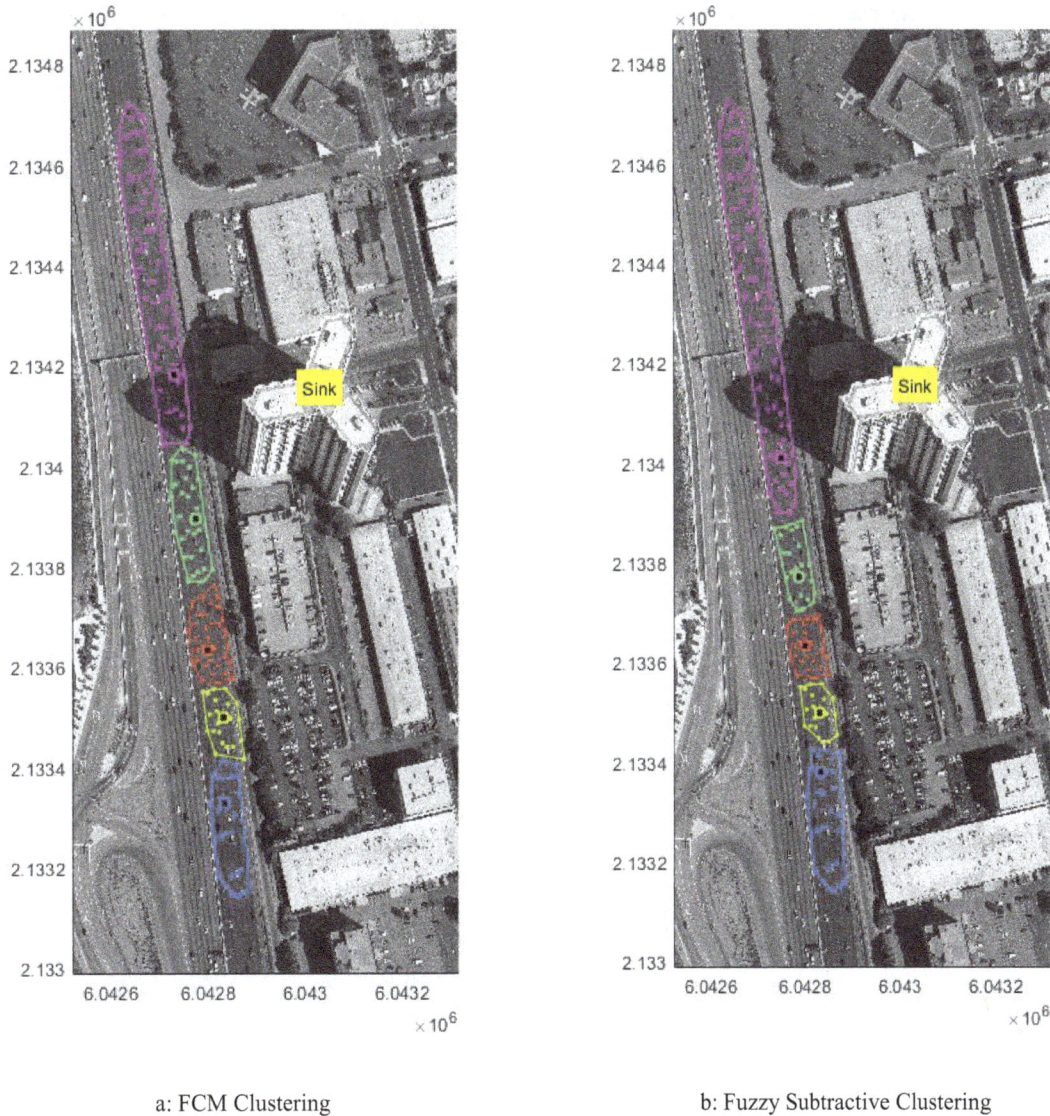

a: FCM Clustering                                    b: Fuzzy Subtractive Clustering

Figure 6. The clustering obtained by applying the FCM and Fuzzy Subtractive algorithms, clusters are shown by different colors and the cluster-heads are shown by bolded points

## 6. CONCLUSION

Traffic monitoring and management in urban intelligent transportation systems can be well carried out using vehicular sensor networks. In a vehicular sensor network, the vehicles can gather the urban traffic information and transmit them to the traffic management center. One of the most challenging problems in these networks is the amount of energy consumed by the vehicles. Taking the advantages of clustering algorithms in urban intelligent transportation systems leads to a significant decrease in the energy consumption. In this research, Fuzzy C-Means and Fuzzy Subtractive algorithms have been used to solve this problem. Results show that the energy consumption is significantly low in techniques using clustering algorithms. Moreover, the results of Fuzzy Subtractive algorithm is better than Fuzzy C-Means in comparison.

It is suggested, for the future works, that a comparison between the two algorithms for different datasets to be drawn. It is also suggested that the performances of these algorithms be compared against other algorithms. Moreover, the optimization methods could be applied on the algorithms and their effect on the results could be explored.

## REFERENCES

Abbasi, A.A. and Younis, M., 2007. A survey on clustering algorithms for wireless sensor networks. Computer communications, 30(14), pp.2826-2841.

Bali, R.S., Kumar, N. and Rodrigues, J.J., 2014. Clustering in vehicular ad hoc networks: taxonomy, challenges and solutions. Vehicular communications, 1(3), pp.134-152.

Bandyopadhyay, S. and Coyle, E.J., 2003, April. An energy efficient hierarchical clustering algorithm for wireless sensor networks. In INFOCOM 2003. Twenty-Second Annual Joint Conference of the IEEE Computer and Communications. IEEE Societies (Vol. 3, pp. 1713-1723). IEEE.

Boyinbode, O., Le, H. and Takizawa, M., 2011. A survey on clustering algorithms for wireless sensor networks. International Journal of Space-Based and Situated Computing, 1(2-3), pp.130-136.

Chaung, K.S., Tzeng, H.L., Chen, S., Wu, J., Chen, T.J., 2006. Fuzzy c-means clustering with spatial information for image segmentation, Computerized Medical Imaging and Graphics, Vol. 30, Issue, 1. Pages 9-15.

Fadaei, M., Abdipour, M., Dindar Rostami, M., 2016. Choosing Proper Cluster Heads to Reduce Energy Consumption in Wireless Sensor Networks Using Gravitational Force Algorithm. International Academic Journal of Science and Engineering 3(6), pp. 24-33.

Hammouda, K. and Karray, F., 2000. A comparative study of data clustering techniques. Fakhreddine Karray University of Waterloo, Ontario, Canada.

Heinzelman, W.B., Chandrakasan, A.P. and Balakrishnan, H., 2002. An application-specific protocol architecture for wireless microsensor networks. IEEE Transactions on wireless communications, 1(4), pp.660-670.

Khan, J.M., Seth, A., 2014 "Performance comparison of FCM and K-means clustering technique", Global Journal of Advanced Engineering Technologies and Sciences, 2, 26-29.

Karthikeyan, A., Sarkar, S., Gupte, A.M. and Srividhya, V., 2013. Selection of cluster head using fuzzy adaptive clustering for energy optimization in wireless sensor network. Journal of Theoretical & Applied Information Technology, 53(1).

Le Borgne, Y.A., Santini, S. and Bontempi, G., 2007. Adaptive model selection for time series prediction in wireless sensor networks. Signal Processing, 87(12), pp.3010-3020.

Lee, W.S., Ahn, T.W. and Song, C., 2016. A Study on Energy Efficiency for Cluster-based Routing Protocol. Journal of the Institute of Electronics and Information Engineers, 53(3), pp.163-169.

Miyamoto, S., Ichihashi, H. and Honda, K., 2008. BasicMethods for c-Means Clustering. In Algorithms for Fuzzy Clustering (pp. 9-42). Springer Berlin Heidelberg.

Pal, N.R. and Chakraborty, D., 2000. Mountain and subtractive clustering method: Improvements and generalizations. International Journal of Intelligent Systems, 15(4), pp. 329-341.

Park, G.Y., Kim, H., Jeong, H.W. and Youn, H.Y., 2013, March. A novel cluster head selection method based on K-means algorithm for energy efficient wireless sensor network. In Advanced Information Networking and Applications Workshops (WAINA), 2013 27th International Conference on (pp. 910-915). IEEE.

Seredynski, M. and Bouvry, P., 2011, October. A survey of vehicular-based cooperative traffic information systems. In 14th International IEEE Conference on Intelligent Transportation Systems (ITSC) (pp. 163-168). IEEE.

Vodopivec, S., Bešter, J. and Kos, A., 2012, July. A survey on clustering algorithms for vehicular ad-hoc networks. In Telecommunications and Signal Processing (TSP), 2012 35th International Conference on (pp. 52-56). IEEE.

Zytoune, O., Fakhri, Y. and Aboutajdine, D., 2010. A novel energy aware clustering technique for routing in wireless sensor networks. Wireless Sensor Network, 2(3), p.233.

# 8

# A NEW HYBRID YIN-YANG-PAIR-PARTICLE SWARM OPTIMIZATION ALGORITHM FOR UNCAPACITATED WAREHOUSE LOCATION PROBLEMS

A. A. Heidari*, O. Kazemizade, F. Hakimpour

School of Surveying and Geospatial Engineering, College of Engineering, University of Tehran, Iran
(as_heidari, kazemizade.omid, fhakimpour)@ut.ac.ir

**KEYWORDS:** Warehouse Location; Optimal; Location Analysis; Optimization; Particle Swarm; Yin-yang-pair Optimization

**ABSTRACT:**

Yin-Yang-pair optimization (YYPO) is one of the latest metaheuristic algorithms (MA) proposed in 2015 that tries to inspire the philosophy of balance between conflicting concepts. Particle swarm optimizer (PSO) is one of the first population-based MA inspired by social behaviors of birds. In spite of PSO, the YYPO is not a nature inspired optimizer. It has a low complexity and starts with only two initial positions and can produce more points with regard to the dimension of target problem. Due to unique advantages of these methodologies and to mitigate the immature convergence and local optima (LO) stagnation problems in PSO, in this work, a continuous hybrid strategy based on the behaviors of PSO and YYPO is proposed to attain the suboptimal solutions of uncapacitated warehouse location (UWL) problems. This efficient hierarchical PSO-based optimizer (PSOYPO) can improve the effectiveness of PSO on spatial optimization tasks such as the family of UWL problems. The performance of the proposed PSOYPO is verified according to some UWL benchmark cases. These test cases have been used in several works to evaluate the efficacy of different MA. Then, the PSOYPO is compared to the standard PSO, genetic algorithm (GA), harmony search (HS), modified HS (OBCHS), and evolutionary simulated annealing (ESA). The experimental results demonstrate that the PSOYPO can reveal a better or competitive efficacy compared to the PSO and other MA.

## 1. INTRODUCTION

The suitability of civic warehouses is influenced by their locations. One of the vital concerns in supply chain management, logistic and location analysis is warehouse location (FL) (Ou-Yang and Ansari, 2017). In recent years, FL problems, which is also termed in the related works as facility location (FL), have been extensively investigated by researchers because of their tactical nature (Aardal et al., 1999; Ou-Yang and Ansari, 2017). There are various mathematical models demonstrating a range of FL tasks, which most of them have a combinatorial nature (Aardal et al., 1999). Consequently, exact algorithms can often be effective for small instances and metaheuristic algorithms (MA) have been utilized in literature as efficient methods to handle larger applied FL tasks.

In the traditional uncapacitated WL (UWL), each warehouse requires an initial cost to be established and a certain cost is associated with the choosing of roads from a client to a warehouse. The objective of the UWL is to determine where to construct the warehouses and which routes to utilize for optimizing the overall costs. In UWL, the decision maker should determine the candidate sites that specify a new warehouse, while some criteria such as total costs to be minimized (Al-Sultan and Al-Fawzan, 1999; Korupolu et al., 2000). The cost includes fixed costs to open plants and depots, and variable cost can be related to the transportation. The other constraints can be the demands of all stores and the warehouse and suppliers capacity limits, which should not be violated. Therefore, the selection of warehouse location will profoundly influence the management planning of plants and organizations. It's worth noting that several applied tasks with no warehouses to obtain, such as portfolio management, machine scheduling, clustering, and computer networks can similarly be addressed according to UWL-based models (Ghosh, 2003; Tcha and Lee, 1984).

Some exact approaches have been developed for solving the UWL. Some of the well-known methods are branch and bound (Klose, 1998), dual approach (DUALLOC) (Erlenkotter, 1978) Lagrangean relaxation (Barcelo et al., 1990), the primal-dual (Körkel, 1989), and linear programming. It can be proved that the UWL is NP-hard (Sevkli and Guner, 2006). Hence, in last 10 years, many researchers tried to develop efficient MA to tackle the UWL. Some of the well-established works can be summarized as tabu search (TS) (Michel and Van Hentenryck, 2004), simulated annealing (SA) (Aydin and Fogarty, 2004), particle swarm optimizer (PSO) (Guner and Sevkli, 2008), and genetic algorithms (GA) (Jaramillo et al., 2002). In 2017, Guo et al proposed a two-stage capacitated FL (TSCFLP) and a hybrid evolutionary algorithm (EA) (Guo et al., 2017). In 2015, Basti and Sevkli used artificial bee colony (ABC) optimizer to realize the p-median UFL problem (Basti and Sevkli, 2015). Moreover, Heidari et al proposed an efficient opposition-based chaotic HS (OBCHS) to tackle the UFL problems and the results revealed that the OBCHS can reveal a better efficacy compared to GA and PSO methods (Heidari et al., 2015a). Ng investigated expanding neighborhood tabu search (ENTS) algorithm for large warehouse location problems in water infrastructure planning (Ng, 2014). Esnaf et al. proposed a fuzzy c-means algorithm with fixed cluster centers for UFL that permits unlabeled data to be assigned to the related clusters centers. In their research, the proposed algorithm is tested on different UFL benchmarks from literature and Turkish fertilizer producer's real data. The algorithm was compared to the PSO and ABC algorithms and the results show that the proposed algorithm have better performance than the other algorithms (Esnaf et al., 2014). In 2017, a hybrid PSO-tabu search optimizer has been proposed to treat the hierarchical FL problem (Ou-Yang and Ansari, 2017).

The Yin-Yang-pair optimization (YYPO) is one of the recent MA that mimics the philosophy of balance among discordant concepts (Punnathanam and Kotecha, 2017).

---

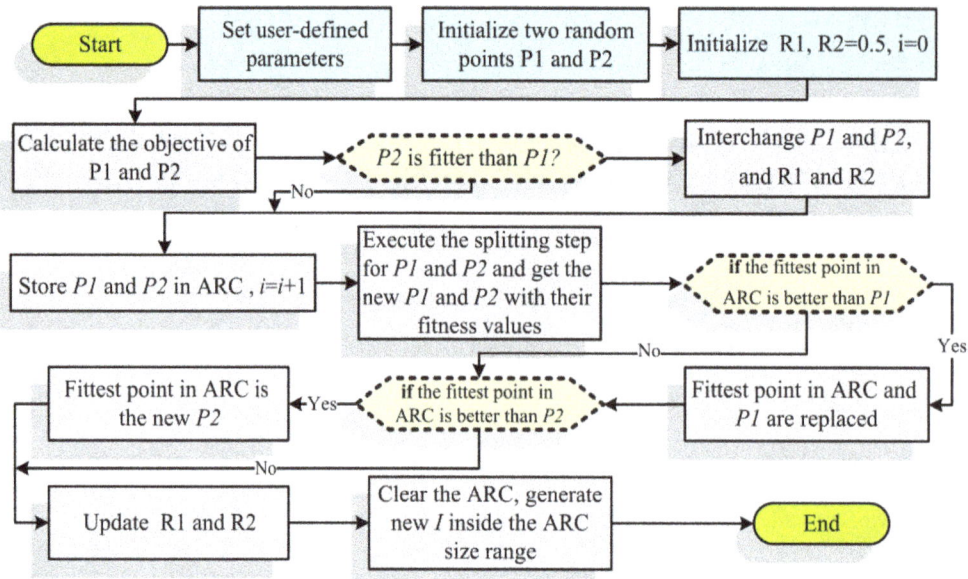

Figure 1. The structure of YYPO algorithm

This algorithm has shown an efficient performance in dealing with several optimization problems. In this paper, a hybrid MA is proposed and substantiated to tackle the UWL problem. For this purpose, the YYPO optimizer is embedded into the well-known PSO algorithm to enhance the efficacy of this algorithm. Then, the proposed PSO-YPO is validated based on several instances of the UWL.

## 2. THE YYPO ALGORITHM

The YYPO optimizer is one of the recent MA inspired by the philosophy of balance among discordant concepts (Punnathanam and Kotecha, 2017). In YYPO, the variables of the tasks should be normalized inside interval (0, 1). This optimizer utilizes two points (P1 and P2) to search the problem landscape. The P1 and P2 are generated in the initial step of YYPO inside the domain of [0, 1] and their fitness are evaluated. The better point is considered as P1 and the other one is assigned as P2. The point P1 plays its role in exploitation phase, while the point P2 tries to highlight the exploration behaviors. The points P1 and P2 act as hubs to sightsee the hypersphere dimensions inside the solution space specified by radii of R1 and R2, correspondingly. These spans have a self-adaptive nature such that R1 has a propensity to every so often decrease and R2 to rise. The YYPO involves two core phases:

- Splitting phase
- Archive phase

### 2.1 Splitting phase

This phase is designed to boost the exploration of the target space. For this purpose, one of the points in consort with its search range is inserted to this step. The splitting phase can create new positions inside the hypersphere at diverse directions. In this step, 2D fresh points are scattered nearby the particular point P and their distances to the P is calculated based on R. Note that D denotes the dimension. The new points can be scattered using two strategies with the same chance: one-way and D-way splitting. In one-way strategy, simply one variable of each P is adjusted. In the D-way method, all variables of each P are updated (Punnathanam and Kotecha, 2017).

One-way splitting: 2D duplicates of the P are kept as S then; one variable of each P in S is adjusted by Eq. (1):

$$S_j^j = S^j + a \times R \tag{1}$$

$$S_{D+j}^j = S^j - a \times R, j = 1,2,3,...,D \tag{2}$$

In aforementioned relations, superscript shows the variable number, the subscript are the P number and $a$ shows a random value inside (0, 1). In D-way splitting step, all variables of P in S should be updated using:

$$\begin{cases} S_k^j = S^j + a\left(R/2^{0.5}\right) & \text{if } B_k^j = 1 \\ S_k^j = S^j + a\left(R/2^{0.5}\right) & \text{else} \end{cases} \tag{3}$$

$$k = 1,2,3,...,2D. \quad j = 1,2,3,...D$$

Here, the binary matrix $B$ is made by randomly choosing 2D distinctive integers between 0 and $2^{D-1}$.

### 2.2 Archive stage

Initialization of this step indicates that $I$ archive (ARC) updates have been completed. Moreover, the archive comprises $2I$ points. The finest point in the ARC, when it is superior to $P1$, is switched with $P1$ (i.e. the former $P1$ is currently in the ARC). Then, the best point in the present ARC, if superior to $P2$, substitutes $P2$. Thus, the best two points will continuously survive sequential repetitions (Punnathanam and Kotecha, 2017). The R1 and R2 are updated as:

$$R1 = R1 - (R1/\alpha) \tag{4}$$

$$R2 = R2 + (R2/\alpha) \tag{5}$$

The $R1$ can shrinkage the volume explored nearby $P1$ in the splitting phase, while $R2$ can expand the volume explored nearby $P2$. In the YYPO, the maximum of R2 is 0.75. Then, the ARC is deflated and a renewed value of $I$ is created inside ($I_{min}$ and $I_{max}$).

The flowchart of YYPO is demonstrated in Figure 1. For more details about the YYPO please refer to (Punnathanam and Kotecha, 2017).

## 3. THE PSO ALGORITHM

The PSO is a well-known population-based MA used in numerous works (Heidari et al., 2017a, b; Heidari and Delavar, 2016; Heidari et al., 2015b; Trelea, 2003). The PSO tries to simulate the idealistic social life of birds (Heidari and Pahlavani, 2017c). The PSO utilizes a swarm of particles (solutions) that can search the target domain to explore and exploit the fittest particle. Temporarily, the search agents all follow the best solution in their searching trajectories. In PSO, particles will track their personal best locations along with the best agent explored thus far. The PSO can be mathematically described as:

$$v_i^{t+1} = \omega v_i^t + c_1\varphi_1(p_i - x_i^t) + c_2\varphi_2(p_g - x_i^t) \qquad (6)$$

$$x_i^{t+1} = x_i^t + v_i^{t+1} \qquad (7)$$

where $v_i^t$ denotes the velocity of agent $i$ at iteration $t$, $w$ shows a weighting factor, $c_j$ is an acceleration factor, $\varphi_1$ and $\varphi_2$ are random values inside (0, 1), $x_i^t$ is the existing location of agent $i$ at iteration $t$, $p_i$ is the *pbest* of solution $i$ at loop $t$, and $p_g$ is the fittest agent so far. A modified PSO (CFM) is also available that utilizes a constriction factor (CF) (Heidari and Ali Abaspour, 2016). In this paper, the CFM version is employed to treat the UWL problem. The main equation of CFM can be described as:

$$v_i(t) = \chi\left[v_i(t) + c_1\varphi_1(p_i - x_i(t)) + c_2\varphi_2(p_g - x_i(t))\right] \qquad (8)$$

$$x_i(t+1) = x_i(t) + v_i(t) \qquad (9)$$

$$\chi = 2/\left|2 - \varphi - (\varphi^2 - 4\varphi)^{0.5}\right|, \ \varphi = c_1 + c_2, \ \varphi > 4, \qquad (10)$$

where $\chi$ denotes the constriction factor.

## 4. THE HIERARCHICAL PSO-YYPO ALGORITHM

The basic PSO can reveal a satisfactory performance in treating different optimization tasks. However, immature convergence to LO may still happen in tackling some larger UWL cases. The reason is that PSO cannot make a stable balance between exploration and exploitation tendencies. In order to mitigate this drawback, an enhanced PSO algorithm is proposed to enrich its efficacy on UWL tasks. For this purpose, the YYPO optimizer can be used an extra step inside the PSO method.

The main structure of combined algorithm (PSOYPO) is similar to the basic PSO. The first difference is that the global best solution of PSO is obtained during iterations via YYPO and inserted into the PSO again for more improvements. The second difference is the hierarchical structure of PSOYPO. In the initial step of PSO, the YYPO is started to generate the best results. Then, the outputs of YYPO are considered as the initial candidate solutions for PSO. The third difference is that the PSO will search the topography of the target problem, then, the YYPO is triggered again improve the local searching capacity of PSO. After that, the basic PSO can continue the search until the last iteration. In PSOYPO, the operators of YYPO have a constructive effect on the precision of solutions based on the two best agents in sorted population. It's worth noting that YYPO is not a population-based optimizer and improves only

two solutions during the searching process. The YYPO is a low complexity MA which starts with two initial positions and produces more points with regard to the dimension of target problem. Hence, when the YYPO wants to be utilized, the population of particles is sorted and only the first and second best agents are inserted as the inputs to the YYPO.

The proposed PSOYPO has the main advantages of both PSO and YYPO. In this work, PSOYPO is used to handle the UWL problem. To learn how the PSO, PSOYPO, and other compared methods are applied to UWL problems, the reader is referred to (Guner and Sevkli, 2008; Sevkli and Guner, 2006).

## 5. UWL PROBLEM

In the UWL, a subset of locations from a given set of potential locations is required for establishing warehouses so as to optimize a given function of these chosen locations while satisfying certain constraints (Basu et al., 2015; Michel and Van Hentenryck, 2004). In this section, the variables of the UWL problem are described.

In UWL problems with $m$ stores and $n$ candidate warehouse sites, $f_j$ is used to represent the cost of opening warehouse $j$ that is fixed and $c_{ij}$ is used to represent the cost of serving customer $i$ from warehouse $j$ or assigning store $i$ to warehouse $j$. We assume that $c_{ij} \geq 0$ for all $i = 1, 2, 3... m$ and $j = 1... n$ and $f_j > 0$ for all $j = 1... n$. A binary variable $y_j$ is used to represent the status of warehouse $j$ in the model. Warehouse $j$ will be opened only if $y_j = 1$ in the solution. A binary variable $x_{ij}$ is used for the road from store $i$ to warehouse $j$ in the model. Store $i$ will be served by warehouse $j$ only if $x_{ij} = 1$ in the solution. However, each $x_{ij}$ can be treated as a continuous variable and will have a binary value in the solution. The solution process of the UWL problem is to find an optimal solution that satisfies the store demand and minimizes the total cost (Esnaf et al., 2014; Sun, 2006). Here, the formally UWL is described as:

$$Minimize \ Z = \sum_{i=1}^{n} f_{ci} \cdot y_i + \sum_{i=1}^{m}\sum_{i=1}^{n} c_{ij} x_{ij} \qquad (11)$$

$$s.t. \ \sum_{i=1}^{n} x_{ij} = 1, \qquad (12)$$

$$0 \leq x_{ij} \leq y_i \in \{0;1\}, \qquad (13)$$

where $i = 1,...,n; j = 1,...,m;$

$x_{ij}$: the quantity provided using facility $i$ to client $j$;

$y_i$: whether facility $i$ is confirmed $(y_i = 1)$ or not $(y_i = 0)$

The UWL problems usually have a mixed integer formulation with binary variables to indicate the locations of warehouses, together with continuous variables that represent system dynamics. This mixed integer nature of UWL problems makes them NP-hard (Ng, 2014; Sevkli and Guner, 2006). The solution methodologies used for these problems concentrate on exact methods and MA. Because the UWL is NP-hard, exact methods are mostly suitable for small-sized UWL cases. In the large-scale cases, MA are used in order to obtain near optimum solutions since finding optimum solutions need much time (Basti and Sevkli, 2015).

In addition, several MA have been applied for tackling UWL problems such as hybrid EA (Guo et al., 2017), ABC (Basti and Sevkli, 2015), an opposition-based chaotic harmony search algorithm (Heidari et al., 2015a), tabu search (Sun, 2006), expanding neighborhood tabu search (Ng, 2014), a simplified artificial fish swarm (Azad et al., 2013), and a modified continuous PSO (Saha et al., 2011). In the next sections, the results of a new effective PSO-based optimizer are reported to investigate the optimum solutions of several UWL problems.

## 6.  RESULTS AND ANALYSIS

In this section, the efficacy of the new PSOYPO is studied carefully. Here, each method is experienced using MATLAB 2013 software. The PSOYPO is utilized to solve 15 well-known UWL benchmark test cases from OR Library (Beasley, 2005). Twelve test problems are somewhat small in size, whiles three test cases are pretty large. The details of tackled benchmark problems are reported in Table 1.

| ID | Name | Type | Size ($m \times n$) | Optimum |
|---|---|---|---|---|
| TF01 | Cap71 | Small | $16 \times 50$ | 932615.8 |
| TF02 | Cap72 | Small | $16 \times 50$ | 977799.4 |
| TF03 | Cap73 | Small | $16 \times 50$ | 1010641 |
| TF04 | Cap74 | Small | $16 \times 50$ | 1034977 |
| TF05 | Cap101 | Small | $25 \times 50$ | 796648.4 |
| TF06 | Cap102 | Small | $25 \times 50$ | 854704.2 |
| TF07 | Cap103 | Small | $25 \times 50$ | 893782.1 |
| TF08 | Cap104 | Small | $25 \times 50$ | 928941.8 |
| TF09 | Cap131 | Small | $50 \times 50$ | 793439.6 |
| TF10 | Cap132 | Small | $50 \times 50$ | 851495.3 |
| TF11 | Cap133 | Small | $50 \times 50$ | 893076.7 |
| TF12 | Cap134 | Small | $50 \times 50$ | 928941.8 |
| TF13 | Cap-A | Large | $100 \times 1000$ | 17156454 |
| TF14 | Cap-B | Large | $100 \times 1000$ | 12979072 |
| TF15 | Cap-C | Large | $100 \times 1000$ | 11505594 |

Table 1. Details of benchmark test cases

Each method is verified based on 30 independent trials. The population size of all methods is equal to the number of warehouses. Note that the recommended settings in related literature are used in this experiment (Heidari et al., 2015a; Sevkli and Guner, 2006). In addition, the algorithms can repeat the process throughout 1.00E+03 iterations. In this research, the mean relative percent error (ARPE) is measured as the performance index. The MRPE is described as:

$$MRPE = \sum_{i=1}^{Z} \left( \frac{(T_i - S) \times 100}{S_i} \right) / Z \qquad (14)$$

where $T_i$ is the $ith$ iteration, $U$ denotes the fittest value and $Z$ represents the iteration number. The other indexes utilized in this study are "Optimum Rate" represented by (HR) and CPU time indicated by (RT). Furthermore, HR is the hit to the optimum rate (HR) ratio that shows the robustness of algorithms. It is truly hard to explore the best positions in every round of search. The higher HR and MRPE value are better than the lower values. When these indexes are higher, it can be recognized that the quality of results is more satisfactory.

The MRPE values of the PSOYPO algorithm are compared to HS, OBCHS, and PSO in Table 2. It can be seen from Table 2 that the proposed PSOYPO can obtain preferable solutions compared to the basic PSO method. It is seen that for problems with bigger sizes, the error rate of all methods has increased. The reason that PSOYPO can obtain results with a better precision is that it has an effective exploitation capacity. Hence, in the case of finding a fruitful region, it can focus on the better positions in the search space. The MPRE values of PSOYPO are very competitive to the OBCHS, but OBCHS still can show

a better efficacy than PSO-based methods for TS13, TS14, and TS15 problems.

| ID | PSO | HS | OBCHS | PSOYPO |
|---|---|---|---|---|
| TS1 | 6.00E-02 | 4.00E-02 | 0.00E+00 | 0.00E+00 |
| TS2 | 8.00E-02 | 9.00E-02 | 0.00E+00 | 0.00E+00 |
| TS3 | 7.00E-02 | 7.00E-02 | 1.00E-02 | 0.00E+00 |
| TS4 | 8.00E-02 | 8.00E-02 | 0.00E+00 | 0.00E+00 |
| TS5 | 1.60E-01 | 1.30E-01 | 0.00E+00 | 2.00E-02 |
| TS6 | 1.70E-01 | 1.00E-01 | 0.00E+00 | 0.00E+00 |
| TS7 | 1.50E-01 | 1.70E-01 | 2.00E-02 | 3.00E-02 |
| TS8 | 1.70E-01 | 1.90E-01 | 0.00E+00 | 0.00E+00 |
| TS9 | 7.90E-01 | 7.60E-01 | 0.00E+00 | 0.00E+00 |
| TS10 | 8.10E-01 | 7.70E-01 | 1.00E-02 | 8.00E-02 |
| TS11 | 7.10E-01 | 7.20E-01 | 0.00E+00 | 0.00E+00 |
| TS12 | 9.10E-01 | 9.40E-01 | 0.00E+00 | 2.00E-02 |
| TS13 | 2.12E+01 | 1.27E+01 | 1.00E-02 | 1.00E-02 |
| TS14 | 9.91E+00 | 4.12E+00 | 1.00E-02 | 2.41E+00 |
| TS15 | 9.62E+00 | 2.41E+00 | 3.00E-02 | 4.52E+00 |

Table 2. Comparison of MRPE values

In Table 3, The HR values of the PSOYPO algorithm are compared to the HS, OBCHS, and PSO algorithms (Also see Figure 2).

| ID | PSO | HS | OBCHS | PSOYPO |
|---|---|---|---|---|
| TS1 | 8.50E-01 | 9.00E-01 | 1.00E+00 | 1.00E+00 |
| TS2 | 7.90E-01 | 8.40E-01 | 1.00E+00 | 1.00E+00 |
| TS3 | 6.50E-01 | 6.80E-01 | 1.00E+00 | 1.00E+00 |
| TS4 | 7.10E-01 | 7.80E-01 | 1.00E+00 | 1.00E+00 |
| TS5 | 5.60E-01 | 4.20E-01 | 1.00E+00 | 1.00E+00 |
| TS6 | 4.60E-01 | 5.40E-01 | 1.00E+00 | 9.80E-01 |
| TS7 | 1.90E-01 | 1.20E-01 | 1.00E+00 | 1.00E+00 |
| TS8 | 6.50E-01 | 7.20E-01 | 9.90E-01 | 9.90E-01 |
| TS9 | 6.00E-02 | 7.00E-02 | 1.00E+00 | 1.00E+00 |
| TS10 | 0.00E+00 | 1.00E-02 | 1.00E+00 | 9.80E-01 |
| TS11 | 0.00E+00 | 0.00E+00 | 1.00E+00 | 1.00E+00 |
| TS12 | 1.00E+00 | 1.40E-01 | 1.00E+00 | 1.00E+00 |
| TS13 | 0.00E+00 | 0.00E+00 | 9.80E-01 | 1.00E+00 |
| TS14 | 0.00E+00 | 0.00E+00 | 9.70E-01 | 9.70E-01 |
| TS15 | 0.00E+00 | 0.00E+00 | 5.00E-01 | 5.60E-01 |

Table 3. Comparison of HR values

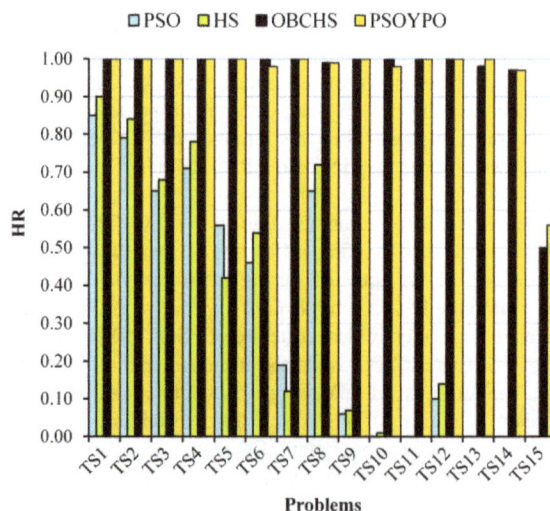

Figure 2. Robustness comparison between the PSO, HS, OBCHS, and PSOYPO algorithms

It can be seen that PSOYPO can obtain better results than PSO in dealing with all problems. The results of the PSOYPO method are very competitive to the OBCHS, but it is observed that the results are better than basic HS. Generally, the PSO cannot show an outstanding effectiveness in comparison with HS, OBCHS, and PSOYPO.

In Table 3, the *RT* values of the PSOYPO algorithm are compared to HS, OBCHS, and PSO. It is seen that the PSOYPO require additional time obtain better results compared to the PSO (see Figures 3 and 4). The reason is that the proposed strategy has a hierarchical structure that increases the time of the process. However, OBCHS is still the fastest method.

| ID | PSO | HS | OBCHS | PSOYPO |
|-----|---------|---------|---------|---------|
| TS1 | 1.30E-01 | 1.20E-01 | 7.00E-02 | 1.50E-01 |
| TS2 | 1.50E-01 | 1.90E-01 | 6.00E-02 | 2.10E-01 |
| TS3 | 2.80E-01 | 2.30E-01 | 1.40E-01 | 3.10E-01 |
| TS4 | 2.30E-01 | 1.05E+00 | 4.50E-01 | 2.60E-01 |
| TS5 | 7.10E-01 | 6.50E-01 | 6.10E-01 | 8.50E-01 |
| TS6 | 8.00E-01 | 9.80E-01 | 8.50E-01 | 9.60E-01 |
| TS7 | 1.13E+00 | 9.90E-01 | 7.80E-01 | 1.23E+00 |
| TS8 | 5.10E-01 | 4.50E-01 | 1.20E-01 | 6.50E-01 |
| TS9 | 4.34E+00 | 4.13E+00 | 2.41E+00 | 5.23E+00 |
| TS10 | 4.62E+00 | 4.14E+00 | 3.98E+00 | 6.62E+00 |
| TS11 | 4.51E+00 | 4.13E+00 | 3.41E+00 | 5.52E+00 |
| TS12 | 4.53E+00 | 3.98E+00 | 2.12E+00 | 4.87E+00 |
| TS13 | 1.47E+01 | 1.47E+01 | 8.12E+00 | 1.56E+01 |
| TS14 | 1.77E+01 | 1.60E+01 | 1.30E+01 | 1.84E+01 |
| TS15 | 2.57E+01 | 2.24E+01 | 1.90E+01 | 2.93E+01 |

Table 4. Comparison of *RT* values

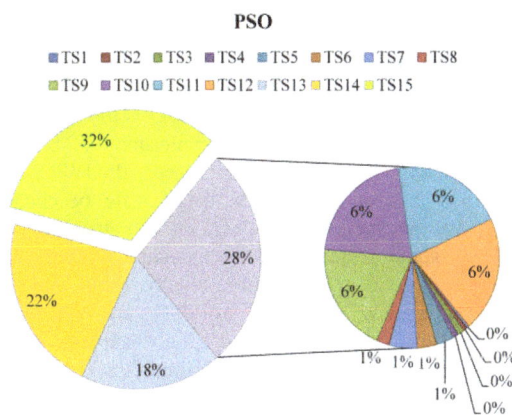

Figure 3. Computational efficacy of the PSO in dealing with TS1 to TS15

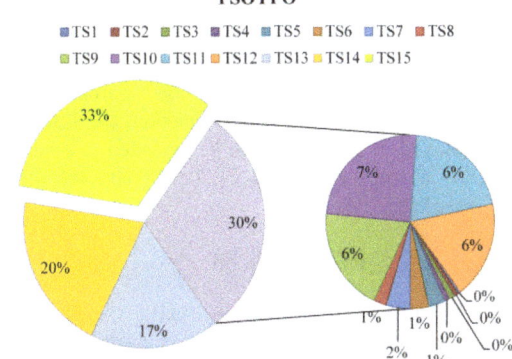

Figure 4. Computational efficacy of the proposed PSOYPO in dealing with TS1 to TS15

Regarding the measured indexes, it is seen that the efficiency of PSOYPO is superior to PSO and other methods. The reason is that it has an enhanced local search potential, which can help it in escaping from LO. To further analysis of the performance of the PSOYPO, it is compared with a GA-based strategy proposed by (Jaramillo et al., 2002) and evolutionary SA (ESA) developed by (Aydin and Fogarty, 2004). Note that the results of GA and ESA are obtained and reported here based on their original works (Sevkli and Guner, 2006). The standard deviation (STD) results of different algorithms for normal scale (NS) and large scales (LS) problems are compared in Tables 5 and Table 6, respectively. In addition, the running time of different methods is tabulated in Table 7 and Table 8. Note that the HS, OBCHS, PSO, and PSOYPO algorithms were tested on the same computer.

| ID | GA | ESA | HS | OBCHS | PSO | PSOYPO |
|------|---------|---------|---------|--------|-------|--------|
| TS1 | 0.00 | 0.00 | 0.00 | 0.00 | 0.00 | 0.00 |
| TS2 | 0.00 | 0.00 | 0.001 | 0.00 | 0.001 | 0.00 |
| TS3 | 0.00033 | 0.00 | 0.00021 | 0.0001 | 0.002 | 0.00 |
| TS4 | 0.00 | 0.00 | 0.00 | 0.00 | 0.002 | 0.002 |
| TS5 | 0.00020 | 0.00 | 0.0001 | 0.00 | 0.005 | 0.003 |
| TS6 | 0.00 | 0.00 | 0.0001 | 0.00 | 0.001 | 0.00 |
| TS7 | 0.00015 | 0.00 | 0.0023 | 0.00 | 0.052 | 0.00 |
| TS8 | 0.00 | 0.00 | 0.00 | 0.00 | 0.014 | 0.01 |
| TS9 | 0.00065 | 0.00008 | 0.0004 | 0.00 | 0.004 | 0.00 |
| TS10 | 0.00 | 0.00 | 0.00 | 0.00 | 0.005 | 0.00 |
| TS11 | 0.00037 | 0.00002 | 0.00075 | 0.00 | 0.002 | 0.00 |
| TS12 | 0.00 | 0.00 | 0.00 | 0.00 | 0.001 | 0.00 |

Table 5. Comparison of STD results for normal scale problems

| ID | GA | ESA | HS | OBCHS | PSO | PSOYPO |
|------|---------|---------|-------|---------|-------|---------|
| TS13 | 0.00 | 0.000 | 0.00 | 0.00 | 0.001 | 0.00 |
| TS14 | 0.00172 | 0.00070 | 0.001 | 0.00002 | 0.001 | 0.00001 |
| TS15 | 0.00131 | 0.00119 | 0.00 | 0.00 | 0.051 | 0.00 |

Table 6. Comparison of STD results for LS problems

From the results in Table 5, it is noticeable that the results of PSOYPO algorithm have a superior accuracy. The STD results are lower than other methods in the majority of test cases. It is seen that for all problems, except TS4, TS5, and TS8, the proposed PSOYPO can show an impressive performance according to the STD results and its results have the least deviation from the optimum. It's also seen that the accuracy of results of the PSO isn't impressive. From Table 6, it is seen that the HS, OBCHS, and PSOYPO algorithms can obtain the best results in solving TS14 test case (see Figure 5). The overall outcomes confirm that the stagnation problems of PSO have been alleviated significantly based on the proposed YYPO-based mechanisms.

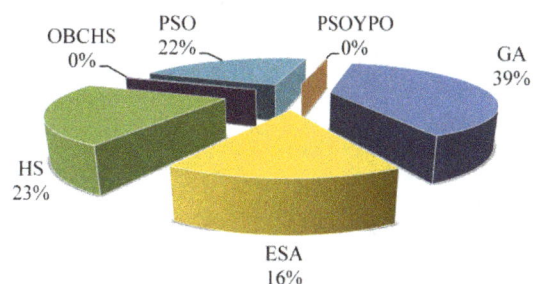

Figure 5. Comparison of STD results for TS14

From Table 8 and Table 9, it is realized that the chaos-embedded OBCHS can outperform the GA, ESA, HS and PSOYPO methods in dealing with TS1 to TS12. However, in many applications, the quality of solutions and efficiency of algorithms is more important than the running time. The UWL is not an exception. Figures 6 and 7 also compare the computational speed of different algorithms for NS and LS problems, respectively. From these illustrations, it can be understood that the proposed PSOYPO can show an acceptable performance in dealing with UWL problems.

| ID | GA | ESA | HS | OBCHS | PSOYPO |
|---|---|---|---|---|---|
| TS1 | 0.287 | 0.041 | 0.042 | 0.014 | 0.552 |
| TS2 | 0.322 | 0.028 | 0.210 | 0.021 | 1.127 |
| TS3 | 0.773 | 0.031 | 0.215 | 0.028 | 2.451 |
| TS4 | 0.200 | 0.018 | 0.041 | 0.014 | 4.445 |
| TS5 | 0.801 | 0.256 | 0.704 | 0.098 | 5.412 |
| TS6 | 0.896 | 0.098 | 0.085 | 0.019 | 6.102 |
| TS7 | 1.371 | 0.119 | 1.012 | 0.064 | 7.921 |
| TS8 | 0.514 | 0.026 | 0.201 | 0.014 | 8.930 |
| TS9 | 6.663 | 2.506 | 4.122 | 0.514 | 9.754 |
| TS10 | 5.274 | 0.446 | 0.492 | 0.197 | 9.851 |
| TS11 | 7.189 | 0.443 | 0.785 | 0.312 | 10.821 |
| TS12 | 2.573 | 0.079 | 1.415 | 0.142 | 10.711 |

Table 7. Comparison of computational performance for normal scale problems

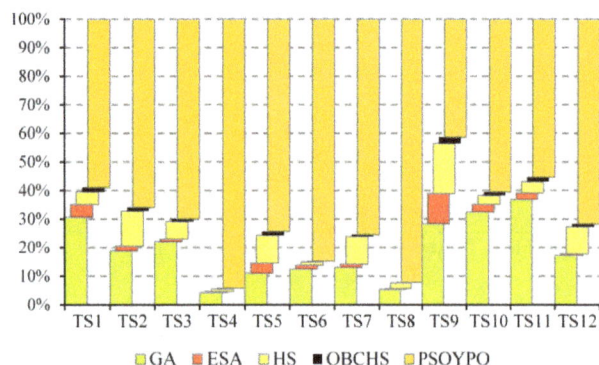

Figure 6. Comparison of time results for NS problems

| ID | GA | ESA | HS | OBCHS | PSOYPO |
|---|---|---|---|---|---|
| TS13 | 184.422 | 17.930 | 47.412 | 15.741 | 19.851 |
| TS14 | 510.445 | 91.937 | 147.52 | 84.112 | 124.874 |
| TS15 | 591.516 | 131.345 | 110.430 | 101.922 | 352.811 |

Table 8. Comparison of computational performance for LS problems

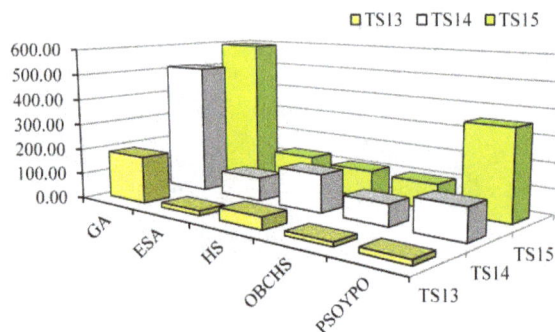

Figure 7. Comparison of time results for LS problems

To validate the efficacy of the PSOYPO, the Wilcoxon rank sum test with 0.05 degree is performed to finish the experiments. The rank sum test can demonstrate the significant improvements in the exploration and exploitation potentials of the proposed PSOYPO over other methods. The final results are tabulated in Table 9. In this Table, + shows that the improvements are significant and p-values are lower than 0.05. The negative values show that the PSOYPO cannot statistically outperform the second method.

| Problem | PSOYPO versus | | | | |
|---|---|---|---|---|---|
| ID | GA | ESA | HS | OBCHS | PSO |
| TS1 | + | + | + | + | + |
| TS2 | + | + | + | + | + |
| TS3 | + | + | + | + | + |
| TS4 | + | + | + | + | + |
| TS5 | - | + | - | + | + |
| TS6 | + | + | + | + | + |
| TS7 | + | - | + | + | + |
| TS8 | + | + | + | - | + |
| TS9 | + | + | + | + | + |
| TS10 | + | + | + | - | + |
| TS11 | + | + | + | - | + |
| TS12 | + | + | + | + | + |

Table 9. Results of the statistical tests

From Table 9, it can be seen that the new results are significantly better than those of PSO algorithm in all cases. However, it is seen that the results are not statistically superior to the OBCHS in solving TS08, TS10, and TS11.

## 7. CONCLUSIONS AND FUTURE DIRECTIONS

In this paper, a new hierarchical PSO-based algorithm is proposed and employed to solve several UWL problems. The Yin-Yang-pair optimization (YYPO) was used to enrich the optimality of the exploited results. The YYPO is the latest MA proposed in 2015 inspiring from the philosophy of balance between conflicting concepts. The proposed PSOYPO was tested on 15 well-known small and large scale benchmark problems. It was observed that the PSOYPO can attain the best solutions throughout a reasonable computational time. The comparative results of this research vividly show that the immature convergence problems of PSO are alleviated, significantly. The New PSOYPO not only exploits better solutions but also it is capable of outperforming PSO in dealing with the UWL tasks.

In addition to utilizing the UWL models and the PSO and YYPO algorithms to real-world tasks, proposing more efficient and competent metaheuristics seems to be a remarkable research direction. Future works in this direction can revolve around the addition of more PSO operators to enhance its efficacy. Proposing new exploration or exploitation approaches can similarly deserve investigation. With adaptation, the PSOYPO technique employed in this paper can be extended to other FL scenarios, such as the capacitated FL (CFL) and the single source CFL problems.

## REFERENCES

Aardal, K., Chudak, F.A., Shmoys, D.B., 1999. A 3-approximation algorithm for the k-level uncapacitated facility location problem. Information Processing Letters 72, pp. 161-167.

Al-Sultan, K., Al-Fawzan, M., 1999. A tabu search approach to the uncapacitated facility location problem. Annals of Operations Research 86, pp. 91-103.

Aydin, M.E., Fogarty, T.C., 2004. A distributed evolutionary simulated annealing algorithm for combinatorial optimisation problems. Journal of heuristics 10, pp. 269-292.

Azad, M., Kalam, A., Rocha, A.M.A., Fernandes, E.M.d.G., 2013. A simplified binary artificial fish swarm algorithm for uncapacitated facility location problems, World Congress on Engineering 2013, WCE 2013. Newswood Limited Publisher, pp. 31-36.

Barcelo, J., Hallefjord, Å., Fernandez, E., Jörnsten, K., 1990. Lagrangean relaxation and constraint generation procedures for capacitated plant location problems with single sourcing. OR Spectrum 12, pp. 79-88.

Basti, M., Sevkli, M., 2015. An artificial bee colony algorithm for the p-median facility location problem. International Journal of Metaheuristics 4, pp. 91-113.

Basu, S., Sharma, M., Ghosh, P.S., 2015. Metaheuristic applications on discrete facility location problems: a survey. Opsearch 52, pp. 530-561.

Erlenkotter, D., 1978. A dual-based procedure for uncapacitated facility location. Operations Research 26, pp. 992-1009.

Esnaf, Ş., Küçükdeniz, T., Tunçbilek, N., 2014. Fuzzy C-Means Algorithm with Fixed Cluster Centers for Uncapacitated Facility Location Problems: Turkish Case Study, Supply Chain Management Under Fuzziness. Springer, pp. 489-516.

Ghosh, D., 2003. Neighborhood search heuristics for the uncapacitated facility location problem. European Journal of Operational Research 150, pp. 150-162.

Guner, A.R., Sevkli, M., 2008. A discrete particle swarm optimization algorithm for uncapacitated facility location problem. Journal of Artificial Evolution and Applications 2008, pp. 1-9.

Guo, P., Cheng, W., Wang, Y., 2017. Hybrid evolutionary algorithm with extreme machine learning fitness function evaluation for two-stage capacitated facility location problems. Expert Systems with Applications 71, pp. 57-68.

Heidari, A.A., Abbaspour, R.A., Jordehi, A.R., 2017a. An efficient chaotic water cycle algorithm for optimization tasks. Neural Computing and Applications 28, pp. 57–85.

Heidari, A.A., Abbaspour, R.A., Jordehi, A.R., 2017b. Gaussian bare-bones water cycle algorithm for optimal reactive power dispatch in electrical power systems. Applied Soft Computing 57, pp. 657–671.

Heidari, A. A., Pahlavani, P., 2017c. An Efficient Modified Grey Wolf Optimizer with Lévy Flight for Optimization Tasks. Applied Soft Computing, 60, pp. 115–134.

Heidari, A.A., Ali Abaspour, R., 2016. Let a CPSO+ algorithm explore your high-quality shortest paths: An effective chaos-enhanced PSO-based strategy, Proceedings of the International Conference on Industrial Engineering and Operations Management, pp. 2125-2132.

Heidari, A.A., Delavar, M.R., 2016. A Modified Genetic Algorithm for Finding Fuzzy Shortest Paths in Uncertain Networks. Int. Arch. Photogramm. Remote Sens. Spatial Inf. Sci. XLI-B2, pp. 299-304.

Heidari, A.A., Kazemizade, O., Abbaspour, R.A., 2015a. OBCHS: An Effective Harmony Search Algorithm with Opposition based Chaos-enhanced Initialization for Solving Uncapacitated Facility Location Problems. Int. Arch. Photogramm. Remote Sens. Spatial Inf. Sci. XL-1/W5, pp. 307-311.

Heidari, A.A., Mirvahabi, S.S., Homayouni, S., 2015b. An Effective Hybrid Support Vector Regression with Chaos-Embedded Biogeography-Based Optimization Strategy for Prediction of Earthquake-Triggered Slope Deformations. Int. Arch. Photogramm. Remote Sens. Spatial Inf. Sci. XL-1/W5, pp. 301-305.

Jaramillo, J.H., Bhadury, J., Batta, R., 2002. On the use of genetic algorithms to solve location problems. Computers & Operations Research 29, pp. 761-779.

Klose, A., 1998. A branch and bound algorithm for an uncapacitated facility location problem with a side constraint. International Transactions in Operational Research 5, pp. 155-168.

Körkel, M., 1989. On the exact solution of large-scale simple plant location problems. European Journal of Operational Research 39, pp. 157-173.

Korupolu, M.R., Plaxton, C.G., Rajaraman, R., 2000. Analysis of a local search heuristic for facility location problems. Journal of algorithms 37, pp. 146-188.

Michel, L., Van Hentenryck, P., 2004. A simple tabu search for warehouse location. European Journal of Operational Research 157, pp. 576-591.

Ng, T.L., 2014. Expanding Neighborhood Tabu Search for facility location problems in water infrastructure planning, 2014 IEEE International Conference on Systems, Man and Cybernetics (SMC). IEEE, pp. 3851-3854.

Ou-Yang, C., Ansari, R., 2017. Applying a hybrid particle swarm optimization_Tabu search algorithm to a facility location case in Jakarta. Journal of Industrial and Production Engineering 34, pp. 199-212.

Punnathanam, V., Kotecha, P., 2017. Multi-objective optimization of Stirling engine systems using Front-based Yin-Yang-Pair Optimization. Energy Conversion and Management 133, pp. 332-348.

Saha, S., Kole, A., Dey, K., 2011. A Modified Continuous Particle Swarm Optimization Algorithm for Uncapacitated Facility Location Problem, Information Technology and Mobile Communication. Springer, pp. 305-311.

Sevkli, M., Guner, A.R., 2006. A continuous particle swarm optimization algorithm for uncapacitated facility location problem, International Workshop on Ant Colony Optimization and Swarm Intelligence. Springer, pp. 316-323.

Sun, M., 2006. Solving the uncapacitated facility location problem using tabu search. Computers & Operations Research 33, pp. 2563-2589.

Tcha, D.-w., Lee, B.-i., 1984. A branch-and-bound algorithm for the multi-level uncapacitated facility location problem. European Journal of Operational Research 18, pp. 35-43.

Trelea, I.C., 2003. The particle swarm optimization algorithm: convergence analysis and parameter selection. Information processing letters 85, pp. 317-325.

# 9

# DEVELOPING A SPATIAL PROCESSING SERVICE FOR AUTOMATIC CALCULATION OF STORM INUNDATION

H. Jafari [a, *], Ali A. Alesheikh [b]

[a] Dept. of GIS, K.N. Toosi University Of Technology, Tehran, Iran  - haniyejafari12@gmail.com
[b] Dept. of GIS, K.N. Toosi University Of Technology, Tehran, Iran  - alesheikh@kntu.ac.ir

**KEY WORDS:** WPS, Storm, Inundation, Web Service, USISM, Hydrological Model

**ABSTRACT:**

With the increase in urbanization, the surface of earth and its climate are changing. These changes resulted in more frequent flooding and storm inundation in urban areas. The challenges of flooding can be addressed through several computational procedures. Due to its numerous advantages, accessible web services can be chosen as a proper format for determining the storm inundation. Web services have facilitated the integration and interactivity of the web applications. Such services made the interaction between machines more feasible. Web services enable the heterogeneous software systems to communicate with each other. A Web Processing Service (WPS) makes it possible to process spatial data with different formats. In this study, we developed a WPS to automatically calculate the amount of storm inundation caused by rainfall in urban areas. The method we used for calculating the storm inundation is based on a simplified hydrologic model which estimates the final status of inundation. The simulation process and water transfer between subcatchments are carried out respectively, without user's interference. The implementation of processing functions in a form of processing web services gives the capability to reuse the services and apply them in other services. As a result, it would avoid creating the duplicate resources.

## 1. INTRODUCTION

### 1.1 Theoretical background

The global changes in climate and the development of urbanization process have increased the frequency, severity, and losses of natural disasters. Flood and storm inundation are introduced as one of the natural disasters, which make many challenges to Hydrologic experts due to the intensity and shortage of time to react (Apel et al., 2009; Bonta, 2004; Cheng, 2010). As the ground is often covered with impermeable materials in the urban areas, runoff speed is high and the infiltration of rainwater into the soil seems to be low. As a result, heavy showers in an urban area can cause flood and inundation (Smith, 2006). For instance, the 40-hour rainfall on the 13th 14th and 15th of April in 2012 in Tehran caused serious storm inundation. Indeed, within the last 50 years, the rainfall mass was unprecedented and the regions 1, 3, 4, 5, 6 and 15 faced with serious problems. In some streets, the water level reached to as much as 30 cm and the movement of people and cars was stopped. Moreover, the floodwater broke into the tunnel of metro's line 4 and thoroughly disrupted the movement of trains in that line. In some stations, the floodwater level reached more than 5 meters and caused damages to electric, telecommunication, control, and navigation systems. Accordingly, in recent years, the need for paying attention to urban flood and storm inundation has been discussed more seriously.

Today, via web, many of applications are presented as accessible services.Web services have made the integration and the interactivity of web applications between the machines more feasible. Web services enable the heterogeneous software systems to communicate with each other. With the application of such technology, all implementation details are hidden from the users' view, who would only need a web browser (Bean, 2008). In GIS world, web services are utilized in order to manage, analyse and disseminate spatial data. Spatial web services enable the use spatial data and spatial processing functions to various organizations (Friis-Christensen, et al., 2007).

Different formats of spatial data can be processed through a Web Processing Service (WPS). In fact, the process, defined by a WPS, can encompass a wide range of simple and complex processes. The service can express the simple calculation of the distance between two points or complicated calculations such as modeling the atmosphere changes on earth as a processing source in a standard way. In the current study, since the WPS standard was considered as the basis of our processing functions the details of developing  such services are elaborated in the following sections.

Many different open source and proprietary software have hitherto been generated to implement WPS services. In fact, some software have been developed specifically for WPS implementation, while other ones have been introduced for different purposes, to which the capability of WPS implementation has been added. Table 1 presents a classification based on the role of different software in regard to WPS implementation and application (Evangelidis, et al., 2014). In Table 1:

Client applications: offer the interface to users, which enable them to search for information related to the capabilities of a WPS servant, send their request and receive the answers to their request. In addition, they create a medium to display the processes output.

---

* Corresponding author

Server software: entails the main core of operations needed to develop a WPS servant via internet. WPS servant serves as an access point, that can manage all client requests and their responses. Technically, WPS server software is integrated into other software such as libraries, plug-ins and desktop spatial processing parts to process the input parameters, dispatched by user request, and reply to them through the outputs.

Libraries: include classes and divisions developed by means of programing languages, and connected to existing software, which create processing divisions.

Desktop GIS software: act as a thick client and supply all processing actions of a WPS through their processing divisions. Software products listed in Table 1 take part in creating and using WPS implicitly or explicitly through the plug-ins, developed for this purpose.

| Name | Client | Server | Library | Desktop GIS | Open Source |
|---|---|---|---|---|---|
| 52° North WPS | | √ | | | √ |
| ArcGIS Server | | √ | | | |
| Deegree WPS | | √ | | | √ |
| disy Cadenza Professional | √ | | | | |
| disy Cadenza Web | √ | √ | | | |
| disy GISterm Professional | √ | | | | |
| disy GISterm Web | √ | √ | | | |
| Erdas Apollo Professional | √ | √ | | | |
| GDAL/OGR | | | | √ | √ |
| GeoMedia SDI Portal | √ | | | | |
| Geoserver | | √ | | | √ |
| Geoshield Project | √ | √ | | | √ |
| GeoTools | | | | √ | √ |
| GNIS Server | √ | √ | | | |
| GRASS GIS | √ | | | √ | √ |
| HSLayers | | | √ | | √ |
| Liquid XML Data Binder | √ | √ | | | √ |
| Maplink Pro | √ | √ | | | √ |
| Netgis Server | | √ | | | |
| OpenGeo Suite | √ | √ | | | √ |
| Openlayers | | | √ | | √ |
| PyWPS | | √ | | | √ |
| QuantumGIS | √ | | | √ | √ |
| Rasdaman | | √ | | | √ |
| uDig | √ | | | | √ |
| WPS.NET | | √ | | | √ |
| WPSint | √ | | | | √ |
| ZOO-Project | | √ | | | √ |

Table 1. Contributions to the implementation of OGC WPS specification standard (Evangelidis, et al., 2014)

Some of the noticeable WPS implementation softwares are briefly introduced as follows.

52°North is regarded as an open source free software, presented in 2004, which was covered by several research and software development unions as well as a wide range of collaborative activities related to software development. The WPS created by 52°North conforms to Open Geospatial Consortium (OGC) standards. It is a Java-based software that is applied as a plug-in (Evangelidis, et al., 2014).

Zoo is recognized as a WPS open source project, which was published in 2009. Zoo develops a WPS on the basis of OGC and provides the developer with a user-friendly framework. The software also makes it possible to develop and chain the services. The main purpose of the software is to collect available open source libraries and apply them in a standard manner. It also aims at simplifying the WPS development by providing a simple method to establish the web services. Zoo offers a capability with respect to managing and chaining of web services, developed by means of different programming languages (Fenoy, et al., 2013).

The PyWPS project began to work in 2006, which aimed to create Grass-based processes applicable to web users. PyWPS can be considered as a translation library, which receives input requests according to OGC standards, sends them to Grass or any other tools (developed by Python), and resend the results (Cepicky & Becchi, 2007).

APOLLO and ArcGIS Server are introduced as two instances of commercial remedies, widely applied in WPS creation. A modeling engine named *Imagine* is planned for APOLLO, that represents the capability of graphical designing of models and complicated spatial algorithms in order to create chained processed models and publish them. Model Builder, a modelling engine similar to APOLLO's, was introduced by ESRI in 2006. ArcGIS Server 10.1 and its following versions support WPS standards. Moreover, the processing tools and models created in ArcGIS can be utilized by any client applicants, supporting WPS standard (Evangelidis, et al., 2014). To date, the functionalities of spatial web services have been applied in many studies in the field of Hydrology. Diaz et al. developed a web-based geoportal for hydrological purposes, which combines processing web services with client applications in order to display spatial data (Diaz et al, 2008). The geoporatl aims at achieving a user-friendly interactive interface for the scholars in the field of hydrology. The server part provides the hydrological models via a library from the distributed spatial processing web services. The client part facilitates the search for catalogue services in order to achieve an appropriate resource. Furthermore, this part interacts with processing web services through the parameters of hydrological models and displays the results to the user. Castronova and his colleagues (2013). designed a modeling service based on WPS standard. Their approach was examined on a hydrological model. They demonstrated how a hydrological model has been implemented and used as a spatial processing service (Castronova et al., 2013). Lou and his team established a distributed, interoperable and standard-based spatial web service, that can automaticaly extract streams network from Digital Elevation Model (DEM). It is worth noting that in this web service, the streams are extracted by inserting the threshold on the flow accumulation map. This service was also implemented consistently with the WPS standard (Luo et al., 2014). Kadlec and his research group built a WPS using 52 North, for analysis of hydrological time series in R. The presented architecture can be used as a model for publishing other time series analysis scripts online (Kadlec et. al, 2016).

In the present study, the spatial web service was applied in order to automatically calculate the amount of storm inundation caused by rainfall in urban areas. The method, adopted for calculating storm inundation, was based on the spatial information systems. Moreover, it is a simplified method of the hydrological models, which calculates the final status of inundation. The simulation process and water transfer between sub catchments are carried out respectively, without the user's interference and automatically, in regard to flow order.

## 2. MATERIALS AND METHODS

### 2.1 Algorithm Of Storm Inundation Calculation

In the current study, Urban storm inundation simulation method (USISM) algorithm was used for storm inundation calculation. The method is based on geographic information systems (GIS)

and that is a kind of simplified distributed hydrologic model. The USISM model does not consider the impact of storm water infrastructures in calculation of hydrological characteristics such as flow direction and flow accumulation. Besides, in this model, curve number method was utilized in order to estimate the runoff, in which the amount of infiltration is calculated according to the soil type. As a result, and due to the lack of land use map, the effect of any land types on the calculation of surface runoff is not taken into account and the curve number is determined to estimate the runoff only based on the percentage of permeable areas. As the urban areas are often covered with impermeable materials, the permeability value has been considered as zero for all surfaces. Figure 2 shows the computational algorithm for USISM model (Zhang & Pan, 2014).

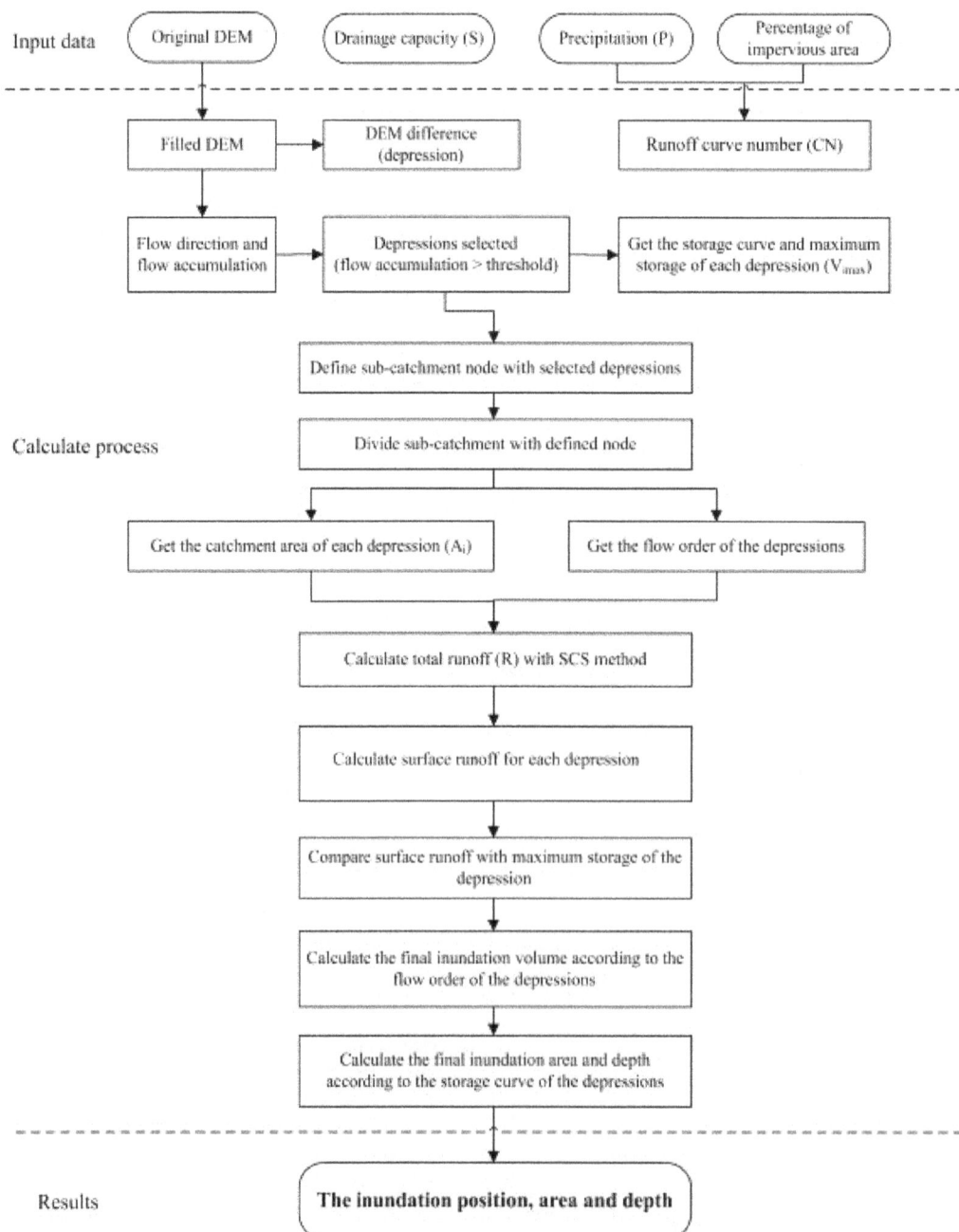

Figure 2. USISM model (Zhang & Pan, 2014)

According to above figure DEM must be filled to find depressions. Depressions are calculate using filled DEM minus original DEM. Flow Direction and Flow Accumulation are calculate using a D8 algorithm. Using threshold for flow accumulation, stream is extract. The depressions on the stream are inundated area potentially. Using ArcGIS 3D analysis tools the maximum storage volume and storage curve of these depressions are obtained. In the next step subcatchment is divide based on flow direction, flow accumulation and depressions outline node. Then subcatchment area is calculated, and the order of the depressions is determined. Using the order of depressions surface runoff volume is calculated. Average velocity of water drainage multiplied by storm duration used for sewer system runoff estimation. In the next step surface runoff volume compared with the maximum storage of the depression. If surface runoff volume is less than the maximum storage, volume of surface runoff is inundation volume. Otherwise, depression is filled and excess water transferred to next depressions. Other than the volume, depth and area of the depression are the final results.

The simulation phases in USISM model are carried out manually by the user, though, in implementation service, the steps are automatically done without the user intervention. Thus, the process of water transfer among the depressions and movement from upstream to downstream catchments, considered as a time consuming computational process in USISM model, is automatically performed in our web processing service.

## 2.2 Implementation of Web Processing Service

In order to implement the automatic computation service of storm inundation, the spatial processing tool was created to transmit surface runoff across the catchments, and calculate the inundation's depth, volume and area using Arcobjects as well as C# programming language in Visual Studio 2015. Using the Model Builder, a combination of spatial processing tools available in ArcGIS 10.2, and the created processing tool, the needed model was established to estimate the storm inundation, which was put as the WPS on ArcGIS for Server 10.2. Flooded depressions, the catchment area related to the research cite, the streams caused by the considered rainfall, flooded points, along with the depth, volume and area were estimated as the outputs of the mentioned services.

## 2.3 Implementation Of Web Application

In order to apply the developed service, a web application was also generated to make it possible to call the created service. The implementation was performed using the ARCGIS API for JAVA SCRIPT. The programming phases were carried out in Visual Studio 2015. In this system, DEM layer, urban passages maps, OSM map and Bing satellite images are accessible to users. Figure 3 illustrates the implemented user interface (UI).

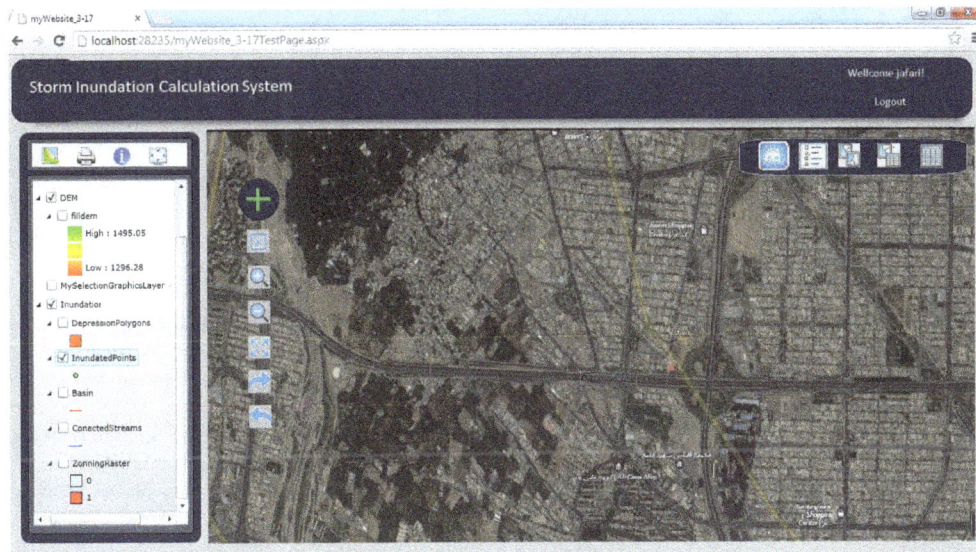

Figure 3. User interface of the developed web application

## 3. RESULTS

In the current study, the district 1 of region 22 of Tehran was selected as the research site. Topographic data used in the research process is DEM, on which the height of the buildings was not considered. The planned speed of water transfer through the sewage system was reported 12 mm per hour. The rainfall data applied in the research process is related to a rainfall event with a return period of 100 years. This event involves a 1-hour rainfall with the amount of 15.74 mm. The data was collected from different departments of Tehran Municipality and National Weather Service. The intensity and duration of rainfall as well as DEM data of related areas were investigated and clicking on run bottom, the direction and accumulation maps of the flow are generated through another service and sent to the storm inundation automatic calculation service. This implemented service runs and output is accessible to the user. The outputs of this processed service consist of flooded depressions, flooded points with the characteristics of their volume, depth and area, the sub catchment of the research site and the streams caused by the rainfall event. Figure 4 and Figure 5 show the maps of flow direction and accumulation of input flow to the automatic calculation service of storm water.

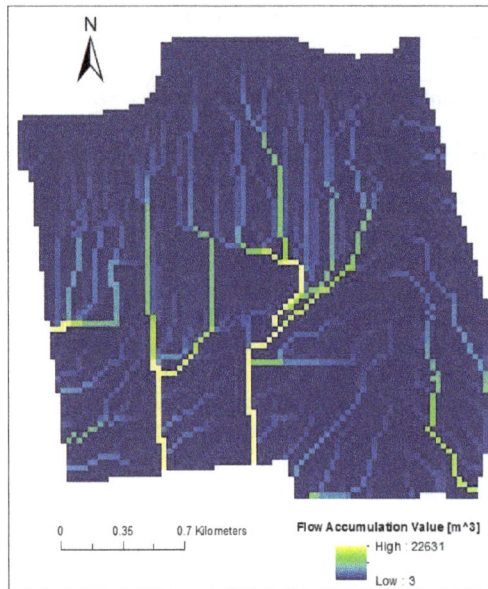

Figure 4. Input flow accumulation map

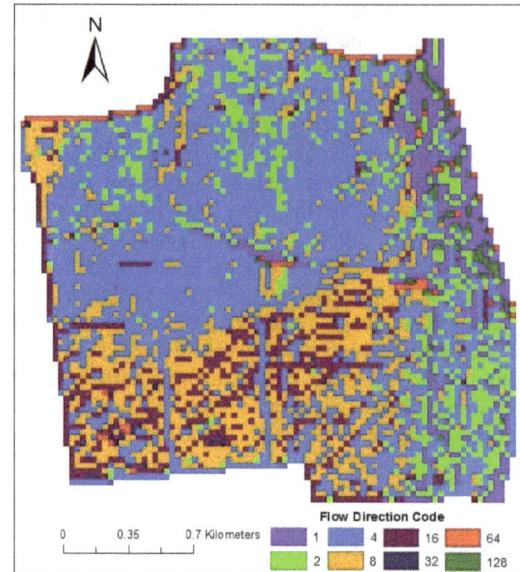

Figure 5. Input flow direction map

Figure 6 illustrates the inundated points, streams, and the sub catchment of the study area calculated by the storm inundation calculation service.

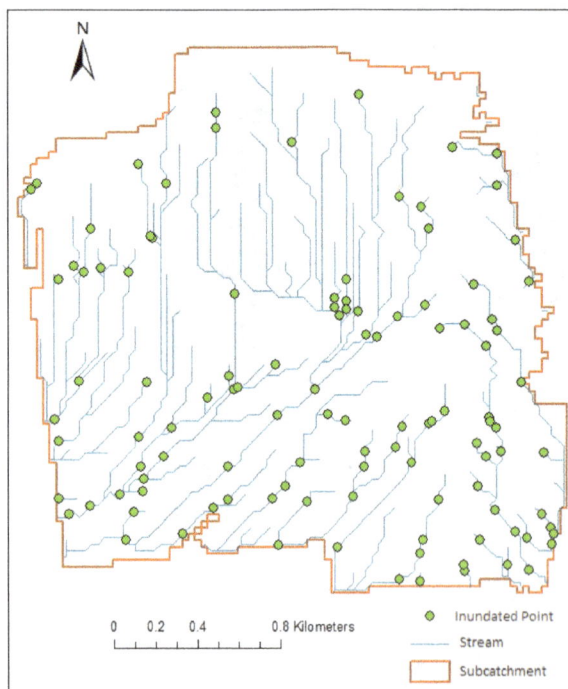

Figure 6. Inundated points, streams, and the sub catchment of the study area calculated by the storm inundation calculation service

In order to evaluate the inundation, the flooding assessment tool bar has been envisaged which includes three tools: spatial query, attribute query and inundation statistics calculation. The spatial query tool enables the user to select his intended areas based on a distinct geometry and consequently, observe its inundation characteristics. On the other hand, the attribute query tool enables the user to enquire according to a special feature and observe the flood characteristics. Besides, the flood statistics tool provides some statistics in regard to the flood status, including a maximum volume of inundation, the maximum area and depth of inundation as well as the total area of flooded regions.

## 4. CONCLUSIONS

In the current study, the automatic calculation process service of storm inundation was developed on the basis of USISM model. In order to examine the implemented service, a web application was also developed. This system made it possible to measure the inundation of a rainfall event. The output of the system can be used for decision-making and carrying out practical measures in minimum possible time.

Processing functions required for automatic calculation of storm inundation were published on a server as a standard WPS service. The results would be sent to the user after receiving rainfall information and required maps, the service runs and result sent to the client. Then, the user observes the results through the UI, who is able to spatially and descriptively query on the map of flooded areas and observe their characteristics. The characteristics consist of volume, depth and area of each inundated point. The implementation of processing functions in a form of processing web services gives the capability to reuse the services and apply them in other services. As a result, it would avoid creating the duplicate resources.

The storm inundation calculation in the developed service is based on some simplifying assumptions. First, the equipment's related to storm water collecting infrastructure was only regarded in determining hydrological characteristics of the research site. Moreover, in calculating inundation amount, the medium capacity has been utilized that was designed for flood collection infrastructure. Indeed, such simplification sometimes overestimates and sometimes underestimates the drainage capacity. In addition, ignoring the spatial distribution of the drainage process makes errors in the simulation. USISM also simulates the final status of storm event, and thus, the intensity and duration of rainfall enjoy such a great importance with respect to the inundation calculation.

## REFERENCES

Apel, H., Aronica, G. T., Kreibich, H. & Theieken, A. H., 2009. Flood risk analyses-how detailed do we need to be?, Natural Hazards. 49, pp. 79-98.

Bean, J., 2008. *SOA and web services interface design: principles, techniques, and standards*. Morgan Kaufmann, USA.

Bonta, J. V., 2004. Development and utility of Huff curves for disaggregating precipitation amounts, *Applied engineering in agriculture*. 20, pp. 641-656.

Castronova, A. M., Goodall, J. L. & Elag, M. M., 2013. Models as web services using the Open Geospatial Consortium (OGC) Web Processing Service (WPS) standard, *Environmental Modelling & Software*. 41, pp. 72-83.

Cepicky, J. & Becchi, L., 2007. Geospatial processing via Internet on remote servers PyWPS, *OSGeo Journal*. 1, pp. 39–42.

Cheng, X. T., 2010. Urban water disasters and strategy of comprehensive control of water disaster, *Journal of Catastrophol*. 25, pp. 10-15.

Diaz, L., Granell, C. & Gould, M., 2008. Case Study: Geospatial Processing Services for Web based Hydrological Applications. In: J. T. Sample, K. Shaw, S. Tu & M. Abdelguerfi, eds. *Geospatial Services and Applications for the Internet*. US: Springer, pp. 31-47.

Evangelidis, K., Ntouros, K., Makridis, S. & Papatheodorou, C., 2014. Geospatial services in the Cloud, *Computers & Geosciences*. 63, pp. 116–122.

Fenoy, G., Bozon, N. & Raghavan, V., 2013. ZOO-Project: the open WPS platform, *Geomatics*. 5(1), pp. 19–24.

Friis-Christensen, A., Ostländer, N., Lutz, M. & Bernard, L., 2007. Designing Service Architectures for Distributed Geoprocessing: Challenges and Future Directions, *Transactions in GIS*. 11, p. 799–818.

Kadlec, J., Ames, D.P., Bayles, M., Seul, M., Hooper, R. and Cummings, B., 2016. Web Based Analysis of Hydrological Time Series in R using Web Processing Services. *8th International Congress on Environmental Modelling and Software*.

Luo, W., Li, x., Molloy, I. & Di, L., 2014. Web Service for extracting stream networks, *GeoJournal*. 79, p. 183–193.

Smith, M. B., 2006. Comment on 'Analysis and modeling of flooding in urban, *Journal of Hydrology*. 317, pp. 355-363.

Zhang, S. & Pan, B., 2014. An urban storm-inundation simulation method based on GIS, Journal of Hydrology. 517, p. 260–268.

# 10

# SPATIO-TEMPORAL PATTERN MINING ON TRAJECTORY DATA USING ARM

S. Khoshahval [a]*, M. Farnaghi [a], M. Taleai [a]

[a] Faculty of Geodesy and Geomatics Engineering, K.N. Toosi University of Technology, Tehran, Iran
(samira_khoshahval@yahoo.com; farnaghi@kntu.ac.ir; taleai@kntu.ac.ir)

**KEY WORDS:** User Trajectory, Association Rule Mining, Location-based Application, Frequent Pattern Mining, Apriori Algorithm

**ABSTRACT:**

Preliminary mobile was considered to be a device to make human connections easier. But today the consumption of this device has been evolved to a platform for gaming, web surfing and GPS-enabled application capabilities. Embedding GPS in handheld devices, altered them to significant trajectory data gathering facilities. Raw GPS trajectory data is a series of points which contains hidden information. For revealing hidden information in traces, trajectory data analysis is needed. One of the most beneficial concealed information in trajectory data is user activity patterns. In each pattern, there are multiple stops and moves which identifies users visited places and tasks. This paper proposes an approach to discover user daily activity patterns from GPS trajectories using association rules. Finding user patterns needs extraction of user's visited places from stops and moves of GPS trajectories. In order to locate stops and moves, we have implemented a place recognition algorithm. After extraction of visited points an advanced association rule mining algorithm, called Apriori was used to extract user activity patterns. This study outlined that there are useful patterns in each trajectory that can be emerged from raw GPS data using association rule mining techniques in order to find out about multiple users' behaviour in a system and can be utilized in various location-based applications.

## 1. INTRODUCTION

Traditionally, the mobile phones were a device of communication; but today thanks to the technology advancements we can use them as tracking gadgets. Mobile devices with GPS (Global Positioning System) receivers are able to capture and record traces of users' movement, including coordinates, timestamps, elevation and other attributes such as velocity and bearing. Owing to GPS-enabled mobiles, huge amounts of user traces are recorded and stored from user daily activities. The need for using hidden information in trajectories developed studies in various fields such as urban planning, transportation, surveillance and security (Gudmundsson, Laube et al. 2011).

Extracting useful information from raw GPS trajectory data is a complex task. Trajectory data is basically a series of points which is not mostly human readable and needs interpretation; hence, trajectory analysis methods are necessary to extract useful information about users' behaviour (Zheng, Zhang et al. 2009).

Users movements make a continuous number of points in each GPS trajectory. Examining trajectory points conveys stops and moves in a route (Zheng, Li et al. 2008). User's stop places make a concentration of points in a specific section of the trace. Loosing GPS signal may cause outlier in a track which implies the necessity of trajectory data pre-processing. Movement analysis calculates several measures such as distance or time between each two points of a trace and uncovers stops and moves which leads to users visited points. If user visits some specific locations permanently, those locations will be considered as user's point of interest (POI) which can be determined using place recognition algorithms (Gong, Sato et al. 2015).

Association rule mining (ARM) is one of the most popular methodologies of data mining which is an admirable apparatus for information discovery in large data itemsets. The major purpose of ARM is to distinguish the most frequent groups of items transpiring together (Zhang and Zhang 2002). Diverse association rule mining algorithms have been proposed by various researchers (Agrawal and Srikant (1994), Agrawal,

Mannila et al. (1996) and Narvekar and Syed (2015) to name a few). Among multiple available algorithms we have selected Apriori which is one of the most famous rule mining algorithms. Apriori advantages are simplicity and its ability to generate rules from frequent itemsets. Furthermore, Apriori divides the problem of finding all association rules into two subproblems which are finding all frequent itemsets and generating rules (Agrawal, Mannila et al. 1996).

This paper aims to depict a solution based on Apriori algorithm to find out user behavioural patterns from user's trajectory data by generating rules. Our solution is composed of two major stages of place recognition and generating rules based on discovered visited places. Place recognition gives us coordinates of user's visited points as stops which are then going to be semantically processed and formed to visited places by adding location name, part of day and some other features to the visited points' coordinates. Rule generation is obtained by taking advantage of Apriori which is a famous ARM algorithm. Eventually, produced rules are interpreted to discover user behavioural patterns.

This paper is divided into 5 sections. Section 2 gives a brief review of the related works. The proposed solution is thoroughly described in section 3. The experimental results are discussed in section 4 and our conclusion is drawn in the final section including the future work.

## 2. RELATED WORKS

Recently trajectory analysis and extracting user behavioural patterns have risen interest of researchers who study moving objects activities. Spaccapietra, Parent et al. (2008) exhibited modelling approaches in order to show the role of trajectories by attaching semantic annotations to bird's migration trajectories. Baglioni, de Macêdo et al. (2009) reached an interpretation of moving objects via enriching trajectories with semantic and geographical information of an ontology.

The related works in the field of user behavioural pattern discovery can be generally divided into place recognition studies

---

\* Corresponding author

and spatial association rule mining studies. In order to discover visited points, researchers put effort on not only semantic trajectory analysis but also finding stops and moves in traces by using time and distance based algorithms. These methods can be divided to three major categories. The studies in the first category are mostly based on time and location. Zheng, Zhang et al. (2009) modelled location histories of user with a graph and proposed a HITS-based model. The studies in the second category focus on indirect parameters such as direction and acceleration which are calculated based on velocity of each GPS points in order to extract user's movement type (see Rehrl, Leitinger et al. (2010) and Bhattacharya, Kulik et al. (2012)). The last category of place recognition algorithms utilizes clustering to group data that shares identical characteristics. In the study of Ashbrook and Starner (2003), a system is designed to cluster GPS data into locations. An improved version of DBSCAN is presented in Gong, Sato et al. (2015) which uses coordinates and time of GPS trajectories to identify activity stop points.

Various approaches have been hypothesized to generate rules using ARM algorithm. Presenting a beneficial ARM algorithm was a concern of a number of researchers. Apriori, the most famous ARM algorithm, was first introduced by Agrawal and Srikant (1994). Apriori finds the frequent itemsets and generate rules based on discovered itemsets. AprioriTID determines the candidate itemsets before starting the pass in database. AprioriHybrid is a combination of Apriori and AprioriTID which transfers between algorithms in different passes which causes extra costs (Agrawal and Srikant 1994). In the following studies, attempts have been made by Han and Pei (2000) to use FP-tree concepts and they presented FP-Growth algorithm which has complexity in calculation. Moreover, other algorithms such as

Eclat and Recursive elimination was introduced by Schmidt-Thieme (2004) and Borgelt (2005), respectively.

All of the named researchers tend to develop more frequent rule mining algorithms but the gained methodologies have rarely been used for spatial pattern mining issues. Mousavi, Hunter et al. (2016) presented semantically interpretation of trajectory patterns using ontologies but researchers mostly have tended to focus on place recognition and spatial association rule mining separately but the purpose of our study is mainly on providing a methodology which uses the results of a place recognition algorithm as inputs for spatial association rule mining in order to discover user patterns. We have used a time and distance based algorithm to discover visited points and Apriori algorithm to generate rules based on extracted places.

## 3. METHOD

The proposed solution is composed of 4 steps of data preparation, visited place extraction, spatiotemporal analysis and association rule mining on extracted visited places (Figure 1). This solution receives trajectory data as input and generate users' behavioural patterns as output. Each step has an output which is going to be the input of next step.

In the first step OSM data and moving object data is gathered and pre-processed. Then, place recognition algorithm extracts visited points. Spatiotemporal analysis is implemented on the visited points to generate visited places' feature table Afterwards. Finally, association rules are generated based on visited places' feature table from which users' behavioural pattern are obtained.

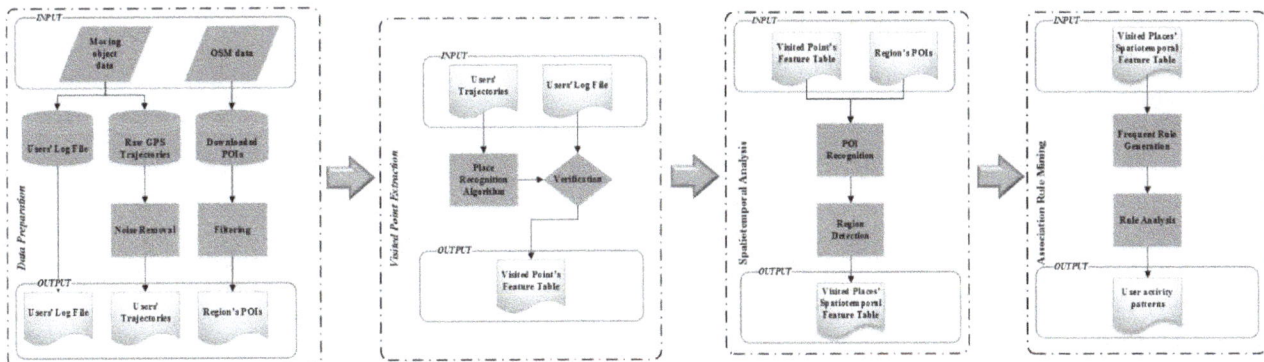

**Figure 1.** Proposed approach

### 3.1 Data preparation

The input data of the solution is composed of three different datasets; including users' Log file, users' trajectories and regions' POIs.

In an attempt to gather needed moving object data, we kept tracks of users for 60 days. Users were asked to use a defined tracking application and start tracking from early morning until night. Moreover, users had to keep a log file of places that they visited each day as a diary to validate the gained results. Loosing GPS signals in some environmental conditions may cause outliers. Therefore, we preprocessed the trajectories in order to eliminate useless points.

In this study, regions' POIs and some base maps of city of Tehran in Iran were downloaded from the Open Street Map (OSM) website. Tehran has 22 districts and 134 regions. Due to public access to OSM there is a mixture of true and false registered POIs for a region and a lot of placemarks without any useful information. The filtering process removes unknown

places of the downloaded POIs. Loosing GPS signal may cause outliers in trajectories and therefore noise removal is necessary to eliminate useless points. Finally, the outputs of this step are available in the format of shape files, KML and tables.

Figure 2a. is a sample trajectory gathered by a user. As it can be seen, there is a concentration of GPS points which is probably a visited place. Figure 2b. is a sample of OSM POIs restricted to a specific region.

**(a)**

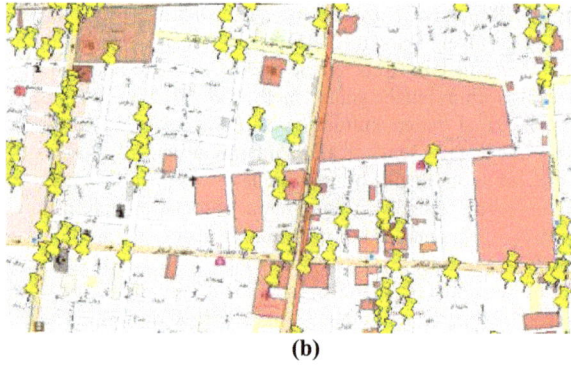

**(b)**

**Figure 2.** (a) User trajectory; (b) Explored POIs for a restricted region of Tehran

## 3.2 Visited point extraction

A trajectory is a record of changes in the position of a moving object during a period of time (Spaccapietra, Parent et al. 2008). In other words, it is a series of points, in the form of equation 1, in which there are features like coordinate, timestamp and etc.

$$\{(t_0, x_0, y_0), (t_1, x_1, y_1), (t_2, x_2, y_2), ..., (t_n, x_n, y_n)\}$$
$$\text{where} \begin{cases} t_i, x_i, y_i \in R, i = 0, 1, 2, ..., N \\ t_0 < t_1 < t_2 < \cdots < t_N \end{cases} \quad (1)$$

A visit point is a location where user stays in the place for a specific time duration (Lv, Chen et al. 2016). In equation 2, P is a set of N visit points that an algorithm can find in a trajectory.

$$P = \{VP_1, VP_2, .., VP_N\} \quad (2)$$

To determine whether the trajectory has any visited point or not, we developed a place recognition algorithm. Figure 3 presents the developed time-based algorithm adopted from Lv, Chen et al. (2016). This algorithm examines the time interval between each two GPS point. In the algorithm, T is a set of GPS trajectories and $\varepsilon_t$ is the time threshold which is considered as a minimum of 5 minutes. $\varepsilon_d$ is defined as the distance threshold which is set to 200 meters. In the first run of the algorithm constant thresholds are applied to parameters and the set of visited points is empty. For each trajectory, the algorithm calculates distance and time interval between points incrementally and compares it with the given thresholds. If the distance is less than the minimum threshold, the point is considered as a potential visited point (Line 2-5) but if it is more than minimum threshold the algorithm checks the time interval (Line 7-9). Eventually, the output of the algorithms is a list of discovered visited points each of which includes 5 attributes of latitude, longitude, elevation, time and day.

---

**Algorithm 1**
Input: GPS traces T, threshold $\varepsilon$
Output: Visited Places VP
1: $\varepsilon_t$ and $\varepsilon_d$ constant, initial set of VPs = Ø
2: **for** trajectory points in T **do**
3: calculate distance interval
4:   **if** measured distance < $\varepsilon_d$ **then**
5: consider Point as a potential VP
6:     **else**
7: calculate time duration
8:     **if** measured time > $\varepsilon$ **then**
9:     add Point to VP set
10: create the table
11:     **else**

---

12:     print no VPs
13: **end**

**Figure 3.** Place recognition algorithm (Lv, Chen et al. 2016)

### 3.3 Spatiotemporal analysis

Taking advantage of GIS tools, we exploited the exported visited points in order to append semantic meanings to the extracted visited points, this entails merging coordinates of visited points to downloaded OSM POIs using proximity tools. Then, visited places are discovered and connected to their location names.

In order to accomplish spatial analysis, we considered human activities and part of day in 8 and 7 separate fields respectively (Figure 4). Each visited place is connected to one or more activity type, therefore user daily activity types are determined. For temporal analysis, we used a division of hours of day like morning until night in the form of early/late morning, early/late afternoon, early/late evening and night. Then, exact hours of each visited place were matched to the correct time interval.

Finally, the output of this step is visited places spatiotemporal feature table which illustrates a list of user POIs including location, day, part of day and region.

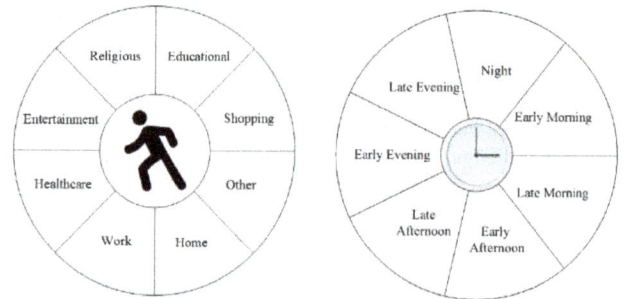

**Figure 4.** Human daily visited places/Part of day

### 3.4 Association rule mining on extracted visited places

Association rule mining is one of the most important data mining techniques and it is a well-known methodology for disclosing considerable relations and patterns among items in large datasets. Basically, association rule mining methods are also called market basket analysis because they enable us to define customers shopping behaviour (Agrawal, Mannila et al. 1996). In market basket analysis, association rules are generated for items with support and confidence larger than the user defined measures. In equation 3, I is an itemset with m items in the database of T with j transactions.

$$I = \{i_1, i_2, ..., i_m\} \quad (3)$$
$$T = \{tr_1, tr_2, ..., tr_j\}$$

There are two major measures of support and confidence which help us to define the frequent itemsets. The support of an association rule, as it is defined by equation 4, is the number of transactions with both items A and B over the whole number of transactions in the database.

$$\text{Support}(A \rightarrow B) = P(A \cap B) \quad (4)$$
$$\approx \frac{\text{NO. transactions with both A and B}}{\text{All transactions in DB}}$$

The confidence of an association rule is shown by equation 5. It is the number of transactions with both items A and B over the number of transactions with A.

$$\text{Confidence}(A \rightarrow B) = P(B|A) \quad (5)$$

$$\approx \frac{\text{NO. transactions with both A and B}}{\text{Transactions with only A}}$$

Apriori firstly introduced by Agrawal, Mannila et al. (1996) is an association rule mining algorithm that finds frequent itemsets of L in database D using support and confidence. This search is level-wise and it goes up the lattice of itemsets. In the first step, the algorithm scans the whole database and eliminates items that are lower than the defined minimum support and then all items are combined together to produce subsets of items for the next scan.

For association rule mining using Apriori algorithm we considered spatial and temporal attributes exported form spatiotemporal analysis step as items of the rule mining database. Thus, we used spatial features such as location name and location category. Moreover, temporal features such as day and part of day were also added to the rule mining database. Then, the frequent items were selected based on defined ARM parameters to generate rules.

### 4. EXPERIMENT AND ANALYSIS

After applying the four steps of the proposed approach on the data, the activity rules were generated. Some of the rules, that were generated for a particular user who was a student with a part-time job can be seen in Tables 1 and Table 2. Table 1 and Table 2 present spatial patterns and spatio-temporal patterns, respectively. Among all generated rules, there was a significant rule by support of 1.2% and confidence of 98% which shows that if the user is at home the next location is work and if the user goes shopping after work the next location is home (Table 1). In table 2, rules with a combination of time and location are shown. For the first rule in table 2, if it is Saturday (the first day of week in Tehran) and part of the day is early in the morning, user is going to have an educational activity such as a class in university by the support of 2.2% and confidence of 87%. For this particular user with a part time job, it can be seen that the second rule (Table 2) shows a type of work activity in the early afternoon. Finally, the user's ending location is home after workplace in the late evening (the last row in Table 2).

| Rules |
|---|
| If *Home* **Then** *Work* |
| If *Work* **Then** *Shopping* **Then** *Home* |

**Table 1.** Spatial user activity patterns

| Rules |
|---|
| If day *Saturday* part of day *Early Morning* **Then** *Educational* |
| If part of day *Early Afternoon* **Then** *Work* |
| If part of day *Late Evening* place *Work* **Then** *Home* |

**Table 2.** Spatio-temporal activity patterns

According to the users' log files, the active user is mostly used to do two different kinds of activities on Fridays which are in the category of shopping and entertainment. Both of these activities happen sporadically. For instance, in one month, the user does shopping activities three times but in the next month, the activity is declined to one. Therefore, the system considers the activities on Friday generally and presents the most frequent one. According to the gained user's behavioral pattern, entertainment was the most frequent event on Fridays early evening (Table 3). Therefore, there would be some differences between our user log file and the systems' presented result for the activities if the frequency of activities were lower than the threshold. This event can be fairly alleviated by reducing the support that is not recommended. Instead of decreasing thresholds it would be better

to increase the period of trajectory gathering process in order to obtain a supplementary database.

| Rules |
|---|
| If day *Friday* part of day *Early Evening* **Then** *Entertainment* |

**Table 3.** The most frequent activity

Although, the resulted if/then rules may appear to be obvious for a known user such as a friend who we are familiar with her daily routine, the rules generated by the proposed solution are gained automatically in a systematic procedure. These rules can be obtained for multiple unknown users with different attributes by the proposed methodology and the activity patterns can be exploited in a location-based service to recommend useful information to users. In such a location based service, where multiple users are producing trajectory data, obtaining behavioural rules is a complex task and it is not obvious for human and analysis is necessary.

### 5. CONCLUSION

In this paper, we proposed a method to reveal user behavioral patterns from GPS trajectory data using place recognition and association rule mining. We applied place recognition algorithm to extract visited points; then, spatiotemporal analysis resulted in visited places feature table as input items to discover spatial association rules. The extracted rules conveyed human readable patterns that provide useful information about user daily activities.

For future work, behavioral patterns can be used in location based services. Understanding a user activity pattern in a system can be a beneficial factor in further usage of location-based services. Having known the user behavioral patterns, a location-based service can provide more accurate recommendations to users. For instance, if the service knows about the history of user activities on Saturday, it can recommend correct places in exact time to the user. Therefore, suggestions of the service would be precise and user-directed.

### REFERENCES

Agrawal, R., H. Mannila, R. Srikant, H. Toivonen and A. I. Verkamo (1996). "Fast Discovery of Association Rules." Advances in knowledge discovery and data mining **12**(1): 307-328.

Agrawal, R. and R. Srikant (1994). Fast algorithms for mining association rules. Proc. 20th int. conf. very large data bases, VLDB.

Ashbrook, D. and T. Starner (2003). "Using GPS to learn significant locations and predict movement across multiple users." Personal and Ubiquitous computing **7**(5): 275-286.

Baglioni, M., J. A. F. de Macêdo, C. Renso, R. Trasarti and M. Wachowicz (2009). Towards semantic interpretation of movement behavior. Advances in GIScience, Springer: 271-288.

Bhattacharya, T., L. Kulik and J. Bailey (2012). Extracting significant places from mobile user GPS trajectories: a bearing change based approach. Proceedings of the 20th International Conference on Advances in Geographic Information Systems, ACM.

Borgelt, C. (2005). Keeping things simple: finding frequent item sets by recursive elimination. Proceedings of the 1st international workshop on open source data mining: frequent pattern mining implementations, ACM.

Gong, L., H. Sato, T. Yamamoto, T. Miwa and T. Morikawa (2015). "Identification of activity stop locations in GPS trajectories by density-based clustering method combined with support vector machines." Journal of Modern Transportation **23**(3): 202-213.

Gudmundsson, J., P. Laube and T. Wolle (2011). Computational movement analysis. Springer handbook of geographic information, Springer: 423-438.

Han, J. and J. Pei (2000). "Mining frequent patterns by pattern-growth: methodology and implications." ACM SIGKDD explorations newsletter **2**(2): 14-20.

Lv, M., L. Chen, Z. Xu, Y. Li and G. Chen (2016). "The discovery of personally semantic places based on trajectory data mining." Neurocomputing **173**: 1142-1153.

Mousavi, A., A. Hunter and M. Akbari (2016). "USING ONTOLOGY BASED SEMANTIC ASSOCIATION RULE MINING IN LOCATION BASED SERVICES."

Narvekar, M. and S. F. Syed (2015). "An Optimized Algorithm for Association Rule Mining Using FP Tree." Procedia Computer Science **45**: 101-110.

Rehrl, K., S. Leitinger, S. Krampe and R. Stumptner (2010). An approach to semantic processing of gps traces. Proceedings of the 1st Workshop on Movement Pattern Analysis, Zurich, Switzerland.

Schmidt-Thieme, L. (2004). Algorithmic Features of Eclat. FIMI.

Spaccapietra, S., C. Parent, M. L. Damiani, J. A. de Macedo, F. Porto and C. Vangenot (2008). "A conceptual view on trajectories." Data & knowledge engineering **65**(1): 126-146.

Zhang, C. and S. Zhang (2002). Association rule mining: models and algorithms, Springer-Verlag.

Zheng, Y., Q. Li, Y. Chen, X. Xie and W.-Y. Ma (2008). Understanding mobility based on GPS data. Proceedings of the 10th international conference on Ubiquitous computing, ACM.

Zheng, Y., L. Zhang, X. Xie and W.-Y. Ma (2009). Mining interesting locations and travel sequences from GPS trajectories. Proceedings of the 18th international conference on World wide web, ACM.

# 11

# COMPARISON of FUZZY-BASED MODELS in LANDSLIDE HAZARD MAPPING

N. Mijani [a], N. Neysani Samani [b,*]

[a] MSc. Student of Remote Sensing and GIS, Department of Remote Sensing and GIS, University of Tehran,
Tehran, Iran - naeim.mijani@ut.ac.ir
[b] Assis. Prof of Remote Sensing and GIS, Department of Remote Sensing and GIS, University of Tehran,
Tehran, Iran - nneysani@ut.ac.ir

**KEY WORDS**: Landslide, Fuzzy-based Models, Quality Sum Index, Accuracy

## ABSTRACT

Landslide is one of the main geomorphic processes which effects on the development of prospect in mountainous areas and causes disastrous accidents. Landslide is an event which has different uncertain criteria such as altitude, slope, aspect, land use, vegetation density, precipitation, distance from the river and distance from the road network. This research aims to compare and evaluate different fuzzy-based models including Fuzzy Analytic Hierarchy Process (Fuzzy-AHP), Fuzzy Gamma and Fuzzy-OR. The main contribution of this paper reveals to the comprehensive criteria causing landslide hazard considering their uncertainties and comparison of different fuzzy-based models. The quantify of evaluation process are calculated by Density Ratio (DR) and Quality Sum (QS). The proposed methodology implemented in Sari, one of the city of Iran which has faced multiple landslide accidents in recent years due to the particular environmental conditions. The achieved results of accuracy assessment based on the quantifier strated that Fuzzy-AHP model has higher accuracy compared to other two models in landslide hazard zonation. Accuracy of zoning obtained from Fuzzy-AHP model is respectively 0.92 and 0.45 based on method Precision (P) and QS indicators. Based on obtained landslide hazard maps, Fuzzy-AHP, Fuzzy Gamma and Fuzzy-OR respectively cover 13, 26 and 35 percent of the study area with a very high risk level. Based on these findings, fuzzy-AHP model has been selected as the most appropriate method of zoning landslide in the city of Sari and the Fuzzy-gamma method with a minor difference is in the second order.

## 1. INTRODUCTION

Slope instability is one of the important natural phenomena. Increasing trend of urbanization and overuse of natural resources has exacerbated this phenomenon (Ercanoglu and Gokceoglu, 2004). Landslides are known as one the most common geological disasters which cause damages and casualties worldwide (Bianchini et al., 2016; Shahabi et al., 2014; Wang et al., 2016). The unplanned urbanization especially in developing countries and wide climate changes through global warming increase the risk of natural hazards. Landslide phenomenon is an important worldwide natural hazard and Iran is no exception (Vakhshoori and Zare, 2016).

Rapid population growth, the expansion of human settlements in mountainous areas, difficulty of predicting the time of occurrence of landslides and having numerous factors involved in this phenomenon reveals the necessity of landslide hazard zonation. Landslide is considered one of the most complicated natural phenomenon which endanger human generations (Nourani et al. 2013).

Landslides are caused due to many factors such as earthquakes, rains and rapid melting of snow (Liu et al., 2011) and are affected by factors such as topography, soil and rock type, fractures and bedding, humidity levels and human activities (Florsheim and Nichols, 2013; Liu et al., 2011). Having data related to number, area and volume of landslides is important for estimation of sensitivity (Guzzetti et al., 1999; Malamud et al., 2004), determination of risk of landslides (Cardinali et al., 2002; Reichenbach et al., 2005) and long term evaluation of slopes due

to the effect of mass movements (Korup ,2005; Imaizumi and Sidle, 2007; Guzzetti et al., 2008).

Landslide susceptibility map is a helpful tool for planning and decision making in landslide hazard managements (Hong et al. 2016; Tsangaratos et al., 2015). Sensitive areas with risk potential can be identified using landslide hazard zonation and landslides or damage caused by those can be prevented to some extent by provision of appropriate strategies and management practices.

Different methods have been provided for landslide hazard zonation such as logistic regression (Atkinson and Massari 2011; Conoscenti et al. 2014), neuro-fuzzy (Tien Bui et al. 2012; Vahidnia et al. 2010), decision trees (Alkhasawneh et al. 2014; Tsangaratos and Ilia 2015) and support vector machines (Dou et al. 2015; Hong et al. 2015; Peng et al. 2014), but none of those have the necessary certainty and provided methods in most cases can be used for specific areas by consideration of necessary reformations.

In this way, it seems that fuzzy-based models could model the uncertain aspect of landslide hazard mapping. Fuzzy theory was presented by Lotfi Zadeh in 1965 includes all theories which use basic concepts of fuzzy sets or membership functions.

Lotfi Zadeh (1965) stated that membership function must be defined for determination of members in one set which means membership value exclude the exact zero and one and it is a value between these two. Zero means that it has no membership in the set and one means that it is fully a member of that set (Zadeh, 1965). Many studies have been carried out in Iran and around the world for landslide hazard zonation using fuzzy logic (Chung and Fabbri 2001; Ercanoglu and Gokceoglu, 2004; Lee 2007; Pradhan, 2010).

---

Thiery et al., (2006) used fuzzy logic to evaluate landslide-prone areas in the northern foothills of the Alps in France and introduced the use of fuzzy logic due to the high accuracy and measurement of outputs with proper definition of fuzzy logic operators and combination of sum and γ operators for generation of landslide map as the best combination.

(Barrile et al., 2016; Akgun et al., 2012; Pourghasemi et al., 2012) some studies used fuzzy membership function to prepare landslide hazard zonation map and introduced fuzzy Logic due to the coordination of data and also for additional flexibility of spatial analysis process as very efficient and useful method in preparation of landslide hazard mapping. Due to the fact that a large part of Iran's area is mountainous, there are many areas susceptible to mass movement occurrences and many researchers are trying to provide different methods for identification and zoning of these natural hazards (Pourghasemi et al., 2012; Ghanavati et al., 2015; Pourghasemi et al., 2016; Vakhshoori and Zare, 2016; Aghdam et al., 2017; Gheshlaghi and Feizizadeh, 2017).

Mattkan et al., (2009) used variables such as geology, pedology, altitude, slope, aspect, and distance from the river, distance from the road, distance from fault, vegetation cover and land use for landslide hazard zoning in Lajim River watershed and calculated the weight of variables affecting the extraction of fuzzy membership functions using the method of relative frequency of landslides. Their results show that Fuzzy Gamma and Fuzzy Ordered Weighted Averaging models have the lowest variation and standard deviation compared to the other models.

With respect to occurrence of numerous landslides in the city of Sari over the past years, The main contribution if this paper reveals to the comprehensive criteria causing landslide hazard considering their uncertainties and comparison of different fuzzy-based models including Fuzzy-AHP, Fuzzy Gamma and Fuzzy-OR and preparing a landslide hazard mapping for Sari city.

## 2. PROPOSED METHOD

In this study, different fuzzy operators including 'AND', 'SUM','PRODUCT' and 'OR' considered and the 'OR' operator is selected, because the other operators classify the target area into either very high or very low susceptible zones that are inconsistent with the physical conditions of the study area. In the case of fuzzy gamma, the success and prediction rates increase for higher gamma in such a way that 0.975 shows the best result. The increasing trend of success and prediction rates of gamma operators along with the increasing of γ value is due to the balanced effect of 'PRODUCT' and 'SUM' operators on its equation (equation 13).

The overall methodology flowchart of the study is shown in figure 1. The methodology consists of three Steps:

**Step (1):** Providing spatial database including landslide conditioning factors and historical land slide locations.

**Step (2):** Landslide hazard mapping using Fuzzy-AHP, Fuzzy-Gamma and Fuzzy-OR approaches.

**Step (3):** Accuracy assessment of the constructed maps using DR, QS and method P parameter.

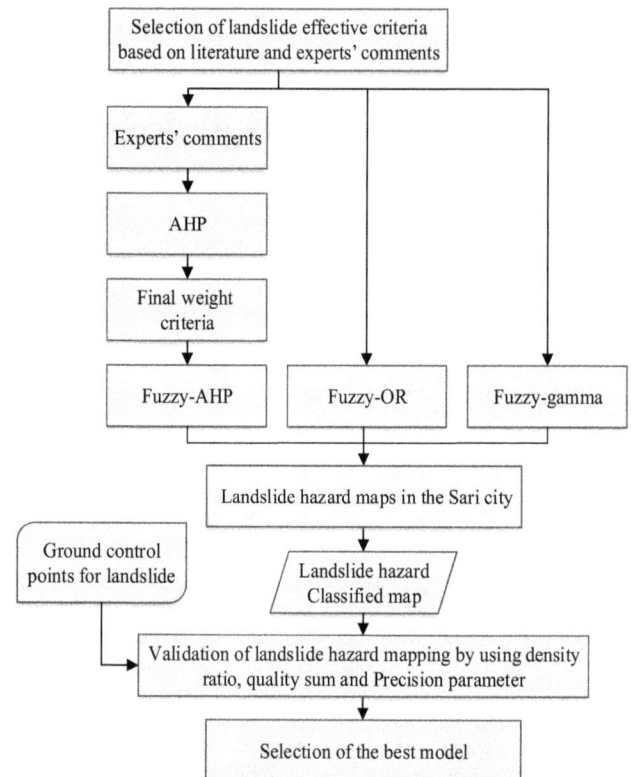

Figure 1. Flowchart of proposed methodology in the study area.

### 2.1 Fuzzy logic and fuzzy-based models

Fuzzy logic has been introduced by Zadeh (1965). Fuzzy Logic Whereas the classical theory of crisp sets can describe only the membership or non-membership of an item to a set, fuzzy logic permits partial membership, which can pose a value from 0 to 1:

$$\mu_A(x) : X \rightarrow [0, 1] \tag{1}$$

in which X refers to the universal set defined in a specific problem and $\mu_A(x)$ the grade of membership for element x in fuzzy set A. The crisp set is a special case of fuzzy sets, in which the membership function for each element takes one of only two values: 0 or 1 (Zadeh, 1965; Samany et al., 2014). To build a fuzzy logic- based model, the proper types of membership function and its parameters should be carefully selected. The process of decomposing a given system input and/or output into fuzzy sets is called fuzzification (Samany et al., 2014). In this study, the "linear" fuzzification algorithms were used. In this study, the fuzzy logic was used for standardized factors in the range of 0–1.

### 2.2 Analytic hierarchy process (AHP)

AHP is one of the most comprehensive methods of multi-criteria decision-making methods (Saaty, 1980) because it provides the capability of formulation of natural complex problems in hierarchy form and it can also consider different qualitative and quantitative criteria (Saaty, 1986). The greatest weight in AHP is related to a layer which has the greatest effect in determination of the objective. In other words, the criteria for weighting each information unit is also based on the greatest effect played by that factor in the layer (Malczwerski, 1999). Based on researches carried out by Saaty and Vargas (1991), a range was suggested for comparison of criteria which includes numerical values from 1 to 9. Each of these numbers show the degree of importance in a way that 1 shows the equal importance and 9 shows the extremely strong importance of a criteria compared to another criteria. The AHP for weighting of criteria consists of four steps (Cay and Uyan, 2013):

1. Creation of hierarchical structure: this step is the most important step in analytic hierarchy process. Hierarchical structure is a graphical representation of a real complex problem on top of which there is the overall objective of problem and criteria, sub-criteria and alternatives are in next levels.

2. Pair-wise comparisons: this model is based on pair-wise weighting model of each one of variables with each other. A pair-wise comparison matrix (n×n) is formed for indicators at this step. i-th row is compared with j-th column in pairwise comparison matrix. Hence, the values in main diagonal are equal to one and each value under the main diagonal is opposite of the value above the main diagonal

3. Preparation of normalized matrix and calculation of weight vector: in this step, the values of each one of comparison matrix columns are initially and pair wisely added to each other and them the value of each element in pairwise

4. comparison matrix is divided by the sum of values in its own column.

Then, the average of elements in each row of normalized matrix is calculated as a result of which weight vector of parameters is created.

5. Calculation of compatibility or incompatibility of weight of values: pairwise comparison matrix (A) should be initially multiplied by weight vector (C) for calculation of compatibility rate in order to obtain a good approximation of $\lambda$max:

$$A \times C = \begin{pmatrix} a_{11} & a_{12} & \cdots & a_{1n} \\ a_{21} & a_{22} & \cdots & a_{2n} \\ \cdots & \cdots & \cdots & \cdots \\ \cdots & \cdots & \cdots & \cdots \\ \cdots & \cdots & \cdots & \cdots \\ a_{n1} & a_{n2} & \cdots & a_{nn} \end{pmatrix} \times \begin{bmatrix} c_1 \\ c_2 \\ c_3 \\ \cdots \\ \cdots \\ c_n \end{bmatrix} = \begin{bmatrix} x_1 \\ x_2 \\ x_3 \\ \cdots \\ \cdots \\ x_n \end{bmatrix} \quad (2)$$

Then, Consistency Index (CI) will be initially calculated as follows for measurement of compatibility rate (CR):

$$CI = \frac{\lambda max - n}{n - 1} \quad (3)$$

In this equation, n is the number of criterion or dimensions of A matrix and $\lambda$max is the biggest eigenvalue of A matrix. Then, compatibility rate is determined as follows:

$$CR = \frac{CI}{RI} \quad (4)$$

System's compatibility is acceptable if inconsistency rate is less than or equal to 0.1 and it would be better for decision maker to review the decision if it is greater than 0.1 (Khan and Samadder, 2015).

## 2.3 Fuzzy AHP

Being one of the multicriteria decision making methods, AHP enables the decision maker to regard a certain hierarchy, make association between options and make a choice. AHP has an approach that makes paired comparison of objective and nonobjective criteria, identifies the priorities among the criteria and consists of the significance of the criteria (Can and Arıkan, 2014). AHP was developed by L. T. Saaty for the first time in 1971 and it has been widely studied in literature (Saaty, 1980). Saaty has proposed the significance scale in which numbers from 1 to 9 are used while the decision maker makes paired

comparison in the application process. However, most of the decisions in real life have uncertain results. In such cases, fuzzy AHP is used instead of AHP. While applying fuzzy AHP, the steps below proposed by Chang should be followed (Chang, 1996):

**Step 1:** $X = \{x_1, x_2, \dots, x_n\}$ being criteria set and $U = \{u_1, u_2, \dots, u_n\}$ being targets set, degree analysis $(g_i)$ is applied for every target by regarding every criterion. M degree analysis value related to the targets is expressed in triangular fuzzy numbers $M_{gi}^1, M_{gi}^2, M_{gi}^3$ as, i = 1,2,...,n and j = 1, 2,...,m. So, $M_{gi}^j$ shows triangular fuzzy number related to j target according to i criteria. For example, $M_{g1}^2$ is triangular fuzzy number related to target-2 according to criteria-1.

**Step 2:** Fuzzy synthetic degree value related to i criterion is stated as;

$$S_i = \sum_{j=1}^{m} M_{gi}^j \otimes \left[ \sum_{i=1}^{n} \sum_{j=1}^{m} M_{gi}^j \right]^{-1} \quad (5)$$

Here, equalities are attained as triangular fuzzy number $(l_i, m_i, u_i)$:

$$\sum_{j=1}^{m} M_{gi}^j = \left( \sum_{j=1}^{m} l_j, \sum_{j=1}^{m} m_j, \sum_{j=1}^{m} u_j \right) \quad (6)$$

$$\left[ \sum_{i=1}^{n} \sum_{j=1}^{m} M_{gi}^j \right]^{-1} = \left( \frac{1}{\sum_{j=1}^{m} u_j}, \frac{1}{\sum_{j=1}^{m} m_j}, \frac{1}{\sum_{j=1}^{m} l_j} \right) \quad (7)$$

**Step 3:** Significance vector is calculated indicated as: $W = (d(A_1), d(A_2), \dots, d(A_n))^T$. W vector is attained by normalizing $W'$ vector. i = 1,2,...,n, is described as:

$$W' = (d'(A_1), d'(A_2), \dots, d'(A_n))^T \quad (8)$$

$$d'(A_i) = min \, V (S_i \geq S_k), \, k = 1, 2, \dots, n \text{ and } k \neq i \quad (9)$$

For the triangular fuzzy numbers $M_1 = (l_1, m_1, u_1)$ and $M_2 = (l_2, m_2, u_2)$, the numbers $M_1$ and $M_2$ should be compared calculating both, $V(M_1 \geq M_2)$ and $V(M_2 \geq M_1)$ values. That is why, $d'(A_i)$ values are calculated according to the equality in number (9) by using the equality in number (10) in order to indicate the likelihood $M_2 \geq M_1$ of $V(M_2 \geq M_1)$ statement.

$$V(M_2 \geq M_1) = \begin{cases} 1, & m_2 \geq m_1 \\ 0, & l_1 \geq u_2 \\ \frac{(l_1 - u_1)}{(m_2 - u_2) - (m_1 - l_1)} & otherwise \end{cases} \quad (10)$$

Elements of W vector are calculated as:

$$d(A_i) = \frac{d'(A_i)}{[d'(A_1) + d'(A_2) + \cdots + d'(A_n)]} \quad (11)$$

$i = 1, 2, \dots, n$

Here, $W'$ vector and W vector is found. The ultimate decision is reached suitably for the hierarchical structure of the AHP approach known with W significance vector which is not fuzzy and calculated from the comparison matrix attained by the triangular fuzzy numbers.

## 2.4 Fuzzy-OR

This operator uses the maximum function in combination and is equal to aggregation and it is calculated as (Chung and Fabbri, 2001):

$$\mu_{OR}(x) = MAX [\mu_A(x), \mu_B(x), \dots, \mu_N(x)] \quad (12)$$

This operator extracts the maximum degree of membership for members which means it extracts the maximum value (weight) of each pixel in all informational layers and provides the final map and that is why this operator considers almost the entire area in landslide hazard zonation in extreme risk class.

## 2.5 Fuzzy-γ

This operator is defined based on multiplication of algebraic fuzzy sum and multiplication and it is calculated as (Chung and Fabbri, 2001):

$$\mu_\gamma(x) = [\mu_{SUM}(x)]^\gamma \times [\mu_{PRODUCT}(x)]^{1-\gamma} \qquad (13)$$

$$\mu_{SUM}(x) = 1 - \prod_{i=1}^{n} \mu_i(x) \quad , \quad \mu_{PRODUCT}(x) = \prod_{i=1}^{n} \mu_i(x)$$

In this equation, $\mu_\gamma(x)$ is a result of fuzzy gamma and $\gamma$ is the parameters determined in the range of zero and one. When $\gamma$ is equal to one, the applied combination is the same fuzzy algebraic sum and when $\gamma$ is equal to zero, combination is equal to fuzzy algebraic multiplication. $\gamma$ varies between zero and one. Fuzzy gamma function of 0.975 has been used in this research.

## 2.6 The effective criteria

The main data layers required for landslide hazard assessment in the study area are shown in Table 1.

Table 1. Main data layers

| Criteria | Description of Criteria |
|---|---|
| Slope | Slope is one of the main factors causing landslides in different areas. The function is straight and linear for fuzzification of slope layer since increases level of slope increases the risk of landslides. |
| Altitude | Altitude has been introduced as one of the factors affecting landslide hazard because it has an important role in controlling of the degree and type of erosion. The type of function for its fuzzification is straight and linear since the increased level of height increases the landslide hazard. |
| Land use | Land uses including forest, agricultural use (rainfed), irrigated agriculture and gardens, built areas, bare land and water-filled areas have been identified in the study area based on carried out evaluations. The type of function for fuzzification of land use layer is reversed and linear. |
| Vegetation cover | The type of function for fuzzification of it is reversed and linear since there is a greater risk of falling in areas with poor vegetation cover which also have high level of slope. |
| Rainfall | Rainfall has a direct relation with landslide hazard and increased rainfall increases the risk of landslide by reduction of shear strength of different levels. Direct linear function has been used for fuzzification of rainfall map. |
| Distance from the road | Construction of roads has the most important role among human activities in creation of new landslides and stimulation of old landslides. Non-normative also is among the causes of landslides in addition to road density. |
| Distance from the river | The type of function for fuzzification of it is reversed and linear since increased distance from the river reduces the risk of landslide and as a result, the score will be less. |
| Aspect | Aspect has been introduced as one of the factors affecting landslide hazard. The function is straight and linear for fuzzification of aspect layer. |

## 2.7 Assessing the accuracy of zoning

In this level, we match the distribution map of landslides in the area and risk zoning maps to evaluate and compare landslide hazard zonation methods using QS and P methods. DR is used for evaluation and comparison of accuracy between zones or levels of risk (Yalcin, 2008).

### 2.7.1 Validation or Quality Sum (QS)

DR is required to be initially calculated for determination of QS which is calculated as following (Gee, 1992):

$$DR = (\frac{S_i}{A_i})/(\sum_i^n S_i/\sum_i^n A_i) \qquad (14)$$

In which $S_i$ is the total area of landslides in each risk level, $A_i$ is i-th level of risk in a zoning map and n is the number of risk levels.

Density of landslides risk is ascending from low levels to high levels of risk in hazard maps which have been prepared properly. A method (map) of zoning in landslides density level with DR= 1 is equal to average density of landslides in the whole area and level with DR of 2 has landslide density two times larger than landslide density of the area.

Thus, better distinction between risk levels using the indicator of DR leads to having risk better accuracy or favorability. QS which is calculated as following shows the validity or favorability of the performance of method for predicting the risk of landslide.

$$QS = \sum_{i=1}^{n}((DR-1)^2 \times s) \qquad (15)$$

In which QS is quality sum, DR is density ration, S is the ratio of risk area to the total area and n is the number of risk classes.

Closeness of deviation of DR values from the average of different zones shows that density of landslides n different classes is close to each other and the level of QS is low and high deviation of DR values from the average of different zones shows that density of landslides are different and as a result, the numerical value of QS will be large. Thus in evaluation of methods, higher value of QS in a method will lead to greater accuracy (favorability) in differentiation.

### 2.7.2 Precision of method (P)

It means the ration of area of landslide area in high and very high risk zones to total area of those zones which is calculated as following (Jade and Sarkar, 1993):

$$P = K_s/S \qquad (16)$$

In which P is the Precision of method in zones with moderate to high risk, Ks is the area of the landslide in zones with moderate to high risk and S is the total area of the landslide region.

## 3. CASE STUDY AREA AND REQUIRED DATA

### 3.1 Case study

City of Sari in located in Mazandaran province of Iran. In longitude of 672590 to 764824 east and latitude of 3981925 to 4077768 north in 39th north zone of UTM (Universal Transverse Mercator). The study area has a particular diversity in terms of climate and simultaneously has four Mediterranean semi-humid, wet and very wet climates. This area also has high topography changes due to being simultaneously located in mountainous and

lowland environmental conditions. The study area has been shown in figure 2.

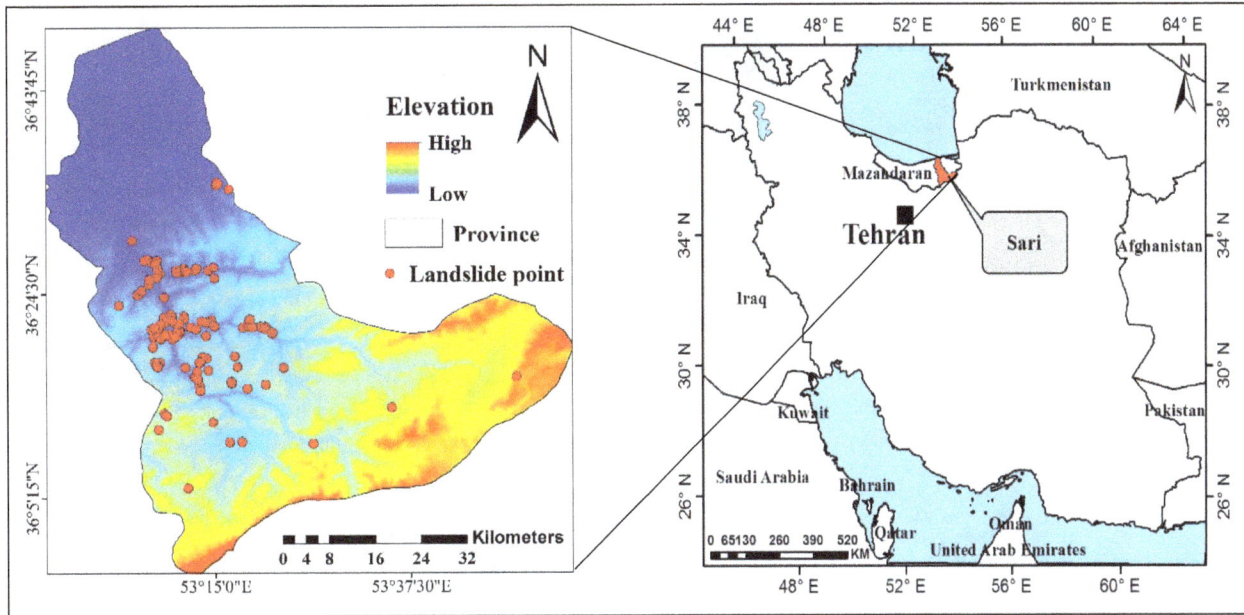

Figure 2. Location of the study area and distribution of landslide data events.

## 3.2 Required data

Initially, distribution data of landslide incidents related to the study area have been prepared from Forest, Rangeland and Watershed organization. Then, these layers were converted into landslide zones using high spatial resolution satellite imagery as well as Google Earth and 1: 100,000 geological map of Sari. This means pointed layer was converted into surface or zoning layer of landslides by determination of distribution of pointed position of landslides on mentioned information sources of area based on this position and the area of occurred landslide as well as its apparent features (cutting area, level of fallen mass, level of dependency to surface displacement of soil) is the dependent variable in the implementation of zoning models as the most important layer used in the present study. Altitude, slope and slope aspect of area were extracted from Aster digital elevation model (DEM). normalized difference vegetation index (NDVI) density of vegetation indicator and Landsat 8 satellite image related to 2015 have been used for preparation of vegetation cover variable. 1:250000 land use map prepared by land use map prepared by Forest Service organization has been used for preparation of land use layer. Shapefile map of isolited lines prepared by Meteorological Agency, Shapefile map of road networks prepared by Roads and Urban Development, Shapefile

map of streams network prepared by Regional Water Authority have been used respectively for preparation of rainfall, road and waterway layers.

## 4. IMPLEMENTATION AND RESULTS

Zoning and risk mapping in this research are based on fuzzy logic. Algorithms of fuzzy functions discussed in this research are linear. The fuzzy membership function tool in ArcGIS 10.4.1 was used to derive membership functions for factors used to derive spatial suitability levels. Maps of various factors have been initially converted to fuzzy maps for landslide hazard zonation using fuzzy linear membership functions. Usage and application of each one of these two functions are done based on two parameters of midpoint and distribution parameter.

Selection of function for fuzzification is based on the nature, importance and relation of each criterion with the selected objective. Since the usage of fuzzy logic model in landslide zoning is based on the analysis of raster (grid), each pixel in each criterion must take a membership value from zero to one based on the ideal function. Fuzzy maps of criteria effective in landslides have been shown in figure 3.

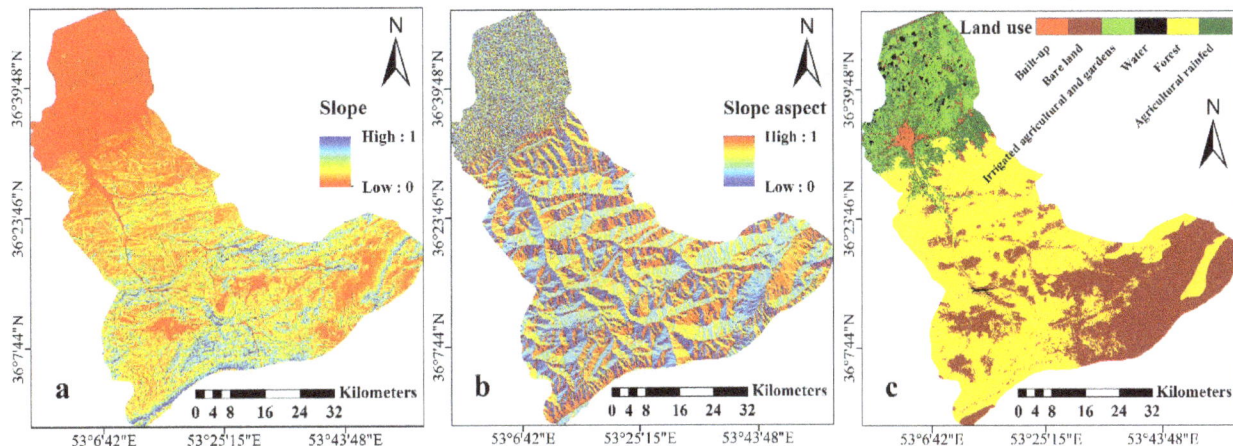

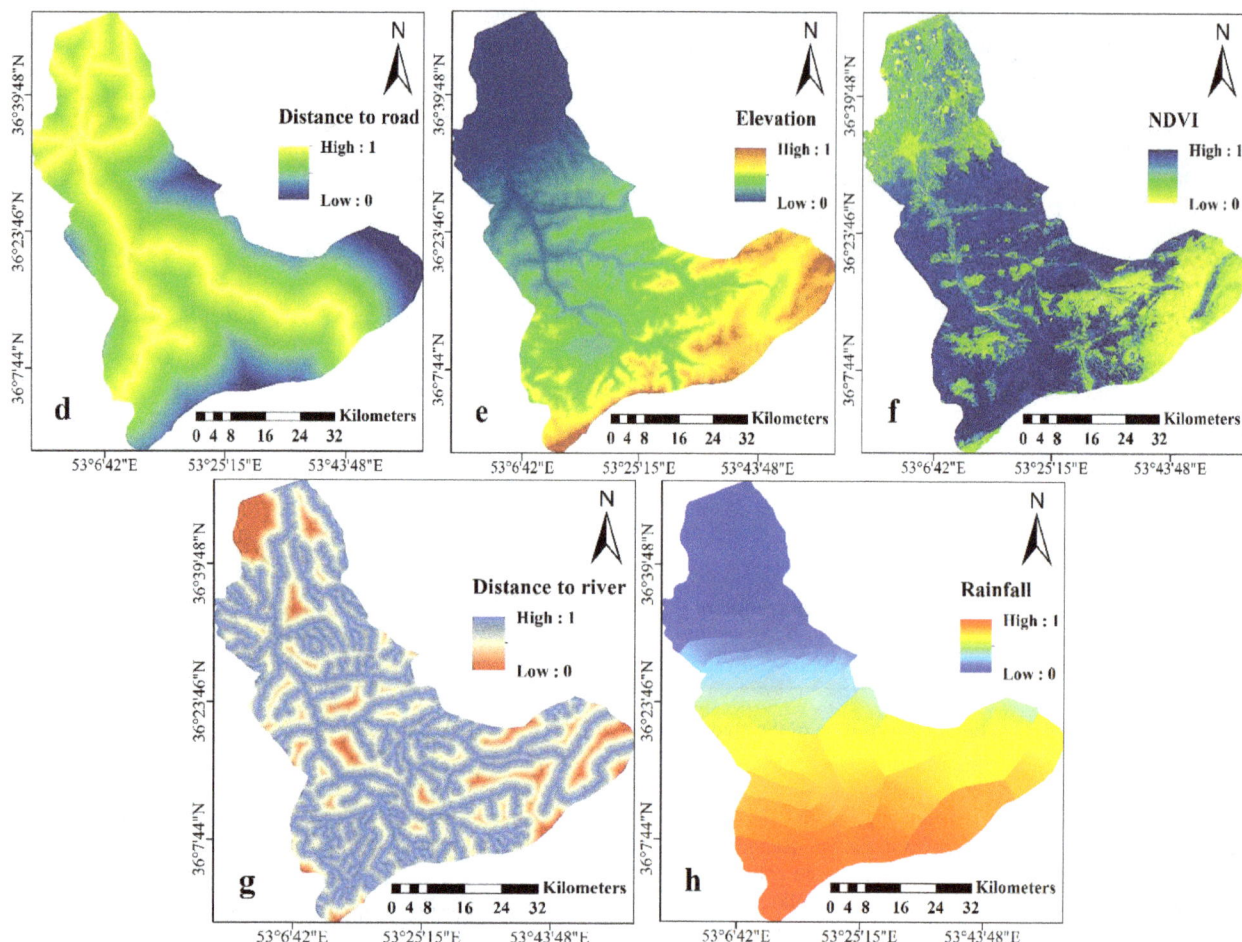

Figure 3. Landslide contributing-factor layers produced for the study area: (a) slope, (b) slope aspect, (c) land use, (d) distance to roads, (e) elevation, (f) NDVI, (g) distance to river, (h) rainfall

After defining membership functions and fuzzification of effective criteria, landslide hazard zonation maps have been overlapped using Fuzzy-AHP model and OR operators and gamma of 0.975 by overlaying the layers of effective classes on each other in landslide with Fuzzy Overlay command. Weights of criteria in Fuzzy-AHP model have been calculated and the results are shown in form of figure 4. In the end, the landslide hazard map of Sari was prepared using each one of these operators and  the results of have been shown in form of figure 5. The area of different classes of risk for Fuzzy-AHP models, gamma of 0.975 and OR operator have  been calculated and their results have been shown in form of figure 6.

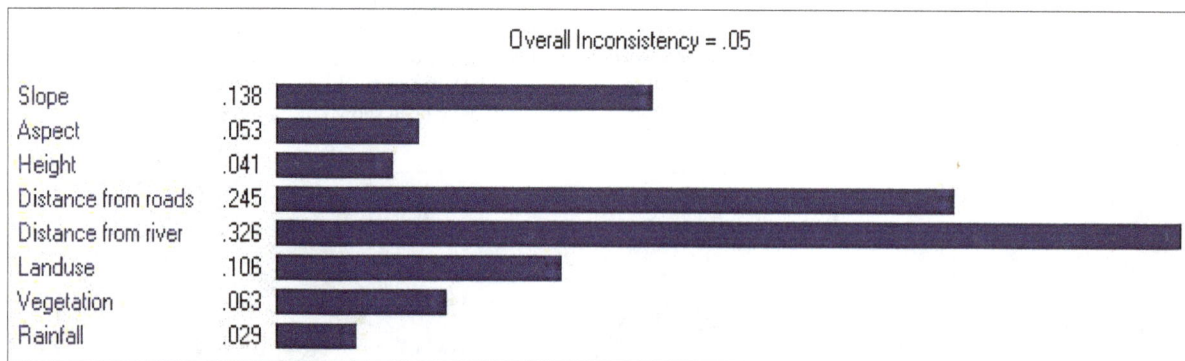

Figure 4. Weight of criteria used in Fuzzy-AHP model for preparation of landslide hazard mapping using AHP method

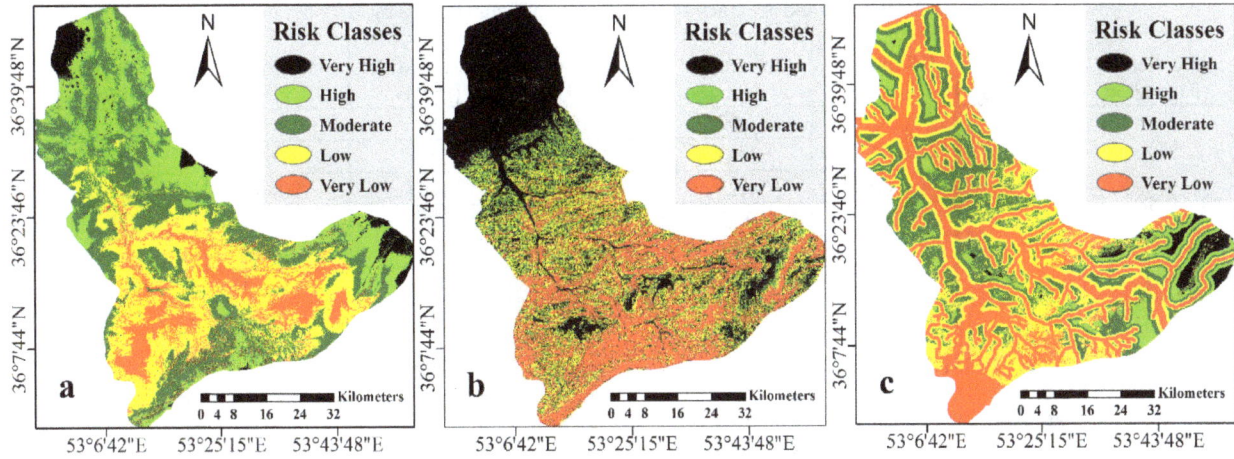

Figure 5. Landslide hazard zonation map using a) Fuzzy-AHP model b) Gamma operator of 0.975 C) OR operator

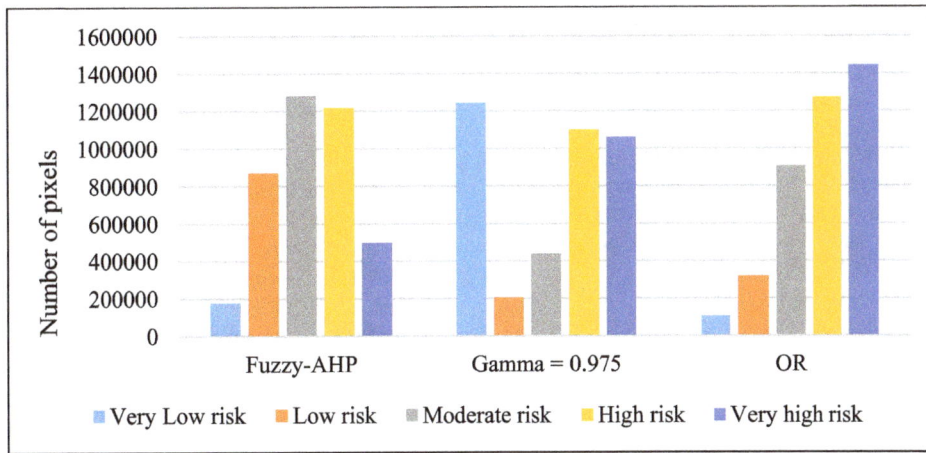

Figure 6. The area of different classes of risk for Fuzzy-AHP models, gamma of 0.975 and OR operator

Validation and accuracy of landslide hazard zonation models for Fuzzy-AHP models, gamma of 0.975 and OR operator have been evaluated using DR, QS and method P parameter quantified and the results have been depicted as tables 2 to 4.

Table 2. Validation and accuracy of landslide hazard zonation using Fuzzy-AHP model

| Precision of (P) method | Quality Sum (QS) | QS in each category | Percentage of area  (S) | Density Ratio (DR) | Area of slide (Pixel) | Area of zone (Pixel) | Danger of slide category |
|---|---|---|---|---|---|---|---|
| 0.92 | 0.45 | 0.04 | 0.04 | 0 | 0 | 177547 | Very Low |
| | | 0.09 | 0.22 | 0.34 | 543 | 870698 | Low |
| | | 0.05 | 0.31 | 0.59 | 1371 | 1281871 | Moderate |
| | | 0.075 | 0.3 | 1.5 | 3288 | 1216427 | High |
| | | 0.2 | 0.13 | 2.23 | 2083 | 497443 | Very High |
| | | | | | 7285 | 4043986 | Sum |

Table 3. Validation and accuracy of landslide hazard zonation using gamma operator of 0.975

| Precision of (P) method | Quality Sum (QS) | QS in each category | Percentage of area  (S) | Density Ratio (DR) | Area of slide (Pixel) | Area of zone (Pixel) | Danger of slide category |
|---|---|---|---|---|---|---|---|
| 0.78 | 0.21 | 0.11 | 0.4 | 0.46 | 1041 | 1243676 | Very Low |
| | | 0.01 | 0.05 | 1.48 | 549 | 204802 | Low |
| | | 0.05 | 0.2 | 1.5 | 1177 | 435605 | Moderate |
| | | 0.04 | 0.27 | 1.4 | 2761 | 1099928 | High |
| | | 0.001 | 0.26 | 0.92 | 1757 | 1059975 | Very High |
| | | | | | 7285 | 4043986 | Sum |

Table 4. Validation and accuracy of landslide hazard zonation using OR operator

| Precision of (P) method | Quality Sum (QS) | QS in each category | Percentage of area (S) | Density Ratio (DR) | Area of slide (Pixel) | Area of zone (Pixel) | Danger of Slide category |
|---|---|---|---|---|---|---|---|
| | | 0.018 | 0.02 | 0.04 | 9 | 106058 | Very Low |
| | | 0.012 | 0.08 | 1.4 | 799 | 316955 | Low |
| 0.88 | 0.11 | 0.002 | 0.23 | 0.89 | 1447 | 905473 | Moderate |
| | | 0.04 | 0.32 | 0.63 | 1457 | 1271074 | High |
| | | 0.04 | 0.35 | 1.37 | 3573 | 1444426 | Very High |
| | | | | | 7285 | 4043986 | Sum |

## 5. DISSCUSION

Hazard zonation map using Fuzzy-AHP model shows that the number of pixels in very high risk classes in terms of landslides in the study area is 497443 which is equal to 13% of total area. Landslides with high risk levels have high risk which is 30% of the area. Each class has the moderate risk of 1281871 pixels which is 31% of area. Sliding pixels of classes with low risk and very low risk are respectively 870698 and 177547 pixels which are equal to 22% and 4% of the area (Table 2). The number of sliding pixels of classes with very high risk, high risk, moderate risk, low risk and very low risk are respectively 1059975, 1099928, 435605, 204802 and 124,676 pixels in 1059975, 1099928, 435605, 204802 and 1243676 with gamma of 0.975 (Table 3). And finally, the number of same pixels for zoning map with the OR operator, are respectively, 1444426, 1271074, 905473, 316955 and 106058 pixels (Table 4).

The results of indicators of QS and method P show that DR of all three used methods are proportionate to increased risk of growing zones in addition to good resolution. Level of QS index which shows comparison and evaluation of methods in comparison with each other has been obtained to be respectively 0.45, 0.21 and 0.11 for Fuzzy- AHP, gamma of 0.975 and OR. The values of method's P are also respectively 0.92, 0.78 and 0.88 for three used operators.

## 6. CONCLUSION

Various methods and many causative factors can be used for landslide hazard mapping production depending on the scale and scope of the study. Comparison of the results of different methods in the same conditions is helpful for assessment of the relative reliability of them, although the reliability of methods is often dissimilar in different conditions.

In this study, Fuzzy-AHP, Fuzzy Gamma and Fuzzy-OR methods were compared at a regional scale, on the city of sari, in the north of Iran considering 8 causative factors. Occurrence of landslides and damages caused by those have become wider and more extensive due to increasing continuation of human changes in nature and use of mountainous areas prone to landslides. Sari has always faced many landslides due to landslides and land use changes. In this research, the validation and accuracy of different landslide hazard zonation have been evaluated and landslide hazard zoning has taken place using landslide distribution map and different algorithms. The results show that using fuzzy-based model is very effective and convenient in reducing and increasing the accuracy of landslide zoning. Also it was determined that the map of fuzzy-AHP model, OR operator and gamma of 0.975 has a high accuracy in landslide hazard zonation. Even though almost the same values were obtained for methods based on method's P which shows the almost the same and there is no limit for intervention of parameters.

Differences in validation and accuracy expresses the priority of method can be caused by things like this: inherent nature of parameters or affecting factors are different in various models and a method which is capable of comparing the priority of effective factors and their weights have greater accuracy and in other words, it will have better compliance with the potential of occurrence of landslides. Hazard zonation map with fuzzy-AHP model shows that the number of pixels in very high risk classes in terms of landslides in the study area is 497443 which is equal to 13% of total area. QS values have been obtained to be respectively 0.45, 0.21 and 0.11 for fuzzy- AHP, gamma of 0.975 and OR. Values of P are also respectively 0.92, 0.78 and 0.88 for three used operators. Based on these findings, fuzzy-AHP model has been selected as the most appropriate method of zoning landslide in the city of Sari and the Fuzzy-gamma method with a minor difference is in the second order.

Nevertheless, in this study landslide hazard mapping produced in a regional scale (small-scale map), and further studies are needed in the landslide hazard mapping production for the slope stability and land use management projects in the larger-scales. Also, more studies for comparing the reliability of other methods in small-scale might be helpful.

## REFERENCES

Aghdam, I. N., Pradhan, B., & Panahi, M. 2017. Landslide susceptibility assessment using a novel hybrid model of statistical bivariate methods (FR and WOE) and adaptive neuro-fuzzy inference system (ANFIS) at southern Zagros Mountains in Iran. *Environmental Earth Sciences*, 76(6), pp. 237.

Akgun, A., Sezer, E. A., Nefeslioglu, H. A., Gokceoglu, C., & Pradhan, B. 2012. An easy-to-use MATLAB program (MamLand) for the assessment of landslide susceptibility using a Mamdani fuzzy algorithm. Computers & Geosciences, 38(1), pp. 23-34.

Althuwaynee, O. F., Pradhan, B., Park, H. J., & Lee, J. H. 2014. A novel ensemble bivariate statistical evidential belief function with knowledge-based analytical hierarchy process and multivariate statistical logistic regression for landslide susceptibility mapping. Catena, 114, pp. 21-36.

Atkinson, P. M., & Massari, R. 2011. Autologistic modelling of susceptibility to landsliding in the Central Apennines, Italy. *Geomorphology*, 130(1), pp. 55-64.

Barrile, V., Cirianni, F., Leonardi, G., & Palamara, R. 2016. A Fuzzy-based Methodology for Landslide Susceptibility Mapping. *Procedia-Social and Behavioral Sciences*, 223, pp. 896-902.

Bianchini, S., Raspini, F., Ciampalini, A., Lagomarsino, D., Bianchi, M., Bellotti, F., & Casagli, N. 2016. Mapping landslide phenomena in landlocked developing countries by means of satellite remote sensing data: the case of Dilijan (Armenia) area. *Geomatics, Natural Hazards and Risk*, pp. 1-17.

Bui, D. T., Pradhan, B., Lofman, O., Revhaug, I., & Dick, O. B. 2012. Landslide susceptibility mapping at Hoa Binh province (Vietnam) using an adaptive neuro-fuzzy inference system and GIS. *Computers & Geosciences*, *45*, pp. 199-211.

Can, ,S., & Arıkan, F. 2014. Bir savunma sanayi firmasında çok kriterli alt yükleniciseçim problemi ve çözümü. Journal of The Faculty of Engineering and Architecture of Gazi University, Cilt 29(4), pp. 645–654.

Cardinali, M., Reichenbach, P., Guzzetti, F., Ardizzone, F., Antonini, G., Galli, M., ... & Salvati, P. 2002. A geomorphological approach to the estimation of landslide hazards and risks in Umbria, Central Italy. *Natural hazards and earth system science*, *2*(1/2), pp. 57-72.

Cay, T., & Uyan, M. 2013. Evaluation of reallocation criteria in land consolidation studies using the Analytic Hierarchy Process (AHP). *Land Use Policy*, *30*(1), pp. 541-548.

Chang, D. Y. 1996. Applications of the extent analysis method on fuzzy AHP. *European journal of operational research*, *95*(3), pp. 649-655.

Chung, C. F., & Fabbri, A. G. 2001. Prediction models for landslide hazard zonation using a fuzzy set approach. *Geomorphology and Environmental Impact Assessment Balkema, Lisse, The Netherlands*, pp. 31-47.

Conoscenti, C., Angileri, S., Cappadonia, C., Rotigliano, E., Agnesi, V., & Märker, M. 2014. Gully erosion susceptibility assessment by means of GIS-based logistic regression: a case of Sicily (Italy). *Geomorphology*, *204*, pp. 399-411.

Dou, J., Yamagishi, H., Pourghasemi, H. R., Yunus, A. P., Song, X., Xu, Y., & Zhu, Z. 2015. An integrated artificial neural network model for the landslide susceptibility assessment of Osado Island, Japan. *Natural Hazards*, *78*(3), pp. 1749-1776.

Ercanoglu, M., & Gokceoglu, C. 2004. Use of fuzzy relations to produce landslide susceptibility map of a landslide prone area (West Black Sea Region, Turkey). *Engineering Geology*, *75*(3), pp. 229-250.

Florsheim, J. L., & Nichols, A. 2013. Landslide area probability density function statistics to assess historical landslide magnitude and frequency in coastal California. *Catena*, *109*, pp. 129-138.

Gee, M.D., 1992. Classification of Landslides Hazard Zonation Methods and a Test of Predictive Capability. In: Bell, Davi, H. (eds.), Proceedings 6th International Symposium on Landslide, pp. 48-56.

Ghanavati, E., Karam, A., & Taghavi, M. E. 2015. Fuzzy logic application in identifying and mapping of landslide hazard: Case study: Taleghan watershed.

Gheshlaghi, H. A., & Feizizadeh, B. 2017. An integrated approach of analytical network process and fuzzy based spatial decision making systems applied to landslide risk mapping. *Journal of African Earth Sciences*.

Guzzetti, F., Ardizzone, F., Cardinali, M., Galli, M., Reichenbach, P., Rossi, M. 2008. Distribution of landslides in the Upper Tiber River basin, central Italy. Geomorphology 96, pp. 105–122.

Guzzetti, F., Carrara, A., Cardinali, M., Reichenbach, P. 1999. Landslide hazard evaluation: a review of current techniques and their application in a multi-scale study. Geomorphology, 31, pp. 181–216.

Hong, H., Chen, W., Xu, C., Youssef, A. M., Pradhan, B., & Tien Bui, D. 2016. Rainfall-induced landslide susceptibility assessment at the Chongren area (China) using frequency ratio, certainty factor, and index of entropy. *Geocarto International*, *32*(2), pp. 139-154.

Hong, H., Pradhan, B., Xu, C., & Bui, D. T. 2015. Spatial prediction of landslide hazard at the Yihuang area (China) using two-class kernel logistic regression, alternating decision tree and support vector machines. *Catena*, *133*, pp. 266-281.

Imaizumi, F., & Sidle, R. C. 2007. Linkage of sediment supply and transport processes in Miyagawa Dam catchment, Japan. *Journal of Geophysical Research: Earth Surface*, *112*(F3).

Jade, S., & Sarkar, S. 1993. Statistical models for slope instability classification. *Engineering Geology*, *36*(1-2), pp. 91-98.

Khan, D., & Samadder, S. R. 2015. A simplified multi-criteria evaluation model for landfill site ranking and selection based on AHP and GIS. *Journal of Environmental Engineering and Landscape Management*, *23*(4), pp. 267-278.

Korup, O. 2005. Geomorphic imprint of landslides on alpine river systems, southwest New Zealand. *Earth Surface Processes and Landforms*, *30*(7), pp. 783-800.

Lee, S. 2007. Application and verification of fuzzy algebraic operators to landslide susceptibility mapping. *Environmental Geology*, *52*(4), pp. 615-623.

Liu, J. P., Zeng, Z. P., Liu, H. Q., & Wang, H. B. 2011. A rough set approach to analyze factors affecting landslide incidence. *Computers & geosciences*, *37*(9), pp. 1311-1317.

Malczewski, J. 1999. *GIS and multicriteria decision analysis*. John Wiley & Sons.

Mttkan, A. A., Sameia, J., Pourali, S.H & Safaei, V.M. 2009. Fuzzy logic models and remote sensing techniques for landslide hazard mapping in the watershed Lajim, Journal of Applied Geology, 5(4), pp. 318-325 (in persion).

Nourani, V., Pradhan, B., Ghaffari, H., & Sharifi, S. S. 2013. Landslide susceptibility mapping at Zonouz Plain, Iran using genetic programming and comparison with frequency ratio, logistic regression, and artificial neural network models. *Natural hazards*, *71*(1), pp. 523-547.

Peng, L., Niu, R., Huang, B., Wu, X., Zhao, Y., & Ye, R. 2014. Landslide susceptibility mapping based on rough set theory and support vector machines: A case of the Three Gorges area, China. *Geomorphology*, *204*, pp. 287-301.

Pourghasemi, H. R., Beheshtirad, M., & Pradhan, B. 2016. A comparative assessment of prediction capabilities of modified analytical hierarchy process (M-AHP) and Mamdani fuzzy logic models using Netcad-GIS for forest fire susceptibility mapping. *Geomatics, Natural Hazards and Risk*, *7*(2), pp. 861-885.

Pourghasemi, H. R., Pradhan, B., & Gokceoglu, C. 2012. Application of fuzzy logic and analytical hierarchy process (AHP) to landslide susceptibility mapping at Haraz watershed, Iran. *Natural hazards*, *63*(2), pp. 965-996.

Pradhan, B., & Lee, S. 2010. Landslide susceptibility assessment and factor effect analysis: backpropagation artificial neural networks and their comparison with frequency ratio and bivariate logistic regression modelling. *Environmental Modelling & Software*, *25*(6), pp. 747-759.

Reichenbach, P., Galli, M., Cardinali, M., Guzzetti, F., & Ardizzone, F. 2005. Geomorphological mapping to assess landslide risk: Concepts, methods and applications in the Umbria region of central Italy. *Landslide Hazard Risk*, pp. 429-468.

Roering, J. J., Kirchner, J. W., & Dietrich, W. E. 2005. Characterizing structural and lithologic controls on deep-seated landsliding: Implications for topographic relief and landscape evolution in the Oregon Coast Range, USA. *Geological Society of America Bulletin*, *117*(5-6), pp. 654-668.

Saaty, T. L. 1980. The Analytical Hierarchy Process, Planning, Priority. *Resource Allocation. RWS Publications, USA.*

Saaty, T. L. 1986. Axiomatic foundation of the analytic hierarchy process. *Management science*, *32*(7), pp. 841-855.

Samany, N. N., Delavar, M. R., Chrisman, N., & Malek, M. R. 2014. FIA 5: A customized Fuzzy Interval Algebra for modeling spatial relevancy in urban context-aware systems. Engineering Applications of Artificial Intelligence, 33, pp. 116-126.

Shahabi, H., Khezri, S., Ahmad, B. B., & Hashim, M. 2014. Landslide susceptibility mapping at central Zab basin, Iran: A comparison between analytical hierarchy process, frequency ratio and logistic regression models. *Catena*, *115*, pp. 55-70.

T.L. Saaty, and L. G. Vargas. 1991 "Prediction, projection and forecasting," Kluwer Academic Publichers, Dordrecht.

Thiery, Y., Malet, J. P., & Maquaire, O. 2006. Test of fuzzy logic rules for landslide susceptibility assessment. In *SAGEO 2006: Information Géographique: observation et localisation, structuration et analyse, representation*, pp. 16-p.

Tsangaratos P, Constantinos L, Dimitrios R, Ioanna I. 2015. Landslide susceptibility assessments using the k-nearest neighbor algorithm and expert knowledge. Case study of the basin of Selinounda river, Achaia County, Greece.

Tsangaratos, P., & Ilia, I. 2016. Landslide susceptibility mapping using a modified decision tree classifier in the Xanthi Perfection, Greece. *Landslides*, *13*(2), pp. 305-320.

Vahidnia, M. H., Alesheikh, A. A., Alimohammadi, A., & Hosseinali, F. 2010. A GIS-based neuro-fuzzy procedure for integrating knowledge and data in landslide susceptibility mapping. *Computers & Geosciences*, *36*(9), pp. 1101-1114.

Vakhshoori, V., & Zare, M. 2016. Landslide susceptibility mapping by comparing weight of evidence, fuzzy logic, and frequency ratio methods. *Geomatics, Natural Hazards and Risk*, *7*(5), pp. 1731-1752.

Wang, L.-J., Guo, M., Sawada, K., Lin, J., Zhang, J. 2016. A comparative study of landslide susceptibility maps using logistic regression, frequency ratio, decision tree, weights of evidence and artificial neural network. Geosci. J. 20, pp. 117–136.

Yalcin, A. 2008. GIS-based Landslide Susceptibility Mapping Using Analytical Hierarchy Process and Bivariate Statistics in Ardesen (Turkey), Comparisons of results and confirmations. Catena, 72, pp. 1-12.

Zadeh, L. A. 1965. Fuzzy sets. *Information and control*, *8*(3), pp. 338-353.

# 12

# A NEW MODEL FOR FUZZY PERSONALIZED ROUTE PLANNING USING FUZZY LINGUISTIC PREFERENCE RELATION

S. Nadi [a], A. H. Houshyaripour [b]

[a] Assistant Professor, Department of Geomatics Engineering, Faculty of Civil and Transportation Engineering, University of Isfahan, Isfahan, Iran - snadi@eng.ui.ac.ir
[b] M.Sc. Remote Sensing, Department of Geomatics Engineering, Faculty of Civil and Transportation Engineering, University of Isfahan, Iran - amirhossein7jm@gmail.com

**KEY WORDS:** Fuzzy Route Planning, Personalized, Uncertainty, Fuzzy AHP, Fuzzy Comparison

**ABSTRACT:**

This paper proposes a new model for personalized route planning under uncertain condition. Personalized routing, involves different sources of uncertainty. These uncertainties can be raised from user's ambiguity about their preferences, imprecise criteria values and modelling process. The proposed model uses Fuzzy Linguistic Preference Relation Analytical Hierarchical Process (FLPRAHP) to analyse user's preferences under uncertainty. Routing is a multi-criteria task especially in transportation networks, where the users wish to optimize their routes based on different criteria. However, due to the lake of knowledge about the preferences of different users and uncertainties available in the criteria values, we propose a new personalized fuzzy routing method based on the fuzzy ranking using center of gravity. The model employed FLPRAHP method to aggregate uncertain criteria values regarding uncertain user's preferences while improve consistency with least possible comparisons. An illustrative example presents the effectiveness and capability of the proposed model to calculate best personalize route under fuzziness and uncertainty.

## 1. INTRODUCTION

Personalized routing, involves different sources of uncertainty. These uncertainties can be raised from user's ambiguity about their preferences, imprecise criteria values and modeling process. To find the best personalized route the model should have the flexibility to consider all these sources of uncertainty. To cope with this problem, integration of routing methods with fuzzy decision theory is an idea that is followed in this article.

During the recent years, a number of researches have been done in personalized routing (Vahidinia et.al., 2008, Khan and Alnuweiri, 2004, Rodriquez and Lazo, 2013). Teodorovic and Kilkuchi (2007) presented a model follows the principles of the classical Clarke-Wright algorithm to develop a set of vehicle routes (Teodorovic and Kilkuchi, 2007). Khan and Alnuweiri (2004) proposed a low-complexity constraint-based routing algorithm for traffic engineering in computer networks that route end-to-end packet flows (Khan and Alnuweiri, 2004). The proposed fuzzy routing algorithm (FRA), modifies the well-known Dijkstra's single-source shortest paths algorithm by including fuzzy membership functions in the path-cost update process (Khan and Alnuweiri, 2004). Zheng and Liu (2004) considered the vehicle routing problem with the travel times as the fuzzy variables. Then they designed a fuzzy optimization model for fuzzy vehicle routing based on the time window (Zheng and Liu, 2004). Boyan and Littman (1994) described the Q-routing algorithm for packet routing in computer networks. They embedded a reinforcement learning into each node of a switching network. In their model each node only used local communication to keep minimal delivery times (Boyan and Littman, 1994).

In this paper we propose a new model for finding the best personalized route based on the fuzzy decision theory, taking into account the uncertainties in either the measurements or preferences to provide users the ability to include their preferences in routing task easily, we propose a to use fuzzy linguistic preference relation AHP method (FLPRAHP). The most challenging issues in including user's preferences in any personalized systems using multi-criteria methods are the number of comparisons, consistency of comparisons and ambiguities in presenting the exact preferences. The FLPRAHP provide a mechanism for improving consistency using least number of comparisons and include user's linguistic phrases about their preferences (Wang and Chen, 2008). The FLPRAHP provides costs of links as fuzzy numbers. To find the best route regarding these fuzzy costs, we proposed a new fuzzy routing method based on incorporation of Center of Gravity Fuzzy Comparison (CGFC) method and dijkstra's algorithm. An illustrative example is then used to present the application of the proposed method.

The paper is organized as follows. In section 2 we present the proposed methodology describing FLPRAHP, center of gravity fuzzy comparison method and their incorporation in Dijkstra's algorithm. In section 3 we detail an illustrative example showing the step by step application of the proposed method. Some conclusions and future directions of the work are included in section 4.

## 2. METODOLOGY

As illustrated in figure 1, the proposed model consists of 2 steps. In the first step, the model provides the basis for capturing users preferences which are always uncertain. The most challenging issues in this step are reducing the number of comparisons between criteria to determine user's preferences and maintaining the comparisons consistent. To cope with these problems, in this step the fuzzy linguistic preference relation method is adopted. This step results in a set of fuzzy numbers represents the cost of each link in the network. The second step uses these fuzzy costs for each link of the network as input values for fuzzy personalized routing engine to solve the best path problem. We design and develop fuzzy personalized routing engine by integrating fuzzy center of gravity ranking method with dijkstra's algorithm.

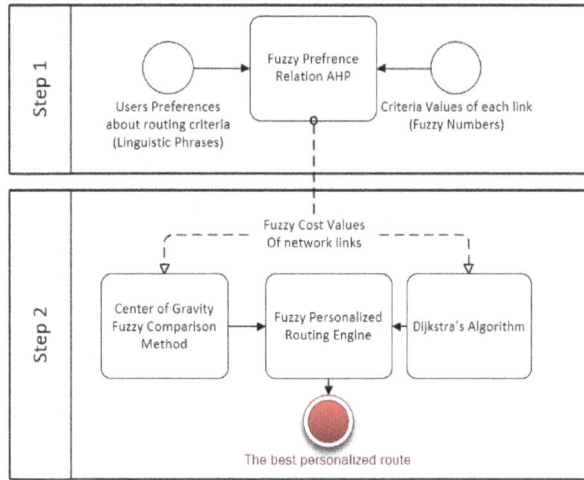

Figure 1. Fuzzy personalized routing model

### 2.1. Conventional Fuzzy AHP

The Analytical Hierarchical Process (AHP) is one of the extensively used multi-criteria decision making methods. Although this method is easy to understand and it can model expert opinions through pairwise comparison, however, the conventional AHP cannot process imprecise or vague information (Chen and Hwang, 1991, Chiclana and Herrera, 1998, Vahidnia et.al, 2008, Zhu, 2014) in conventional AHP decision makers compare criteria using crisp judgments. However, in the real situations most experts can just use their judgments regarding criteria relative meaning which are usually vague. It is the essence of the AHP that human judgments, and not just the underlying information, can be used in performing the evaluations. To model the ambiguity in judgments and also uncertainty in criteria values fuzzy extensions of AHP has been introduced. However, the most challenging issue of these methods are to maintain the comparisons consistence. The FLPRAHP provides a method to capture the experts' preferences about criteria using fuzzy linguistic phrases and calculates importance weight of each criterion using least possible number comparison while maintains consistency. The steps of the conventional Fuzzy AHP are as follows:

Step 1: Hierarchical structure construction by placing the goal of the desired problem on the top level of the hierarchical structure, the evaluation criteria on the middle levels and the alternatives on the bottom level.

Step 2: Constructing the fuzzy judgment matrix $\tilde{A}$. The fuzzy judgment matrix $\tilde{A}$ in equation 1 is a pairwise comparison of criteria that is constructed by assigning linguistic terms to the pairwise comparisons by asking which one of two criteria is more important.

$$\tilde{A} = \begin{bmatrix} \tilde{1} & \tilde{a}_{12} & \cdots & \tilde{a}_{1n} \\ \tilde{a}_{21} & \tilde{1} & & \tilde{a}_{2n} \\ \vdots & \vdots & \ddots & \vdots \\ \tilde{a}_{n1} & \tilde{a}_{n2} & \cdots & \tilde{1} \end{bmatrix} = \begin{bmatrix} \tilde{1} & \tilde{a}_{12} & \cdots & \tilde{a}_{1n} \\ \tilde{a}_{12}^{-1} & \tilde{1} & & \tilde{a}_{2n} \\ \vdots & \vdots & \ddots & \vdots \\ \tilde{a}_{1n}^{-1} & \tilde{a}_{2n}^{-1} & \cdots & \tilde{1} \end{bmatrix} \quad (1)$$

where $\tilde{a}_{ij}$ is the fuzzy number from table 1 resulted by comparing $i$th and $j$th criteria.

Table 1. Membership function of linguistic scales

| Fuzzy numbers | Linguistic scales | Membership function |
|---|---|---|
| $\tilde{1}$ | Equally important | (1,1,3) |
| $\tilde{3}$ | Weakly important | (1,3,5) |
| $\tilde{5}$ | Essentially important | (3,5,7) |
| $\tilde{7}$ | Very strong important | (5,7,9) |
| $\tilde{9}$ | Absolutely important | (7,9,9) |
| $\tilde{1}^{-1}, \tilde{3}^{-1}, \tilde{5}^{-1}, \tilde{7}^{-1}, \tilde{9}^{-1}$ | Relative less important | ... |

Step 3: Calculating fuzzy weights of each criterion. The fuzzy weights of each criterion are calculated using equation 2 (Ekel et.al., 2006).

$$\tilde{w}_i = \frac{\tilde{r}_i}{\tilde{r}_1 \oplus \dots \oplus \tilde{r}_n} \quad (2)$$

$$\tilde{r}_i = [\tilde{a}_{i1} \otimes \tilde{a}_{i2} \otimes \dots \otimes \tilde{a}_{in}]^{\frac{1}{n}} \quad \forall i = 1,2,\dots,n$$

Where $\tilde{w}_i$ is the importance weights of $i$th criterion.

Step 4: Hierarchical layer sequencing. The final fuzzy weight value of each alternative is calculated by hierarchical layer sequencing using equation 3.

$$\tilde{U}_i = \sum_{j=1}^{n} \tilde{w}_j \cdot \tilde{r}_{ij} \quad . \quad \tilde{U}_i = (l.m.u) \quad (3)$$

Where $\tilde{r}_{ij}$ is the fuzzy value of the $j$th criterion, $\tilde{U}_i$ is a fuzzy number shows the final score of $i$th criterion.

Step 5: Ranking alternatives.

To prepare alternative for ranking at the final step, one approach is defuzzification which transform fuzzy numbers to crisp ones. Equation 4 shows one of the simplest but least useful methods named weighted fuzzy mean.

$$X(\tilde{U}_i) = (l + m + u)/3 \quad (4)$$

Where l, m and u are lower, mid and upper band of fuzzy number $\tilde{U}_i$ and $X(\tilde{U}_i)$ is fuzzy mean of $\tilde{U}_i$ which can be used to determine the optimum alternative.

In this paper we propose to use fuzzy ranking method of center of gravity that uses more knowledge from fuzzy numbers to compare them. Furthermore, when the number of criteria become large the number of comparisons as well as maintaining them consistent would be a challenging issue especially in personalized systems where public users are supposed to make comparisons. Here we propose to use FLPRAHP methods in personalized systems which solve these problems. Following we explain the FLPRAHP and Fuzzy Ranking methods in sections 2.2 and 2.3 respectively.

### 2.2. Fuzzy Linguistic preference relation AHP

In the second step of conventional fuzzy AHP described in section 2.1, the amount of comparison can be reduced using the relationship between elements of the matrix $\tilde{C}$ (Berredo et.al., 2005, Ekel et.al., 2006, Wang and Chen, 2008). Given that the fuzzy positive matrix $\tilde{A} = (\tilde{a}_{ij})$ is reciprocal which means that $\tilde{a}_{ji} = \tilde{a}_{ij}^{-1}$ where $\tilde{a}_{ij} \in [1/9.9]$ the fuzzy preference relation matrix $\tilde{P} = (\tilde{p}_{ij})$ where $\tilde{p}_{ij} \in [0.1]$ can be calculated using transformation in equation 5 (Ekel et.al., 2006).

$$\tilde{p}_{ij} = \frac{1}{2}(1 + log_9 a_{ij}) \quad . \quad \tilde{p}_{ij} = (p_{ij}^L.p_{ij}^M.p_{ij}^R) \quad (5)$$

Where $p_{ij}^L$, $p_{ij}^M$ and $p_{ij}^R$ are left, mid and right band of fuzzy number $\tilde{p}_{ij}$.

For $\tilde{A} = (\tilde{a}_{ij})$ being consistent, $\tilde{a}_{ik}$ should be equal to $\tilde{a}_{ij} \otimes \tilde{a}_{jk}$. Taking logarithm on both sides equation 6 yields (Ekel et.al., 2006).

$$\tilde{a}_{ij} \otimes \tilde{a}_{jk} \cong \tilde{a}_{ik} \quad (6)$$

$$log_9 a_{ij} \oplus log_9 a_{jk} = log_9 a_{ik}$$
$$log_9 a_{ij} \oplus log_9 a_{jk} \ominus log_9 a_{ik} = 0$$
$$log_9 a_{ij} \oplus log_9 a_{jk} \oplus log_9 a_{ki} = 0$$
$$\frac{1}{2}(1 + log_9 a_{ij}) \oplus \frac{1}{2}(1 + log_9 a_{jk}) \oplus \frac{1}{2}(1 + log_9 a_{ki}) = \frac{3}{2}$$

Substituting $\frac{1}{2}(1 + log_9 a_{ij})$ from equation 5 the relationship between elements of matrix $\tilde{A}$ can be determined as detailed in equation 7 (Ekel et.al., 2006).

$$\tilde{p}_{ij} \oplus \tilde{p}_{jk} \oplus \tilde{p}_{ki} = \frac{3}{2}$$
$$p_{ij}^L + p_{jk}^L + p_{ki}^R = \frac{3}{2}$$
$$p_{ij}^M + p_{jk}^M + p_{ki}^M = \frac{3}{2} \qquad (7)$$
$$p_{ij}^R + p_{jk}^R + p_{ki}^L = \frac{3}{2}$$

These relationships for more than three criteria are as equation 8 (Ekel et.al., 2006).

$$p_{i(i+1)}^L + p_{(i+1)(i+2)}^L + \cdots + p_{(j-1)j}^L + p_{ji}^R = \frac{(j-i+1)}{2}$$
$$p_{i(i+1)}^M + p_{(i+1)(i+2)}^M + \cdots + p_{(j-1)j}^M + p_{ji}^M = \frac{(j-i+1)}{2} \qquad (8)$$
$$p_{i(i+1)}^R + p_{(i+1)(i+2)}^R + \cdots + p_{(j-1)j}^R + p_{ji}^L = \frac{(j-i+1)}{2}$$

By using these relationships, the required comparisons for n criteria will be reduced from n(n-1)/2 to just n-1 comparisons while the consistency is maintained.

### 2.3. Centre of Gravity Fuzzy Ranking Method

There are many ways to compare two fuzzy numbers. Center of gravity is one of the most common and useful techniques (Chen and Chen, 2009, Chen and Chen, 2007, Chan and Qi, 2002, Phani and Shankar, 2011).

Step 1: Considering $\tilde{A}_i = (a^L . a^{M1} . a^{M2} . a^R . w_{\tilde{A}_i})$ as a generalized trapezoidal fuzzy number where $a^L$, $a^{M1}$, $a^{M2}$ and $a^R$ are real numbers and $0 < w_{\tilde{A}_i} \leq 1$ is its maximum membership value. In this step the standardized trapezoidal fuzzy number must be calculated using 9.

$$\tilde{A}_i^* = \left( \frac{a^L}{k}, \frac{a^{M1}}{k}, \frac{a^{M2}}{k}, \frac{a^R}{k}; w_{\tilde{A}_i} \right) \qquad (9)$$
$$= (a^{L*}, a^{M1*}, a^{M2*}, a^{R*}; w_{\tilde{A}_i})$$

Where $k = max(a^L . a^{M1} . a^{M2} . a^R, 1)$ and $0 \leq a^{L*} \leq a^{M1*} \leq a^{M2*} \leq a^{R*} \leq 1$.

Step 2: Computing the Centre of gravity $(x_{\tilde{A}_i}^*, y_{\tilde{A}_i}^*)$ using equation 10 (Chen and Chen, 2009).

$$y_{\tilde{A}_i}^* = \begin{cases} \dfrac{w_{\tilde{A}_i}\left( \dfrac{a^{M2*} - a^{M1*}}{a^{R*} - a^{L*}} + 2 \right)}{6} & . \quad if\ a^{L*} \neq a^{R*} \\ \dfrac{w_{\tilde{A}_i}}{2} & . \quad if\ a^{L*} = a^{R*} \end{cases} \qquad (10)$$
$$x_{\tilde{A}_i}^* = \frac{y_{\tilde{A}_i}^*(a^{M1*} + a^{M2*}) + (a^{M1*} + a^{R*})(w_{\tilde{A}_i} - y_{\tilde{A}_i}^*)}{2w_{\tilde{A}_i}}$$

Step 3: Computing the mean and standard deviation (Chen and Chen, 2009)

$$\bar{x}_{\tilde{A}_i} = \frac{a^{L*} + a^{M1*} + a^{M2*} + a^{R*}}{4} \qquad (11)$$

$$S_{\tilde{A}_i} = \sqrt{\frac{\sum(x_j - \bar{x}_{\tilde{A}_i})^2}{4-1}} \quad . \quad x_j = \{a^{L*}, a^{M1*}, a^{M2*}, a^{R*}\}$$

Step 4: Computing rank of the standard fuzzy number using equation 12 (Chen and Chen, 2009).

$$Rank(\tilde{A}_i^*) = x_{\tilde{A}_i}^* + \left( w_{\tilde{A}_i} - y_{\tilde{A}_i}^* \right)^{S_{\tilde{A}_i}} \times \left( y_{\tilde{A}_i}^* + 0.5 \right)^{1 - w_{\tilde{A}_i}} \qquad (12)$$

This rank then could be used to rank fuzzy numbers. This approach can also be used for triangular fuzzy numbers by considering $a^{M1} = a^{M2}$ in the equations.

### 2.4. Fuzzy Personalized Routing Engine

The route planning algorithms determine a path through a network from an origin to a destination. For determination of this path, the corresponding minimization problem, over an impedance function, has to be solved. This paper proposes that the impedance of each link should be calculated according to the users' preferences using fuzzy linguistic preference relation AHP method as described in section 2.2. The route planning in a network with non-negative crisp costs of links can be solved easily using well-known existing approaches such as Dijkstra's labeling algorithm. However, when the costs of links are in the form of fuzzy numbers, as in the case in this paper, a fuzzy routing approach is required. We propose to adopt Dijkstra's algorithm for fuzzy numbers based on the fuzzy center of gravity ranking method which is described in section 2.3. The advantage of this approach over other existing approaches, which work only for acyclic, layered graphs, is that it is usable for any general directed graphs, including transportation networks.

Figure 2 illustrates the proposed algorithm for calculating the fuzzy personalized least cost route. In this approach, a directed graph G=(V,A) is defined with node set V, arc set A, start node (origin) o∈V, target node (destination) d∈V and $cost\_of\_\widetilde{travelling}[u.v] = (L_{uv}.M_{uv}.R_{uv})$ is fuzzy cost of travelling from node u∈V to node v∈V which share a link and $0 \leq L_{uv} \leq M_{uv} \leq U_{uv} \leq 1$.

```
set G(V,A) = the graph with a set of vertex, V, and a set of arcs, A
set O = the origin
for each vertex pairs (u, v) in V sharing a link
    cost_of_travelling [u.v]=FLPRAHP (criterion₁, criterion₂, ..., criterionₙ)
for each vertex v in V
    cost[v] = infinity
    previous[v] = undefined
end for
cost[o] = 0̃
Q = the set of all nodes in V
while Q is not empty
    u = vertex in Q with smallest fuzzy cost (ranked by fuzzy center of gravity)
    remove u from Q
    for each neighbour v of u
        alt = cost[u]⊕cost_of_travelling[u.v]
        cgr[alt] = rank of alt using fuzzy centre of gravity
        cgr[cost[v]] = rank of cost[v] using fuzzy centre of gravity
        if cgr[alt] < cgr[cost[v]]
            cost[v] = alt
            previous[v] = u
    end for
end while
set d = the destination
theRoute=Trace_back[previous(D)]
return: cost[d] & theRoute
```

Figure 2. The proposed algorithm for determining personalized routing algorithm

### 3. NUMERICAL EXAMPLE

Figure 3 shows the network to be evaluated and Table 2 illustrates the attributes of each link. The personalized fuzzy route between the origin, A, and the destination, F, are required for different decision strategies. In this example we supposed that the user is interested to find the best route regarding the "distance", "traffic volume" and "quality of road".

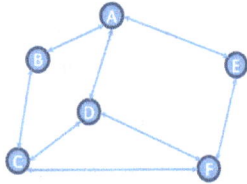

Figure 3. Sample network to be evaluated

Table 2. Fuzzy values of different criteria each link

| link | Road Distance | Road Quality | Traffic Volume |
|------|---------------|--------------|----------------|
| AB | $\tilde{3} = (1.3.5)$ | $\tilde{1} = (-1.1.3)$ | $\tilde{5} = (3.5.7)$ |
| AD | $\tilde{4} = (2.4.6)$ | $\tilde{2} = (0.2.4)$ | $\tilde{4} = (2.4.6)$ |
| AE | $\tilde{5} = (3.5.7)$ | $\tilde{2} = (0.2.4)$ | $\tilde{6} = (4.6.8)$ |
| BC | $\tilde{3} = (1.3.5)$ | $\tilde{1} = (-1.1.3)$ | $\tilde{4} = (2.4.6)$ |
| CD | $\tilde{1} = (-1.1.3)$ | $\tilde{0} = (-2.0.2)$ | $\tilde{3} = (1.3.5)$ |
| CF | $\tilde{6} = (4.6.8)$ | $\tilde{3} = (1.3.5)$ | $\tilde{1} = (-1.1.3)$ |
| DF | $\tilde{2} = (0.2.4)$ | $\tilde{1} = (-1.1.3)$ | $\tilde{2} = (0.2.4)$ |
| EF | $\tilde{3} = (1.3.5)$ | $\tilde{2} = (0.2.4)$ | $\tilde{2} = (0.2.4)$ |

Regarding the method described in section 2, the following steps have been followed.

**Step 1.** Determining users' preferences:

Regarding the hierarchical tree illustrated in figure 4 there are 3 criteria namely "traffic volume", "road distance" and "road quality" in which we first should determine user preferences regarding them.

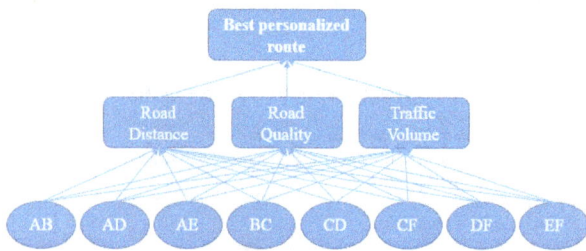

Figure 4. AHP hierarchical structure

Therefore, the first step goes through the process of completing the pair-wise comparison matrix and determining the user preferences using fuzzy linguistic preference relation AHP explained in section 2.2. To do this, the user just completed the first row of pair-wise comparison matrix based on linguistic preference relation in table 1. Other comparisons can be obtained using equation 8. Table 3 shows the comparisons by user.

Table 3. Fuzzy pair-wised comparison

| A | Distance | HVT | Traffic |
|---|----------|-----|---------|
| Distance | $\tilde{1}$ | $\tilde{3}$ | $\tilde{5}$ |
| HVT | ? | $\tilde{1}$ | ? |
| Traffic | ? | ? | $\tilde{1}$ |

Fuzzy preference relation value $P$ is calculated based on the equation 5 as showed in table 4.

Table 4. Fuzzy preference relation of comparisons

| P | Distance | HVT | Traffic |
|---|----------|-----|---------|
| Distance | $\tilde{1}$ | (0.5 0.75 0.87) | (0.75 0.87 0.94) |
| HVT | (0.13 0.25 0.50) | $\tilde{1}$ | (0.57 0.62 0.75) |
| Traffic | (0.06 0.13 0.25) | (0.25 0.38 0.43) | $\tilde{1}$ |

Having this matrix in hand, the weights of each criterion is calculated using equation 2 as follow.

$r(road\ distance) = [0.72 \quad 0.87 \quad 1.35]$
$r(road\ quality) = [0.42 \quad 0.54 \quad 1.04]$
$r(traffic\ volume) = [0.25 \quad 0.37 \quad 0.69]$
$w_1 = w(road\ distance) = [0.23 \quad 0.49 \quad 0.97]$
$w_2 = w(road\ quality) = [0.14 \quad 0.30 \quad 0.75]$
$w_3 = w(traffic\ volume) = [0.08 \quad 0.21 \quad 0.50]$

**Step 2.** Calculating overall cost of each link:

By aggregating users' preferences regarding each criteria and their values for each link the overall cost of each link is calculated as follow.

$cost\_of\_travelling[A.B] = \tilde{3} \otimes w_1 \oplus \tilde{1} \otimes w_2 \oplus \tilde{5} \otimes w_3$
$\qquad = [0.78 \quad 2.82 \quad 8.38] = \overline{2.82}$

$cost\_of\_travelling[A.E] = \tilde{5} \otimes w_1 \oplus \tilde{2} \otimes w_2 \oplus \tilde{6} \otimes w_3$
$\qquad = [1.46 \quad 4.31 \quad 11.57] = \overline{4.31}$

$cost\_of\_travelling[A.D] = \tilde{4} \otimes w_1 \oplus \tilde{2} \otimes w_2 \oplus \tilde{4} \otimes w_3$
$\qquad = [1.07 \quad 3.40 \quad 9.60] = \overline{3.40}$

$cost\_of\_travelling[B.C] = \tilde{3} \otimes w_1 \oplus \tilde{1} \otimes w_2 \oplus \tilde{4} \otimes w_3$
$\qquad = [0.70 \quad 2.61 \quad 7.88] = \overline{2.61}$

$cost\_of\_travelling[D.C] = \tilde{1} \otimes w_1 \oplus \tilde{0} \otimes w_2 \oplus \tilde{3} \otimes w_3$
$\qquad = [0.16 \quad 1.12 \quad 3.94] = \overline{1.12}$

$cost\_of\_travelling[E.F] = \tilde{3} \otimes w_1 \oplus \tilde{2} \otimes w_2 \oplus \tilde{2} \otimes w_3$
$\qquad = [0.68 \quad 2.49 \quad 7.63] = \overline{2.49}$

$cost\_of\_travelling[D.F] = \tilde{2} \otimes w_1 \oplus \tilde{1} \otimes w_2 \oplus \tilde{2} \otimes w_3$
$\qquad = [0.31 \quad 1.70 \quad 5.91] = \overline{1.70}$

$cost\_of\_travelling[C.F] = \tilde{6} \otimes w_1 \oplus \tilde{3} \otimes w_2 \oplus \tilde{1} \otimes w_3$
$\qquad = [1.43 \quad 4.05 \quad 10.79] = \overline{4.04}$

Figure 5 illustrates the resulted network with overall cost of each link.

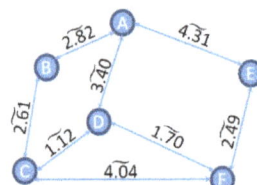

Figure 5. overall fuzzy cost of each link

**Step 3.** Finding the best path using fuzzy personalized routing method:

In the final step, fuzzy personalized routing engine, which has explained in section 2.4, uses fuzzy cost of travelling on each link to find the best route. The data structure in Table 5 shows the steps for determining best route.

Table 5. the proposed data structure to determine optimum personalized route

| Vertex \ loop | Cost (v) | | | | | | Previous vertex | | | | | | removed | | | | | |
|---------------|----|----|----|----|----|----|------|------|------|------|------|------|------|------|------|------|------|------|
| | l1 | l2 | l3 | l4 | l5 | L6 | l1 | l2 | l3 | l4 | l5 | L6 | l1 | l2 | l3 | l4 | l5 | L6 |
| A | $\tilde{0}$ | - | - | - | - | - | - | - | - | - | - | - | ✓ | - | - | - | - | - |
| B | ∞ | $\overline{2.82}$ | - | - | - | - | Null | A | - | - | - | - | | ✓ | - | - | - | - |
| C | ∞ | ∞ | $\overline{5.43}$ | $\overline{4.52}$ | $\overline{4.52}$ | - | Null | Null | B | D | D | - | | | | | ✓ | - |

| D | ∞ | $\overline{3.40}$ | $\overline{3.40}$ | - | - | - | Null | A | A | - | - | - | | | ✓ | - | - | - |
|---|---|---|---|---|---|---|---|---|---|---|---|---|---|---|---|---|---|---|
| E | ∞ | $\overline{4.31}$ | $\overline{4.31}$ | $\overline{4.31}$ | - | - | Null | A | A | A | - | - | | | | ✓ | - | - |
| F | ∞ | ∞ | ∞ | $\overline{5.10}$ | $\overline{6.8}$ | $\overline{8.56}$ | Null | Null | Null | D | E | C | | | | | | ✓ |

In this table, l1, l2, …, l6 columns preserve the results of the steps in routing algorithm which are illustrated in figure 2. For the sample network illustrated in figure 3 the algorithm determined the best path in 6 loops in this table. Table 6 shows the required comparisons for each loop using fuzzy center of gravity.

Table 6. required comparisons in each loop

| Loop number | Fuzzy numbers | $x^*$ | $y^*$ | $\bar{x}$ | $s$ | Rank |
|---|---|---|---|---|---|---|
| l2 | [0.78  2.82  8.38]=$\overline{2.82}$ | 3.93 | 0.37 | 3.99 | 3.28 | 4.157 |
|  | [1.07  3.40  9.60]=$\overline{3.40}$ | 4.61 | 0.38 | 4.69 | 3.68 | 4.785 |
|  | [1.46  4.31  11.57]=$\overline{4.31}$ | 5.65 | 0.39 | 5.78 | 4.34 | 5.764 |
| l3 | [1.48  5.43  16.26]=$\overline{5.43}$ | 7.61 | 0.36 | 7.72 | 6.39 | 7.671 |
|  | [1.07  3.40  9.60]=$\overline{3.40}$ | 4.61 | 0.38 | 4.69 | 3.68 | 4.784 |
|  | [1.46  4.31  11.57]=$\overline{4.31}$ | 5.65 | 0.39 | 5.78 | 4.34 | 5.764 |
| l4 | [1.23  4.52  13.54]=$\overline{4.52}$ | 6.34 | 0.36 | 6.43 | 5.32 | 6.430 |
|  | [1.46  4.31  11.57]=$\overline{4.31}$ | 5.65 | 0.39 | 5.78 | 4.34 | 5.764 |
|  | [1.38  5.1  15.51]=$\overline{5.1}$ | 7.25 | 0.36 | 7.33 | 6.12 | 7.317 |
| l5 | [1.23  4.52  13.54]=$\overline{4.52}$ | 6.34 | 0.36 | 6.43 | 5.32 | 6.430 |
|  | [2.14  6.8  19.20]=$\overline{6.80}$ | 9.22 | 0.38 | 9.38 | 7.35 | 9.247 |

To find the path to any vertex in the network, one can find the minimum cost value in its row and trace back using the previous vertex named in the same loop in "Previous vertex" columns. For example for f as the destination minimum cost belongs to loop 4 with $\overline{5.10}$ which is related to its previous vertex D in loop 4. Then, for vertex D the minimum cost belongs to loop 3 with $\overline{3.40}$ which is related to its previous vertex A in this loop. Therefore, the best path to F would be A→D→F. it should be explained that the next best paths to F with $\overline{6.8}$ and $\overline{8.56}$ cost values can be determined by tracing back from loops 5 and 6 respectively.

## 4. CONCLUSION AND FUTURE DIRECTIONS

Personalized route planning algorithms use different criteria of the network and aggregates them regarding users' preferences to determine overall cost for each link. These cost then are used to determine the best route. In this paper we emphasize that in this process there are two kinds of uncertainties. The first one is in the criteria values e.g. traffic, quality of the road and so on. Another one is from the ambiguity in determining users' preferences. Furthermore, the number of comparison between criteria to determine users' preferences as well as maintaining the consistency of comparisons are other important challenging issues. In this paper we propose a multi-criteria personalized routing model based on the fuzzy linguistic preference relation AHP model and modify Dijkstra's routing algorithm using fuzzy center of gravity ranking method to cope with this issues. Finally, we provide an illustrative example which shows the capability of the proposed model to capture the mentioned uncertainties.

## REFERENCES

Berredo R. C., Ekel P. Y. and R. M. Palhares (2005). "Fuzzy preference relations in models of decision making", Nonlinear Analysis Theory, Methods & Applications 63(5-7): 735–741.

Boyan J. A. and M. L. Littman (1994) "Packet routing in dynamically changing networks: A reinforcement learning approach", In Cowan, J. D., Tesauro, G., and Alspector, J. (eds.), Advances in Neural Information Processing Systems 6(NIPS): 8 pages, Morgan Kaufmann.

Chan F. T. S. and H. J. Qi (2002). "A fuzzy basis channel-spanning performance measurement method for supply chain management", Journal of Engineering Manufacture 216(8): 1155–1167.

Chen S. J. and S. M. Chen (2009). "Fuzzy Risk Analysis Based on the Ranking of Generalized Trapezoidal Fuzzy Numbers", Applied Intelligence 26(1): 1-11.

Chen, S. M. and J. H. Chen (2007). "Fuzzy Risk Analysis Based on Ranking Generalized Fuzzy Numbers with Different Heights and Different Spreads", Expert Systems with Applications 36(3): 6833-6842.

Chen, S. J. and C. L. Hwang (1991). "Fuzzy Multiple Attribute Decision making", Springer.

Chiclana F. and E. Herrera-Viedma (1998). "Integrating three representation models in fuzzy multipurpose decision making based on fuzzy preference relations", Fuzzy Sets and Systems 97(1): 33–48.

Ekel P.Y., Silva M.R., Schuffner Neto F. and R.M. Palhares (2006). "Fuzzy preference modeling and its application to multiobjective decision making", Computers and Mathematics with Applications 52(1-2): 179–196.

Khan J. A. and H. M. Alnuweiri (2004). "A fuzzy constraint-based routing algorithm for traffic engineering", In Proc. of the Global Telecommunications Conference, GLOBECOM '04 3:1366- 1372, IEEE.

Phani Bushan Rao P. and N. Ravi Shankar (2011). "Ranking Fuzzy Numbers with a Distance Method using Circumcenter of Centroids and an Index of Modality", Advances in Fuzzy Systems Vol. 2011, 7pages.

Rodriguez-Puente R. and M. S. Lazo-Cortes (2013). "Algorithm for shortest path search in Geographic Information Systems by using reduced graphs", SpringerPlus 2(291): 13pages.

Teodorovic D. and S. Kikuchi (2007). "Application of fuzzy sets theory to the saving based vehicle routing algorithm", Civil Engneering Systems 8, 1991(2): 87-93.

Vahidniaa M.H., Alesheikh A., Alimohamadi A. and A. Basiri (2008). "Fuzzy analytical hierarchy process in GIS application", The International Archives of the Photogrammetry Remote Sensing and Spatial Information Sciences Vol. XXXVII part B2.

Wang T. C. and Y.H. Chen (2008). "Applying fuzzy linguistic preference relations to the improvement of consistency of fuzzy AHP", Information Sciences 178(19): 3755–3765.

Zheng Y. and B. Liu (2006). "Fuzzy vehicle routing model with credibility measure and its hybrid intelligent algorithm" Applied Mathematics and Computation 176(2): 673-683.

Zhu K. (2014). "Fuzzy analytic hierarchy process: Fallacy of the popular methods" European Journal of Operational Research 236(1): 209-217.

# 13

# TOWARDS A CLOUD BASED SMART TRAFFIC MANAGEMENT FRAMEWORK

M. M. Rahimi [a], F. Hakimpour [b]*

[a] MSc student of GIS, Department of Surveying and Geomatics Engineering, College of Engineering, University of Tehran
rahimi.masoud@ut.ac.ir
[b] Assistant Professor, Department of Surveying and Geomatics Engineering, College of Engineering, University of Tehran
fhakimpour@ut.ac.ir

**KEY WORDS:** Traffic Management, Big Data, Cloud Computing, Hadoop

**ABSTRACT:**

Traffic big data has brought many opportunities for traffic management applications. However several challenges like heterogeneity, storage, management, processing and analysis of traffic big data may hinder their efficient and real-time applications. All these challenges call for well-adapted distributed framework for smart traffic management that can efficiently handle big traffic data integration, indexing, query processing, mining and analysis. In this paper, we present a novel, distributed, scalable and efficient framework for traffic management applications. The proposed cloud computing based framework can answer technical challenges for efficient and real-time storage, management, process and analyse of traffic big data. For evaluation of the framework, we have used OpenStreetMap (OSM) real trajectories and road network on a distributed environment. Our evaluation results indicate that speed of data importing to this framework exceeds 8000 records per second when the size of datasets is near to 5 million. We also evaluate performance of data retrieval in our proposed framework. The data retrieval speed exceeds 15000 records per second when the size of datasets is near to 5 million. We have also evaluated scalability and performance of our proposed framework using parallelisation of a critical pre-analysis in transportation applications. The results show that proposed framework achieves considerable performance and efficiency in traffic management applications.

## 1. INTRODUCTION

Nowadays, one of the most important challenges in transportation systems is traffic congestion. According to recent statistics, transportation has the second place in greenhouse gas emission factors ranking of USA (STATISTICS, 2015). In 2014, traffic congestion costs 6.9 billion hours of citizenry and 3.1 billion gallons of fuel, 160 billion dollars loss to US economy (STATISTICS, 2015). Infrastructure improvement is an expensive solution to traffic congestion challenge.

A Traffic Management System (TMS) as one of the most important components of Intelligent Transportation System (ITS) offers capabilities that can potentially be used to reduce road traffic congestion, improve response time to incidents and ensure better travel experience for commuters.

Some of the most important services of TMS are vehicle routing to shorten commuter journey, traffic prediction that enables early detection of bottlenecks, parking management that ensure optimal usage of parking spots and interact with routing and prediction services for improved control of traffic flow and finally infotainment services that provide useful information for both drivers and passengers (Djahel et al., 2015).

In recent years, researchers have shown great interest in using advances in wireless sensing and communication technologies along with novel techniques and methodologies in TMSs to make them more efficient. With the emerging of new sources of traffic data like Wireless Sensor Networks (WSNs), Machine to Machine communication (M2M), Floating Car Data (FCD), mobile sensing and social media new opportunities for different transportation applications has been created. These big data can be used for real-time road management and decision-making, data mining and knowledge extraction. In order to exploit this big data potential in TMS applications, traditional frameworks face some technical challenges like:

- Data heterogeneity: these data are from different heterogeneous sources in different structures.
- Data management and storage: when a dataset outgrows the storage capacity of a single physical machine, it is necessary to partition it across multiple separate machines.
- Data processing and analysis: with this tremendous valuable big data, real-time query processing, analysis and data mining in traditional frameworks is a time-consuming inefficient task.

All these problems call for well-adapted distributed framework for TMS that can efficiently handle big traffic data integration, indexing, query processing, mining and analysis.

Cloud computing is a useful, scalable and cost-effective solution to answer traffic big data challenges in TMS. Hadoop (Apache) as a popular cloud computing framework on commodity hardware provides high availability and scalability along with fault tolerance for real-time traffic management applications.

In this paper, we attempt to facilitate storage, management, processing and analysis of traffic heterogeneous big data by using a state-of-the-art distributed framework for smart traffic management system. The main requirements of the proposed framework are:

- Low-latency data storage and access: the framework should contain an efficient and flexible tool for data gathering and management for massive traffic data with high performance and low latency.

---
\* Corresponding author

- Fast data processing, analysis and data mining: the framework should contain fast powerful calculation engines to support different type of query processing, traffic modeling, analysis and data mining.
- Easy Implementation for different applications: the programming paradigm for implementation of different calculation models should be easy for implementation and development.
- Fault Tolerance, Elasticity and High Availability: consistency of increasing size of traffic data, adapting to unpredictable changes and high uptime to support real-time applications along with fault tolerance is some of the most important features in a distributed system.

This research is a major step to development of an efficient and real-time traffic management system. The rest of the paper is organized as follows: First we review major related works to our framework, then we will propose our framework architecture and finally, we will evaluate the framework performance and scalability.

## 2. RELATED WORKS

Hadoop is a popular open-source software framework for cloud computing which consists of two main modules. Hadoop Distributed File System (HDFS) and a parallel processing framework named MapReduce. HDFS is a distributed file system. In HDFS, every single file partitions across several separate machines including a master node and some slave nodes. The master node should store file system metadata while the data stores on slave nodes. MapReduce is a parallel programming model for large-scale data processing. MapReduce can be written in various programming languages like Java, Ruby and Python. Based on HDFS, HBase (Apache) is developed as a No-SQL distributed database. In HBase real-time and random big data read and write is done in a fast and easy manner. Hive (Hadoop; Thusoo et al., 2009) is a data warehouse software which facilitates reading, writing, and managing large datasets residing in distributed storage using SQL commands. Pig (Hadoop) is a platform for large-scale datasets analysis programs which contains a high level programming language "PigLatin" along an infrastructure for evaluating these programs. Mahout (Hadoop) is a distributed machine learning and data mining tool. The aim of mahout is build an environment for quickly creating scalable performant machine learning applications. Oozie (Hortonworks) is a workflow scheduler. Oozie is a web-based java program that uses for scheduling apache Hadoop jobs. Flume is distributed, reliable and available service for efficiently collecting, aggregating, and moving large amounts of streaming event data. Sqoop (Apache) is an open source tool used for integration between Hadoop framework and structured relational databases (e.g. Oracle).

Some of the most critical challenges of using distributed systems in spatial applications are storing, management and processing of spatial big data. Hadoop-GIS (Aji et al., 2013) is high-performance scalable system for spatial big data storage on HDFS which supports spatiotemporal queries using SQL like commands. Similarly, SpatialHadoop (Eldawy and Mokbel, 2015) is developed as an efficient framework for spatial queries which employs a two level spatial index structure and some basic spatial functionalities.

In recent years, several efforts have been made on development of an efficient traffic management system in both academic and commercial research societies. In commercial society, *Trafficware* in Houston government and *TransCore* in Washington D.C. are trying to provide an advanced system for enabling the Traffic management personnel to manage the congestion more effectively. Besides, as a public traffic management platform, *Waze* is an android-based application that uses crowd sourcing data and methods. The users have the app running in the background while travelling to their destination, thereby passively contributing traffic data and other incident data. In academic society, there are some studies on traffic management in literature. Xiao et al. (Xiao et al., 2015) proposed a spatial data oriented platform "DriveNet" for freeways performance analysis. The aim of DriveNet is traffic data integration, sharing, analysis and visualization. In this platform a central web server processes user's requests using HTTP(s) protocol. Then the server obtained data from different sources and process and analyse them. This analysis is done using internal and external modules like R servers. The main drawback of this platform is using central server and lack of fault tolerance, elasticity and high availability. With the aim of finding patterns in traffic data, (Khazaei et al., 2015) developed a cloud based big data analytic platform. (Xiong et al., 2016) discussed different aspects in the design of ITS including future trends and current ITS development considerations. (Sekar et al., 2017) has developed a data mining framework for traffic congestion prediction and route planning using hybrid clustering techniques.

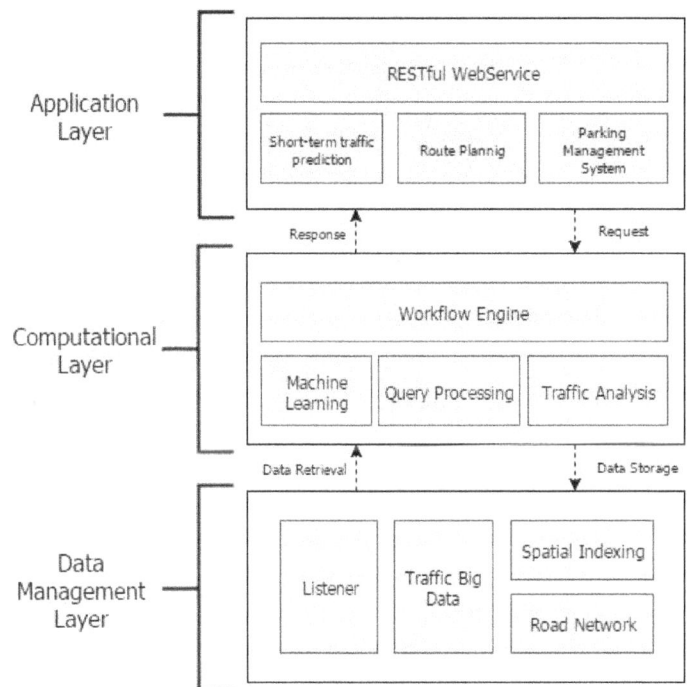

Figure 1. Proposed framework architecture

(Yu et al., 2013) has developed a traffic big data mining system based on Software as a Service (SaaS) architecture. It is observed that most of the efforts on development of an efficient smart traffic management system suffer from the lack of comprehensiveness.

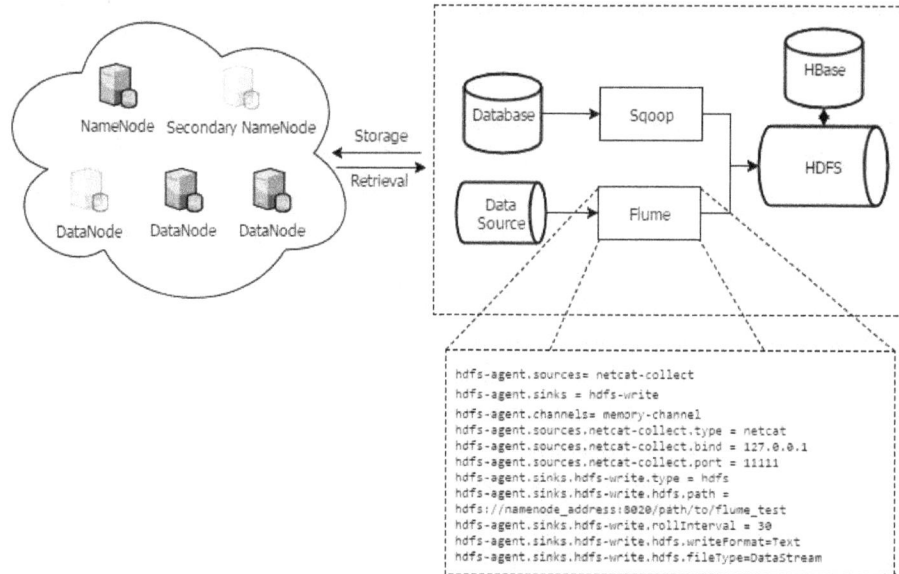

```
hdfs-agent.sources= netcat-collect
hdfs-agent.sinks = hdfs-write
hdfs-agent.channels= memory-channel
hdfs-agent.sources.netcat-collect.type = netcat
hdfs-agent.sources.netcat-collect.bind = 127.0.0.1
hdfs-agent.sources.netcat-collect.port = 11111
hdfs-agent.sinks.hdfs-write.type = hdfs
hdfs-agent.sinks.hdfs-write.hdfs.path =
hdfs://namenode_address:8020/path/to/flume_test
hdfs-agent.sinks.hdfs-write.rollInterval = 30
hdfs-agent.sinks.hdfs-write.hdfs.writeFormat=Text
hdfs-agent.sinks.hdfs-write.hdfs.fileType=DataStream
```

Figure 2. Data management layer

An efficient smart traffic management system requires data integration, management, storage and analysis as long as data mining information and knowledge discovery techniques. Thus, the proposed framework is a new idea to fill in the gaps we had found these systems to make a comprehensive framework. In this paper, we focus more on data integration, data analysis in traffic management applications.

## 3. THE FRAMEWORK ARCHITECTURE

In this section, we have proposed a framework for traffic management system. The main goal of the framework is answering to traditional TMS drawbacks that has discussed earlier. The proposed framework is using Hadoop cloud computing technologies including HDFS, MapReduce paradigm and some of the other members of Hadoop ecosystem. As shown in Figure 1, the proposed framework consists of three layers architecture. Data from different sources imports to data management layer which low-latency data storage and access. Then the computational layer is used to provide different computation paradigms to ensure TMS applications requirements. The upper layer is application layer that provides system interaction with end users using web services or graphical interfaces.

### 3.1 Data Management Layer

Traffic Management needs applicable data about current traffic status and traffic flow characteristics. These heterogeneous data should be collected from different sources in different formats in TMS for further applications. As the result, the cornerstone of traffic management is data management layer. The data management layer aims to integrate all kinds of data into a distributed file system to fully exploit their potential. Data management layer should also interact with different databases to take advantage of their capabilities. Figure 2 represents data management layer architecture.

Generally, there are two kind of data source in the proposed framework. One of them are road networks which represents system functional area. Besides road networks, continuous stream of real-time traffic data for further processing is also another kind of available data in TMS which are collected from different sources and in different formats. Table 1 represents different traffic data types and their characteristics in TMS.

The generated Input data of TMS from different sources sends to a central server. Server by flume agent collects the data and stores them in HDFS. In the data manager layer Sqoop tool is used for transferring data between HDFS and structured relational databases that is a critical requirement in development of an adequate TMS.

One of the most important spatial data in TMS is roads network. A roads network is a directed graph G(V,E,M) where V is a set of vertices which are main points or intersection of roads, E is a set of edges which are vectors connecting two vertices and M is a set of allowed movements.

In a distributed framework, storage and indexing of roads network are important challenges. To answer these challenges and due to static characteristic of roads network, in the proposed framework we have used HBase database. HBase is a distributed, column-based, No-SQL database that uses HDFS as its underlying storage. HBase provides read/write access of individual rows and batch operations for reading and writing data in bulk. The storage unit in HBase is cell which defined uniquely by row-key, column family name, column name and version. By default, a cell version is a timestamp auto-assigned by HBase at the time of cell insertion. On a physical level, every column family data continuously writes on the disk and this data will be sorted by its row key, column name and version. In this framework, we have stored roads network in two HBase tables.

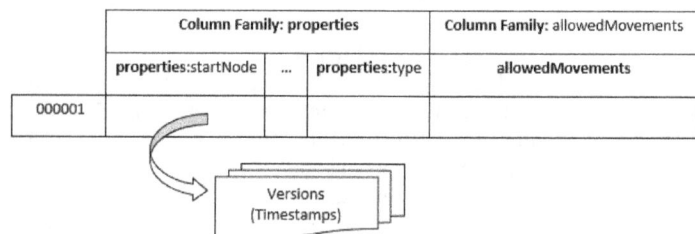

Figure 3. Road nework data model

In the first table, vertices and their properties are stored using "nodeID" as row key. In the second table, edges and properties are stored in different column families and an "edgeID" is used as row key. For each edge, a set of allowed movements is pre-computed and their IDs are stored separated by semicolons. Figure 3 represents road network data model.

| Data Source Class | Data Source | Data Structure | Descriptions |
|---|---|---|---|
| Infrastructural Sensors | Pneumatic Road Tube | Structured | Volume/count, speed and vehicle classification |
| | Magnetic sensors | Structured | Volume/count, speed and vehicle classification |
| | Inductive Loops | Structured | Volume/count, speed, vehicle classification and occupancy |
| | Active infrared | Structured | Volume/count, speed and vehicle classification |
| | Passive infrared | Structured | Volume/count, speed, vehicle classification and occupancy |
| | Microwave radar | Structured | Volume/count, speed, vehicle classification and occupancy |
| | Ultrasonic | Structured | Volume/count |
| | Acoustic sensor | Structured | Volume/count, speed, vehicle classification and occupancy |
| | Bluetooth scanning | Structured | Travel time prediction |
| | V2X (V2V or V2I) | Structured | Travel time prediction |
| Floating Car Data | Vehicle-based | Structured | Vehicle location, speed and heading |
| | Mobile-based | Structured | Mobile location, speed and heading |
| Web and social networks | Web and social networks | Unstructured | Collective information from the public |
| Video surveillance | CCTV | Unstructured | BLOB videos |
| Media | Media | Semi-structured | Audio / Textual Reports |

Table 1. Different traffic data and their properties in TMSs

In HBase, data is stored by key and is randomly accessible based on key. In addition, HBase does not have capabilities for secondary indexing. Therefore, with the aim of fast and efficient storage and retrieval of spatial road network, a secondary spatial index is needed. At present several spatial index algorithm has been proposed in literature (Finkel and Bentley, 1974; Guttman, 1984). However, integration between this methods and Hadoop environment faces several challenges. While Hadoop is using functional programming, all of these methods need procedural approach. Besides, in Hadoop once a file is written on HDFS, it cannot be modified later.

Considering this limitations, in this paper we adopt regular grid spatial indexing method for road network. We have divided the

geographical region of map in k different regions. For each rectangular region, we find any edges in data table that are either connected to intersect the rectangle. We build another table in HBase as shown in Figure 4 that has a column family for edges. Each row in this table represents a region in the index. All "edgeID"s are separated by semicolons.

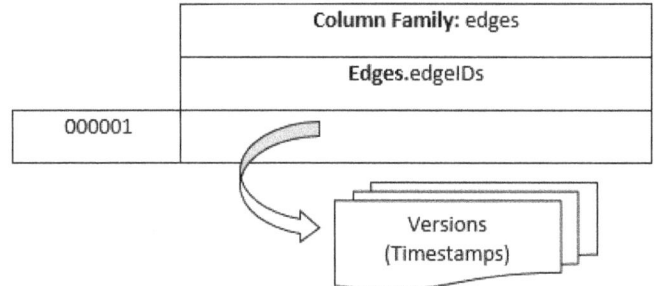

Figure 4. Spatial index data model

## 3.2 Computational Layer

Storage and management of various data in TMS is an important challenge in development of an adequate framework. However, this data should be used for extraction of valuable information and traffic parameters. An adequate TMS should have capable of running data analysis or mining the data using statistical approaches or machine learning methods. By using of big data technologies, in the proposed computational layer we have developed a computational engine for every traffic analysis needed for TMS services (Figure 5).

As mentioned before, one of the basic requirements of an adequate TMS is easy implementation for different applications. Therefore, in the central core of the computational layer, we have used Hadoop MapReduce engine. MapReduce paradigm due to easy programming, automatic load balancing and scalability is a suitable approach for using in a TMS framework. Generally, in a MapReduce program, map function is used for creating intermediate data. Then the intermediate data aggregated using reduce function to generate the final result. By this simple architecture, parallel functions can be automatically managed and load balance significantly reduces compared to traditional platforms.

There are some other modules in the computational layer. Hive and Pig are two open-source data warehousing and query processing tools. Hive prepares an easy read and write interface for distributed storage by using SQL-like language named HiveQL. In Hive, a command-line tool and a JDBC driver for users interaction is included. Hive converts HiveQL queries to MapReduce programs and then runs them on Hadoop. It is clear that MapReduce and Hive final results are the same but MapReduce coding and debugging is more complicated than simple SQL-like commands. Generally, Hive provides higher level of abstraction compared to MapReduce. However like every other high level abstraction, using hive needs additional computation which leads to lower performance of the system (Thusoo et al., 2009). Similarly, Pig also provides higher level of abstraction compared to MapReduce. Pig is an analytical platform for big datasets that contains a high level programming language "PigLatin" for data analysis programming along an infrastructure for evaluating this programs. PigLatin is a simple, expandable and optimize programming language.

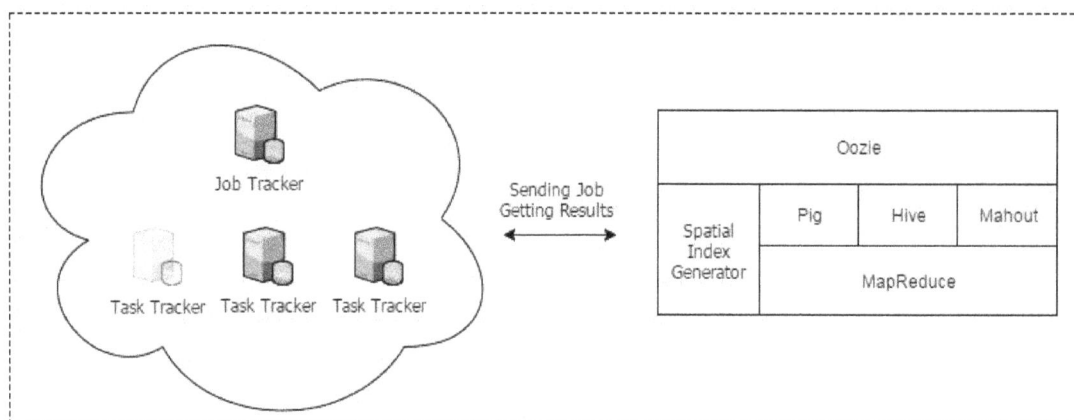

Figure 5. Computational layer

As mentioned earlier, in some complicated traffic management analysis data mining and machine learning tools are needed. One of the most important open-source projects in Hadoop ecosystem is Mahout. Mahout is a distributed framework for machine learning. Mahout supports various algorithms like classification, clustering and filtering in a parallel manner. The main goal of Mahout is development of a scalable machine-learning library. In this work, we have used Oozie as the workflow engine.

### 3.3 Application Layer

Some of the most important services of TMS are vehicle routing to shorten commuter journey, traffic prediction that enables early detection of bottlenecks, parking management that ensure optimal usage of parking spots and interact with routing and prediction services for improved control of traffic flow and finally infotainment services that provide useful information for both drivers and passengers (Djahel et al., 2015). In order to make an interaction between user and system, a set of RESTful web services are designed. Various functions for simple and flexible interaction of user with this web services has been considered.

## 4. FREAMEWORK EVALUATION

We have setup our framework on a Grid5000's Nancy cluster using 3 nodes. The master node is responsible for data processing along with cluster management. The other slave nodes are only responsible for data processing. The commodity environment used for this evaluation is shown in Table 2. Since our work is still under development, we have only implemented a prototype of our proposed framework and parts of data management and computational layer components have been used in this evaluation.

We have evaluated data management layer performance by importing a real OpenStreetMap[1] (OSM) trajectory datasets as a sample traffic data to our framework. Several datasets with different sizes are selected to evaluate the performance. As shown in Fig.6 (a), the speed of records import to data management layer exceeds 8000 records per second when the size of datasets is near to 5 million.    We also evaluate performance of data retrieval in our proposed framework. As shown in Fig.6 (b) the retrieval performance is faster than the import process. The data retrieval speed exceeds 15000 records per second when the size of datasets is near to 5 million.

| Hadoop Version | 2.7.1 |
|---|---|
| HBase Version | 1.1.2 |
| Flume Version | 1.5.2 |
| Cluster Model | Dell PowerEdge R730 |
| Nodes Processors | Intel(R) Xeon(R) CPU E5-2620 v3 @ 2.40GHz |
| Memory | 64GB RAM |
| Network | 10 Gigabit Ethernet DA/SFP+ |

Table 2. Evaluation environment

For evaluation of computational layer, we have parallelized a critical pre-analysis in traffic management application. In this use case, we have evaluated scalability and performance of our proposed framework capabilities for traffic management applications. FCD is one of the most important data sources in traffic management applications that represents Spatio-temporal trajectory of a vehicle. Due to FCD nature, there are a lot of noises in this data. Therefore, FCD has limited accuracy and cannot be matched on the road network by itself. To face this issue, an important traffic analysis named map matching has been developed. In this study we have used (Liu et al., 2012) method.

Our analysis is conducted on the OSM real trajectories and road network that has been used for data management and computational layers evaluation. Fig.7 (a) represents scalability evaluation of our map matching implementation. As shown in Fig.7 (a), with the growth of computation nodes, processing time has decreased significantly. We also evaluate map matching speed using different processing nodes. As shown in Fig.7 (b), by the increase of computational nodes, matching speed shows a significant growth.

---

[1] www.openstreetmap.org

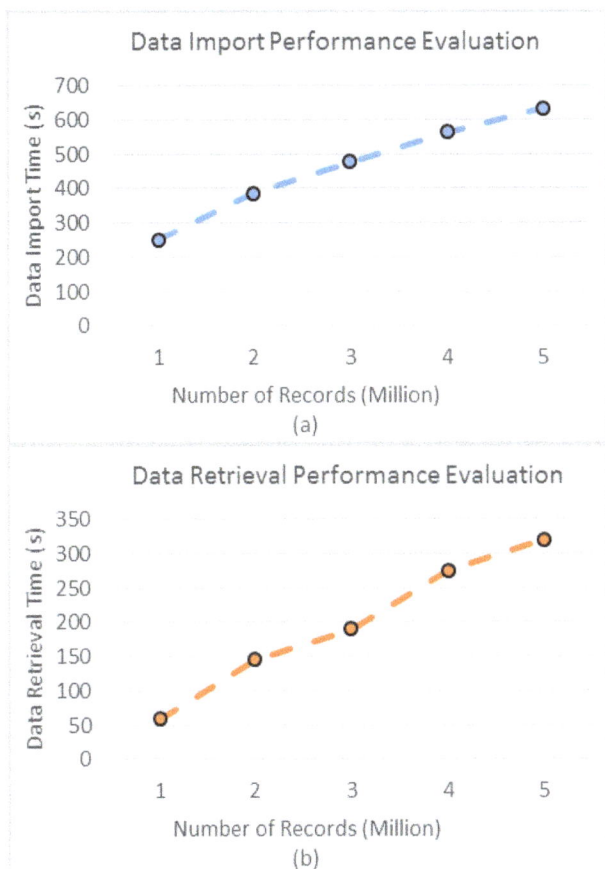

Figure 6. Data management layer evaluation

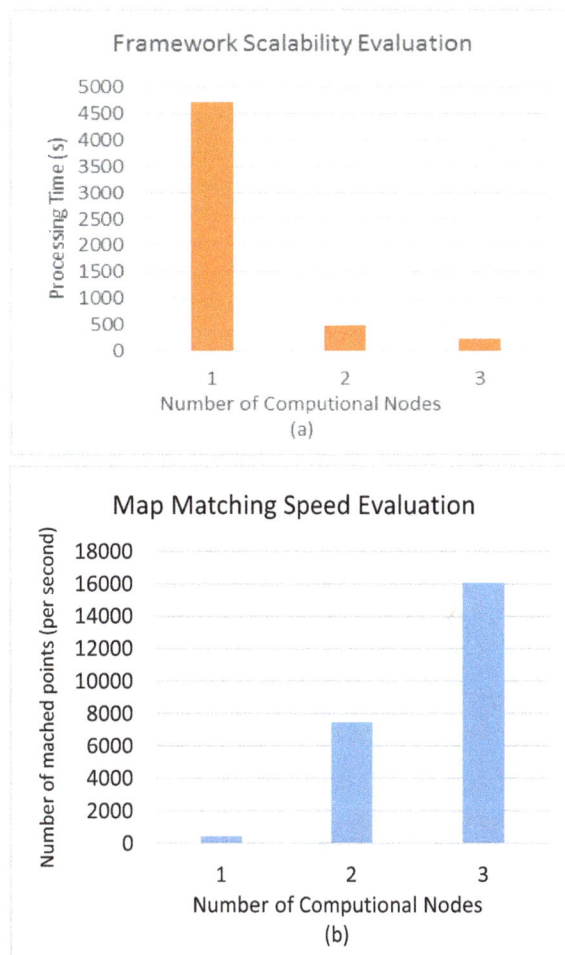

Figure 7. Computational layer evaluation

## 5. CONCLUSION AND FUTURE WORK

Traffic Big data has initiated great opportunities for traffic management applications. Beside such opportunity, several technical challenges have appeared. This paper takes a major step towards development of an efficient and real-time traffic management system using cloud computing technologies. The proposed distributed three layer framework contains several open-source components and libraries for supporting traffic management applications. Our evaluation results show that the proposed framework has efficiently stored, managed and analysed traffic big data. Using Hadoop ecosystem and MapReduce paradigm, proposed approach provides high availability and scalability along with fault tolerance for real-time traffic management applications. In our future work, we plan to improve our framework by considering data mining and knowledge discovery techniques, improving our spatial indexing method and concentrate more on our application layer to support more application functionalities.

## ACKNOWLEDGEMENTS

Experiments presented in this paper were carried out using the Grid'5000 testbed, supported by a scientific interest group hosted by Inria and including CNRS, RENATER and several Universities as well as other organizations[2].

[2] https://www.grid5000.fr

## REFERENCES

Aji, A., Wang, F., Vo, H., Lee, R., Liu, Q., Zhang, X., Saltz, J., 2013. Hadoop GIS: a high performance spatial data warehousing system over mapreduce. Proceedings of the VLDB Endowment 6, 1009-1020.

Apache, Apache Sqoop.

Apache, Welcome to Apache Flume.

Apache, Welcome to Apache Hadoop!

Apache, Welcome to Apache HBase™.

Djahel, S., Doolan, R., Muntean, G.-M., Murphy, J., 2015. A communications-oriented perspective on traffic management systems for smart cities: challenges and innovative approaches. IEEE Communications Surveys & Tutorials 17, 125-151.

Eldawy, A., Mokbel, M.F., 2015. Spatialhadoop: A mapreduce framework for spatial data, 2015 IEEE 31st International Conference on Data Engineering. IEEE, pp. 1352-1363.

Finkel, R.A., Bentley, J.L., 1974. Quad trees a data structure for retrieval on composite keys. Acta informatica 4, 1-9.

Guttman, A., 1984. R-trees: a dynamic index structure for spatial searching. ACM.

Hadoop, APACHE HIVE TM.

Hadoop, Welcome to Apache Pig!

Hadoop, What is Apache Mahout? Apache.

Hortonworks, Apache Oozie.

Khazaei, H., Zareian, S., Veleda, R., Litoiu, M., 2015. Sipresk: A Big Data Analytic Platform for Smart Transportation, EAI International Conference on Big Data and Analytics for Smart Cities.

Liu, S., Liu, C., Luo, Q., Ni, L.M., Krishnan, R., 2012. Calibrating large scale vehicle trajectory data, 2012 IEEE 13th International Conference on Mobile Data Management. IEEE, pp. 222-231.

Sekar, E.V., Anuradha, J., Arya, A., Balusamy, B., Chang, V., 2017. A framework for smart traffic management using hybrid clustering techniques. Cluster Computing, 1-16.

STATISTICS, B.O.T., 2015. Transportation Statistics Annual Report, in: Transportation, U.S.D.o. (Ed.).

Thusoo, A., Sarma, J.S., Jain, N., Shao, Z., Chakka, P., Anthony, S., Liu, H., Wyckoff, P., Murthy, R., 2009. Hive: a warehousing solution over a map-reduce framework. Proceedings of the VLDB Endowment 2, 1626-1629.

Xiao, S., Liu, X.C., Wang, Y., 2015. Data-Driven Geospatial-Enabled Transportation Platform for Freeway Performance Analysis. IEEE Intelligent Transportation Systems Magazine 7, 10-21.

Xiong, G., Zhu, F., Dong, X., Fan, H., Hu, B., Kong, Q., Kang, W., Teng, T., 2016. A kind of novel ITS based on space-air-ground big-data. IEEE Intelligent Transportation Systems Magazine 8, 10-22.

Yu, J., Jiang, F., Zhu, T., 2013. Rtic-c: A big data system for massive traffic information mining, Cloud Computing and Big Data (CloudCom-Asia), 2013 International Conference on. IEEE, pp. 395-402.

# 14

# PROVIDING THE FIRE RISK MAP IN FOREST AREA USING A GEOGRAPHICALLY WEIGHTED REGRESSION MODEL WITH GAUSSIN KERNEL AND MODIS IMAGES, A CASE STUDY: GOLESTAN PROVINCE

Ali Shah-Heydari pour [a], Parham Pahlavani [a,*], Behnaz Bigdeli [b]

[a] School of Surveying and Geospatial Engineering, College of Engineering, University of Tehran, Tehran, Iran
[b] School of Civil Engineering, Shahrood University of Technology, Shahrood, Iran.
(shahheydary1372, pahlavani@ut.ac.ir; bigdeli@shahroodut.ac.ir)

KEYWORDS: Forest Fire, Geographically Weighted Regression, Fire Risk Map, Golestan Forest, Gaussian kernel

**ABSTRACT:**

According to the industrialization of cities and the apparent increase in pollutants and greenhouse gases, the importance of forests as the natural lungs of the earth is felt more than ever to clean these pollutants. Annually, a large part of the forests is destroyed due to the lack of timely action during the fire. Knowledge about areas with a high-risk of fire and equipping these areas by constructing access routes and allocating the fire-fighting equipment can help to eliminate the destruction of the forest. In this research, the fire risk of region was forecasted and the risk map of that was provided using MODIS images by applying geographically weighted regression model with Gaussian kernel and ordinary least squares over the effective parameters in forest fire including distance from residential areas, distance from the river, distance from the road, height, slope, aspect, soil type, land use, average temperature, wind speed, and rainfall. After the evaluation, it was found that the geographically weighted regression model with Gaussian kernel forecasted 93.4% of the all fire points properly, however the ordinary least squares method could forecast properly only 66% of the fire points.

## 1. INTRODUCTION

Forest fire is a major problem for forests across the world. The existence of geographic information about the fire occurrence is a necessary matter to deal with the fire. Fire causes the spatial pattern of vegetation and the process of human life to change as well as it can change global ecosystem due to its direct relation with carbon cycle and atmosphere composition.

To reduce the damages caused by forest fire including destruction of natural resources, environment and life of forest creatures, we should assess fire dangers. Due to the destructive nature of fire, the researches related to real fire occurrences are impractical. As an alternative, scholars use simulation models that are based on field studies to better understand the behavior and effects of fire occurrence.

The policies related to fire-fighting can be divided into two groups: preventive and operational activities. Since prevention is always better than cure, we can prevent the fire and its ruinous consequences by forecasting the areas that have a high risk of fire and performing preventive activities like access restriction and resource allocation. Due to the limited forest resources in Iran, it is very important to conserve these resources in our country.

So far, various models have been presented in the field of the fire in forest areas. In one of this researches, the risk map of fire occurrence was provided using geographic information system and remote sensing for part of the Spanish coasts (Chuvieco and Congalton, 1989). During last years, fire occurred only in 3.47%

of low-dangerous regions from the perspective of this article. In 2003, a new model was developed using artificial neural network to forecast the risk of fire occurrence in a region located at north-western Spain. This study divided regions into four regions based on fire occurrence probability (Amparo and Oscar, 2003). In a research, the fire risk map of Paveh region in Kermanshah was provided by weighting different layers like vegetation, height, and slope using Analytic Hierarchy Process (AHP) (Mohammadi et al., 2010). Another model was developed to forecast fire occurrence in Boreal forests of northern china using the fire information between 1965 and 2009. In another research, the relationship between the human and natural factors of fire occurrence was studied based on the occurrence pattern (Xiaowei et al., 2014). In this model, the risk of fire occurrence was evaluated using meta-heuristic method of artificial neural network and statistical method of logistic regression .According to the results of this study, the artificial neural network method is significantly more accurate than the statistical method and logistic regression. The accuracy of the statistical method is dependent upon the number of sampled points that the fire has been occurred (Jafari and Mohammadzadeh, 2016). In another study, a new model was developed to calculate the risk of fire occurrence in Espirito Santo region. This model was created statistically as well as for each effective parameter in the fire occurrence, a weight was considered based on the parameter effectiveness (Fernando Coelho et al., 2016).

The aim of this study is to determine the risk of fire occurrence in the forest regions of Golestan province using the MODIS images by applying a geographically weighted regression with Gaussian kernel. Since this method takes into account the main

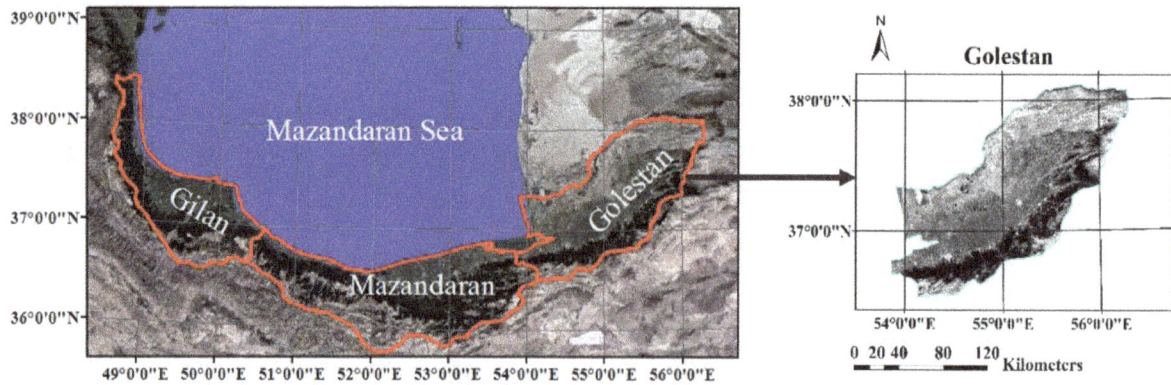

**Figure 1.** The study area is Golestan province located in the northern part of Iran, the northern slopes of the Alborz mountain range and in south coastal plain of Mazandaran sea

features of spatial data, i.e. spatial non-stationary and spatial autocorrelation, it is an appropriate method to model this problem. In this regard, the effective biophysical and human parameters in fire occurrence have been used in the process of modeling.

## 2.    MATERIAL AND METHODS

### 2.1.  The study region

Since the largest number of forest fires during the year belong to Golestan Province, this province was chosen as the study region. Golestan forest is one of the most important forest regions of Iran that is located at eastern Golestan Province and west North-Khorasan Province. This forest is among the tourist regions of Iran and several roads pass through it, therefore human factors play a crucial role in fire occurrence in this region.

The study region of this research (Figure 1) is located in the range of 36° - 38° North and 53° - 56° East and its area is about 20205 km$^2$.

### 2.2.  Data

Different satellite and field data including the following ones was used to model the risk of fire occurrence:

**Satellite data**: includes the information of the fire occurrences, soil type, land use map and Digital Elevation Model (DEM) (Figure 2). Since 68% of fires have been occurred in June, July and August, the fires occurred in these three months were extracted and used in this article. 430 fire points between 2011 and 2016 were extracted from MODIS images. Ground cover map was prepared by NOAA images with resolution of 1 km. The slope and aspect maps were created using the DEM of the region.

**Field data**: include the information of meteorology, rivers (Figure 3), roads and residential regions (Figure 4) of study region. The meteorological information namely temperature, wind, and monthly rainfall were extracted for all training and testing points through interpolating the information of the existing synoptic stations in the region. The information related to residential regions, rivers, and roads was received from organization of Golestan Province Natural Resources and the

layers of distance from residential regions, rivers, and roads were calculated using ArcGIS software.

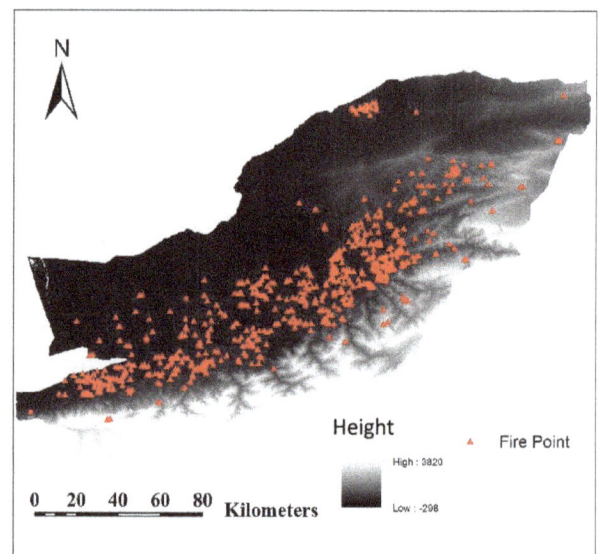

**Figure 2**. DEM of study area with spatial distribution of fire location

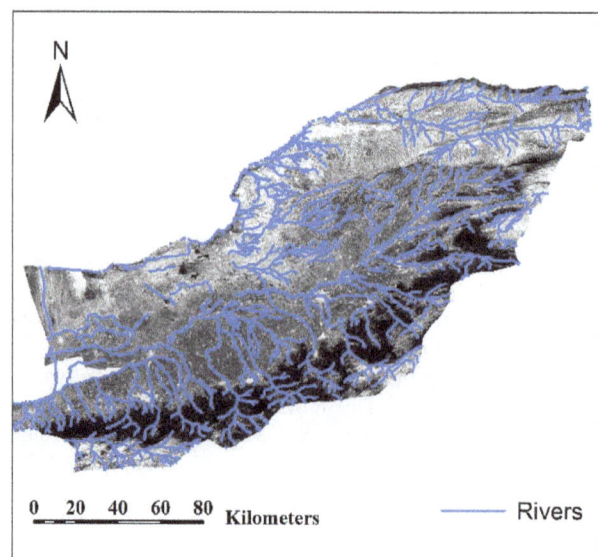

**Figure 3**. Rivers data

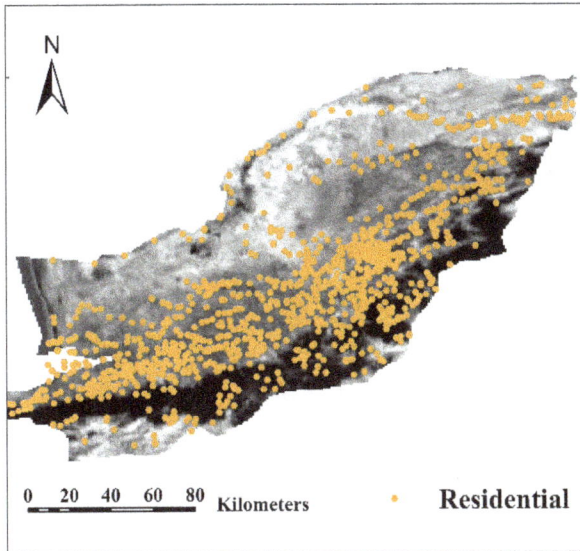

**Figure 4**. Residential region data

## 2.3. Geographically weighted regression

Spatial data has specific features that two of them are spatial autocorrelation and spatial non-stationary. The first is based on Tyler law and represents the inverse relationship of dependencies with distance and the second represents the variation of spatial autocorrelation in space and heterogeneous environment. The ordinary least squares (OLS) method cannot be adapted with these two features because in this method, it is assumed that the data are completely independent from each other and the environment is homogenous. Hence, OLS provides a set of answers without taking into consideration the spatial dependencies of all points of region. For this reason, GWR method was presented by Brunsdon in 1998. In this method, the temporal dependencies of observations are considered as weighted matrices and due to heterogeneous environment and the existence of spatial non-stationary, regression coefficients are calculated locally and separately for each point. The general relationship of GWR is as follows (Brunsdon, 1988):

$$y = \beta_o(u,v) + \sum_{j=1}^{p} \beta_j(u,v)X_j + \varepsilon \tag{1}$$

where $y$ is a dependent variable, $X_j$ is the $j^{th}$ independent variable, $p$ is the number of independent variables, $\varepsilon$ is the remaining of model and $\beta_j$ is the regression coefficients that are a function of the position of observational points $(u,v)$. The determination of geographical weights in GWR is very important. Hitherto, various kernels have been presented for this purpose. One of the kernels that has demonstrated high performance is Gaussian kernel that is represented in equation (2) (McMillen, 1988):

$$W_{ij} = \varphi\left(\frac{d_{ij}}{\sigma h}\right) \tag{2}$$

where $W_{ij}$ is the geographical weight related to $i^{th}$ observation in $i^{th}$ point, $\varphi$ is the standard normal distribution function, $d_{ij}$ is the Euclidean distance of the points $i$ and $j$, $\sigma$ is the standard deviation of $d_{ij}$ values for each point and $h$ is the parameter of bandwidth.

The important step in the determination of geographical weights is to choose appropriate bandwidth. Because if this parameter is too large, the results of GWR tend to OLS and if it is too small the variance of results increase significantly (Charlton et al., 2001)

There are different methods to determine optimal bandwidth. One of those is cross validation method that its function is as follows (Brunsdon, 1988):

$$\sum_{i=1}^{n}[y_i - \tilde{y}_i(h)]^2 \tag{3}$$

Where $\pi$ is the number of observations, $y_i$ is $i^{th}$ observation and $\tilde{y}_i$ is the estimated value of $i^{th}$ observation using other observations that itself is a function of bandwidth parameter and each bandwidth which minimizes this function is considered as optimal bandwidth.

The output of GWR includes several parameters that among them, the determination coefficient of $R^2$ is employed to measure the quality of model fitness (Charlton et al., 2001). This parameter is obtained using equation (4) (McMillen, 1988):

$$R^2 = 1 - \left(\frac{SS_E}{SS_T}\right) \tag{4}$$

$$SS_E = \sum_{i=1}^{n}[y_i - \hat{y}_i]^2 \tag{5}$$

$$SS_T = \sum_{i=1}^{n}[y_i - \bar{y}]^2 \tag{6}$$

Where $n$ is the number of observations, $y_i$ is $i^{th}$ observation, $\hat{y}_i$ is the estimated value of $i^{th}$ observation and $\bar{y}$ is the average of observations.

In addition, the Root Mean Square Error (RMSE) and Normalized Root Mean Square Error (NRMSE) values of model's residues which are obtained using following equations are used to measure the distribution of these residues.

$$RMSE = \sqrt{\frac{1}{n}\sum_{i=1}^{n}(y - \hat{y})^2} \tag{7}$$

Where $n$ is the number of data, $y$ is the real value of dependent variable and $\hat{y}$ is the estimated value of dependent variable.

## 3. RESULT AND DISCUSSION

The programming environment of MATLAB has been used to implement and model the problem. To model the risk of fire occurrence, the points are divided into two groups namely training and testing points. The training points are a combination of fire points (400 points) and non-fire points (325 points) in the period of 2011-2015. In addition, the fire points of 2016 (30 points) were extracted to evaluate the model. The fire points have a value of 1 and the non-fire points have a value of zero.

The factors (Table 1) used in this article were divided into three classes namely human, climate, and field factors. The factors were selected which have more important and available.

According to the spatial positions of training and testing points, the related factors were linked to each of them. Then, the effective parameters in fire occurrence were normalized to improve the model performance. The modeling was conducted once by OLS and again by GWR with Gaussian kernel.

| | Field Factors | Human Factors | Climate Factors |
|---|---|---|---|
| 1 | Height (m) | Distance from rivers (m) | Average temperature |
| 2 | Slope | Distance from roads (m) | Maximum wind speed (m/s) |
| 3 | Aspect | Distance from residential regions (m) | Rainfall (mm) |
| 4 | Soil Type | | |
| 5 | Land use | | |

**Table 1.** Effective Factors

In the process of modeling by OLS method, first, the coefficients of used factors in the model were extracted. The parameters' coefficients show the effectiveness of that parameter in the fire occurrence. After calculating the coefficients, the risk of fire occurrence in these points related to 2016 were calculated having the factors related to the testing points. Then, the risk value of all points were interpolated using inverse distance weighting (IDW) interpolation method that is a distance-based weighted method. Finally, the risk map was provided using the interpolated values (Figure 5). The real fire points of 2016 were also placed on the map for visual evaluation of this method. The calculated parameters of this model are listed in Table 2.

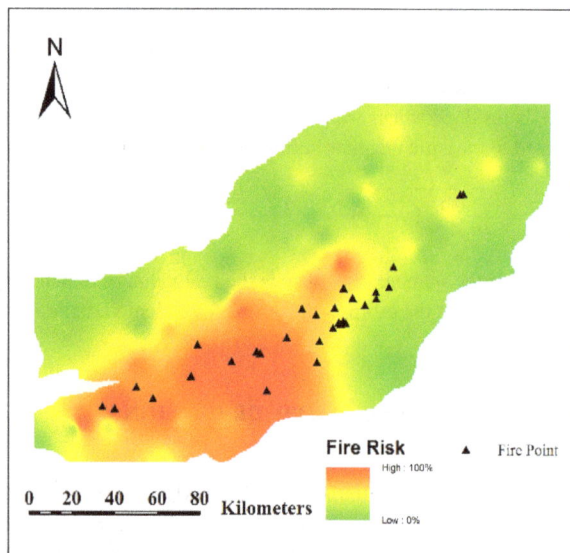

**Figure 5.** Fire risk map provided by OLS method

| Variable | $R^2$ | RMSE |
|---|---|---|
| Value | 0.6642 | 0.2903 |

**Table 2.** OLS parameters

In GWR model, the general process is similar to that of OLS, however it is closer to the reality of problem by considering spatial non-stationary. In this model, we train the problem process to the model using Gaussian kernel and having the testing points. Then, we forecast the risk of fire occurrence for testing points by using coefficients and vicinities (Figure 6). The

real fire points of 2016 were also placed on the map for visual evaluation of this method.

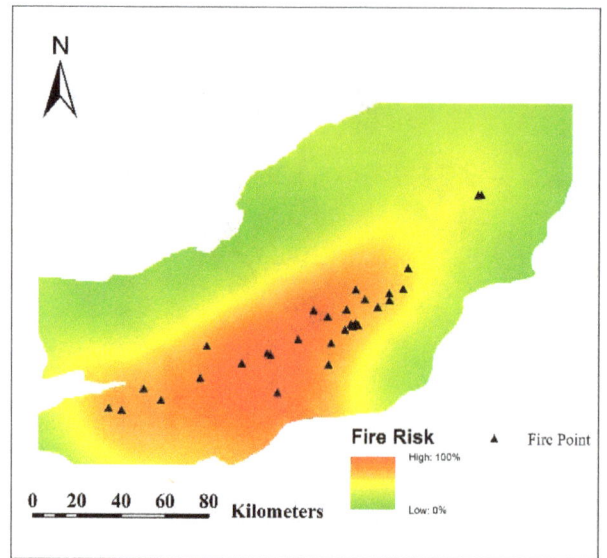

**Figure 6.** Fire risk map provided by GWR method

A set of parameters were calculated according to the features of model (Table 3). Bandwidth shows the used vicinities for forecasting in testing points. The value of $R^2$ shows the fitness of function into reality. In fact, the more the closer this value to 1, the more accurate the fitness implementation.

| Variable | Bandwidth | $R^2$ | RMSE |
|---|---|---|---|
| Value | 0.500 | 0.7941 | 0.2345 |

**Table 3.** GWR parameters

In GWR model, 28 points out of 30 fire points (in 2016) were in the range with fire-occurrence risk more than 70% (93.4% of all fire points) whereas in OLS model, only 20 points were in the range with fire-occurrence risk more than 70% (66% of all fire points). On the other hand, the resulted fitness of GWR method was more accurate than that of OLS method ($R^2$ is closer to 1 in GWR method); in addition, the error of GWR was less (less RMSE in GWR method). The function of outputs shows that GWR function is approached to reality acceptably and the fires were in high-risk forecasted range mostly.

## 4.    CONCLUSION

Today, fire occurrence has become an important issue due to the destruction of the large parts of the earth's forests. One of the solutions of this issue is prevention and timely reaction. Forecasting fire occurrence is very effective to perform preventive and timely actions. Therefore, the geographically weighted regression model was considered as main model. The advantage of this model compared to ordinary least squares method is to consider spatial data features. The obtained results from modeling Eleven effective parameters in fire occurrence with two methods, ordinary least squares and geographically weighted regression with Gaussian kernel, show this advantage. As mentioned, GWR method forecasted up to 93.4% of fire points in high-risk range whereas OLS method forecasted only 66% of fire points in the high-risk range. It is suggested that forecasting fire occurrence be done by using smart methods like neural network in future studies and its results be compared with regression-based methods to achieve the best forecast.

# 5. REFERNCES

A. Mahdavi, S. R. Fallah Shamsi, R. Nazari, 2011. Forests and rangelands' wildfire risk zoning using GIS and AHP techniques. Caspian J. Env. Sci. 2012, Vol. 10 No.1 pp. 43~52.

Adab, H., Nokhandan, M., Miza Bayati, R. and Adabi Firouzjani, A., 2008. Mapping fire risk in forests of Mazandaran province using Molgan Precautionary Index and GIS. Abstracts of 1st International Conference on Climate Change and Botany in Caspian Ecosystems, Iran,: 178-189.

Akbari, D., Amini, J. and Saadat Seresht, M., 2007. Presenting a rapid model for mapping forest fire. Abstracts of 2nd Disaster Management Conference, Iran, 12- 26 January 2007: 7p.

Almeida, R., 1994. Forest fire risk areas and definition of the prevention priority planning action using GIS. EGIS foundation, Available: http://libraries.maine.edu/Spatial/gisweb/spatdb/egis/eg94193.html, Accessed 10 April 2013.

Amparo, A.B., Oscar, F.R. , 2003. An intelligent system for forest fire risk prediction and fire fighting management in Galicia, Expert Systems with Applications 25, pp. 545–554.

Chou, Y.H., Minnich, R.A., Chase, R.A., 1993. Mapping probability of fire occurrence in San Jacinto Mountains, California, USA. Environ. Manage Vol. 17, pp. 129-140.

Chuvieco, E. and Congalton, R.G., 1989. Application of remote sensing and geographic information systems to forest fire hazard mapping. Remote Sensing of Environment, Vol. 29, pp. 147–159.

C. Brunsdon, S. Fotheringham and M. Charlton, 1998 ,"Geographically weighted regression – modelling spatial non-stationarity", The Statistician, Vol.47, No.3, PP. 431-443,.

D. P. McMillen and J. F. McDonald, 1998 ,"Locally weighted maximum likelihood estimation: Monte Carlo evidence and an application", presented at the Regional Science Association International meetings, Santa Fe, NM,.

Díaz-Delgado, R., Lloret, F., Pons, X., 2004. Spatial patterns of fire occurrence in Catalonia, NE Spain. Landsc. Ecol. 19, pp. 731-745.

Fernando Coelho Eugenio , Alexandre Rosa dos Santos,2016. Applying GIS to develop a model for forest fire risk: A case study in Espírito Santo, Brazil. Journal of Environmental Management, Vol. 173 , pp. 65-71.

Jafari Goldarag, Y., Mohammadzadeh, A. , 2016. Fire Risk Assessment Using Neural Network and Logistic Regression. Indian Soc Remote Sens.

Liu, Z., Yang, J., Chang, Y., Weisberg, P.J., He, H.S., 2012. Spatial patterns and drivers of fire occurrence and its future trend under climate change in a boreal forest of Northeast China. Glob. Change Biol. 18, pp. 2041-2056.

M. Charlton and A. S. Fotheringham, "Geographically Weighted Regression", White Paper. Kildare, Ireland: National Centre for Geocomputation, National University of Ireland, Maynooth, 11 , 2001.

Mohammadi, F., Shabanian, N., Pourhashemi, M. and Fatehi, P., 2010. Risk zone mapping of forest fire using GIS and AHP in a part of Paveh Forests.

Neeraj, S., & Hussin, Y.A., 1996. Spatial modeling for forest fire hazard prediction, management and control in Corbett national park, India, Remote sensing and computer technology for natural resources assessment (pp. 185– 192). University of Joensuu.

Renard, Q., Raphael, P., Ramesh, B.R., Kodandapani, N., 2012. Environmental susceptibility model for predicting forest fire occurrence in the Western Ghats of India. Int. J. Wildland Fire 21, pp. 368-379.

Salamati, H., Mostafa Lou, H., Mastoori, A., and Honardoust, F., 2011. Assessment and mapping forest fire risk using GIS in Golestan province forests Abstracts of 1st International Conference on Wildfire in Natural Resources Lands, Iran, 26-28 Oct. 2011, 10 p.

Xiaowei, L., Gang, Z., 2014. A comparison of forest fire indices for predicting fire risk in contrasting climates in China. Nat Hazards 70, pp. 1339–1356.

Yang, J., Healy, H.S., Shifley, S.R., Gustafson, E.J., 2007. Spatial patterns of modern period human-caused fire occurrence in the Missouri Ozark Highlands. For. Sci. 53, pp. 1-15.

# 15

# EXPERIMENTAL COMPARISON BETWEEN MAHONEY AND COMPLEMENTARY SENSOR FUSION ALGORITHM FOR ATTITUDE DETERMINATION BY RAW SENSOR DATA OF XSENS IMU ON BUOY

A. Jouybari[a] *, A. A. Ardalan[a], M-H. Rezvani[b]

[a] University of Tehran, School of Surveying and Geospatial Engineering, Tehran, Iran – (a.jouybari, ardalan) @ut.ac.ir
[b] University of Tasmania, School of Land and Food, Hobart, Tasmania, Australia - Mohammadhadi.Rezvani@utas.edu.au

KEY WORDS: Xsens Kalman Filter, Mahoney Filter, Complementary Filter, Integration, Raw Data, IMU

ABSTRACT:

The accurate measurement of platform orientation plays a critical role in a range of applications including marine, aerospace, robotics, navigation, human motion analysis, and machine interaction. We used Mahoney filter, Complementary filter and Xsens Kalman filter for achieving Euler angle of a dynamic platform by integration of gyroscope, accelerometer, and magnetometer measurements. The field test has been performed in Kish Island using an IMU sensor (Xsens MTi-G-700) that installed onboard a buoy so as to provide raw data of gyroscopes, accelerometers, magnetometer measurements about 25 minutes. These raw data were used to calculate the Euler angles by Mahoney filter and Complementary filter, while the Euler angles collected by XSense IMU sensor become the reference of the Euler angle estimations. We then compared Euler angles which calculated by Mahoney Filter and Complementary Filter with reference to the Euler angles recorded by the XSense IMU sensor. The standard deviations of the differences between the Mahoney Filter, Complementary Filter Euler angles and XSense IMU sensor Euler angles were about 0.5644, 0.3872, 0.4990 degrees and 0.6349, 0.2621, 2.3778 degrees for roll, pitch, and heading, respectively, so the numerical result assert that Mahoney filter is precise for roll and heading angles determination and Complementary filter is precise only for pitch determination, it should be noted that heading angle determination by Complementary filter has more error than Mahoney filter.

## 1. INTRODUCTION

Different kinds of technologies enable the measurement of orientation, inertial based sensory systems have the advantage of being completely self-contained such that the measurement is independent of motion and environment or location. An IMU (Inertial Measurement Unit) contains gyroscopes and accelerometers enabling the tracking of rotational and transfer movements. In order to measure in three dimensions, tri-axis sensors consisting of 3 mutually orthogonal sensitive axes are required. A MARG (Magnetic, Angular Rate, and Gravity) sensor is a combination of IMU along with tri-axis magnetic sensor. An IMU alone can only measure an attitude relative to the direction of gravity which is sufficient for many applications (Euston et al., 2007; Luinge et al., 2004). MARG systems or AHRS (Attitude and Heading Reference Systems) are able to provide a complete measurement of orientation relative to the direction of gravity and the earth's magnetic field.

A gyroscope measures angular velocity which, sensor orientation will be computed over the time if initial conditions are known (Bortz, 1971; Ignagni, 1990). Precision gyroscopes are really expensive and grave for most applications while low accuracy MEMS (Micro Electrical Mechanical System) devices are used in a majority of applications (Yazdi et al. 1998). Accumulating error will occur in computed orientation because

of the integration of gyroscope measurement errors. Therefore, gyroscope by itself can not present a complete measurement of orientation. The accelerometer measures the earth's gravitational and magnetometer measures magnetic fields thus, beside a gyroscope they create an absolute reference of orientation. However, these sensors are likely to be subject to high levels of noise; for example, the measured direction of gravity will corrupt by the noise due to the motion of the platform. The task of an orientation filter is to compute a single estimate of orientation through the optimal fusion of gyroscope, accelerometer and magnetometer measurements.

These days The Kalman filter (Kalman, 1960) plays important role in majority of orientation filter algorithms (Foxlin, 1996; Luinge et al., 1999; Marins, 2001) and commercial inertial orientation sensors. Different commercial inertial systems have used Kalman-based algorithm; for example, Xsens (Xsens Technologies, 2009), micro-strain (MicroStrain, 2009), VectorNav (VectorNav, 2009), Intersense (InterSense, 2008), PNI (PNI sensor corporation) and Crossbow (Crossbow, 2007). The Kalman-based algorithms for orientation determination from sensor's raw data have a number of disadvantages, however, the widespread use of Kalman-based algorithm has emphasised that they have good accuracy and their effectiveness. Implementation of Kalman-based algorithm can

---

* Corresponding author

be really complicated (Kallapur et al., 2009; Barshan and Durrant-Whyte, 1995; Foxlin, 1996; Luinge et al., 1999; Marins et al., 2001; Sabatini, 2006; Luinge and Veltink, 2006). (Mahony et al., 2008) developed the complementary filter which is shown to be an efficient and effective solution; however, performance is only validated for an IMU.

We used Mahoney and Complementary Filter for orientation determination from raw data that has been achieved by accelerometer, gyroscope, and magnetometer accelerometer. Their performances are benchmarked against an existing commercial filter (Xsens Kalman Filter (XKF3i)).

## 2.  MAIN BODY

### 2.1  The Complementary Filter

When looking for the best way to make use of a IMU-sensor, thus combine the accelerometer and gyroscope data, a lot of people get fooled into using the very powerful but complex Kalman filter. However, the Kalman filter is great, there are 2 big problems with it that make it hard to use: Very complex to understand and Very hard.

Complementary Filter is extremely easy to understand, and even easier to implement. Most IMU's have 6 DOF (Degrees of Freedom). This means that there are 3 accelerometers, and 3 gyrosocopes inside the unit. IMU will be able to measure the precise position and orientation of the object it is attached to. This because an object in free space has 6DOF. So if we can measure them all, we know everything. The sensor data is not good enough to be used in this way.

We will use both the accelerometer and gyroscope data for the same purpose: obtaining the attitude of the object. The gyroscope can do this by integrating the angular velocity over time. To obtain the attitude with the accelerometer, we are going to determine the position of the gravity vector (g-force) which is always visible on the accelerometer. This can easily be done by using an atan2 function. In both these cases, there is a big problem, which makes the data very hard to use without filter.

The problem with accelerometers:
As an accelerometer measures all forces that are working on the object, it will also see a lot more than just the gravity vector. Every small force working on the object will disturb our measurement completely. If we are working on an actuated system, then the forces that drive the system will be visible on the sensor as well. The accelerometer data is reliable only on the long term, so a "low pass" filter has to be used.

The problem with gyroscopes:
It is possible to obtain the angular position by use of a gyroscope. It is very easy to obtain an accurate measurement that was not susceptible to external forces. The less good news was that, because of the integration over time, the measurement has the tendency to drift, not returning to zero when the system went back to its original position. The gyroscope data is reliable only on the short term, as it starts to drift on the long term.

The complementary filter gives us a "best of both worlds" kind of deal. On the short term, we use the data from the gyroscope, because it is very precise and not susceptible to external forces.

On the long term, we use the data from the accelerometer, as it does not drift. In its most simple form, the filter looks as follows:

$$angle = 0.98 \times (angle + gyroData \times dt) + 0.02 \times (accData)$$

The gyroscope data is integrated every timestep with the current angle value. After this it is combined with the low-pass data from the accelerometer (already processed with atan2). The constants (0.98 and 0.02) have to add up to 1 but can of course be changed to tune the filter properly. It is very easy to compare Complementary Filter with Kalman filter.

The Complementary filter algorithm is designed in a way that has to be repeated in an infinite loop. Every iteration the pitch and roll angle values are updated with the new gyroscope values by means of integration over time. The filter then checks if the magnitude of the force seen by the accelerometer has a reasonable value that could be the real g-force vector. If the value is too small or too big, we know for sure that it is a disturbance we don't need to take into account. Afterwards, it will update the pitch and roll angles with the accelerometer data by taking 98% of the current value, and adding 2% of the angle calculated by the accelerometer. This will ensure that the measurement won't drift, but that it will be very accurate on the short term (Jan, 2013).

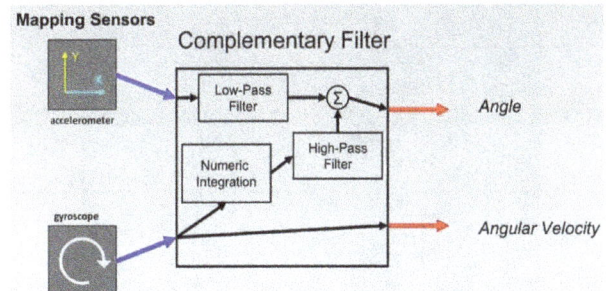

Fig. 1: Complementary filter process schematic (SegBot, 2014)

### 2.2  Xsens Kalman Filter (XKF3i)

The orientation of the IMU sensor (Xsens MTi-G-700) is computed by Xsens Kalman Filter. XKF3i uses signals of the rate gyroscopes, accelerometers and magnetometers to compute a statistical optimal 3D orientation estimate of high accuracy with no drift for both static and dynamic movements. XKF3 is a proven sensor fusion algorithm, which can be found in various products from Xsens and partner products.

The design of the XKF3i algorithm can be summarized as a sensor fusion algorithm where the measurement of gravity (by the 3D accelerometers) and Earth magnetic north (by the 3D magnetometers) compensate for otherwise slowly, but unlimited, increasing (drift) errors from the integration of rate of turn data (angular velocity from the rate gyros). This type of drift compensation is often called attitude and heading referencing and such a system is referred to as an Attitude and Heading Reference System (AHRS) (MTi User Manual, 2015).

### 2.3  Study area

A study area was selected in Southern IRAN, Kish Island in Persian Gulf with Coordinates: 26°32′N 53°58′E (Fig. 2).

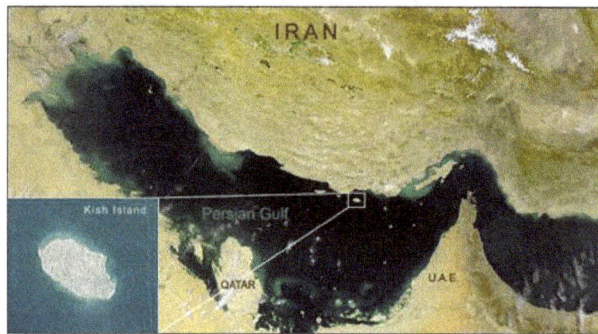

Fig. 2: Kish Island location in Iran.

## 2.4 Data sets

A field test and data acquisition have done in June 2016 in Kish Island beach. As we can see in (fig. 3) a lightweight buoy with the onboard inertial Xsens sensor used (fig. 4). The inertial sensor needs electrical power supply during the data acquisition, therefore, a boat used for putting a battery on it and to restrain the buoy.

Fig. 3: Lightweight buoy with IMU

Fig. 4: Xsens IMU Sensor

IMU data acquired with 8 HZ data rate during 25 minutes. Despite accelerometer (fig. 5), gyroscope (fig. 6), and magnetometer's data (fig. 7), attitude data which uses Xsens Kalman Filter for computation, also acquired.

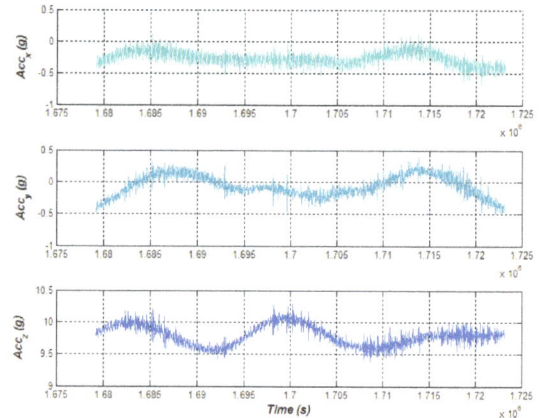

Fig. 5: tri-axis accelerometer data

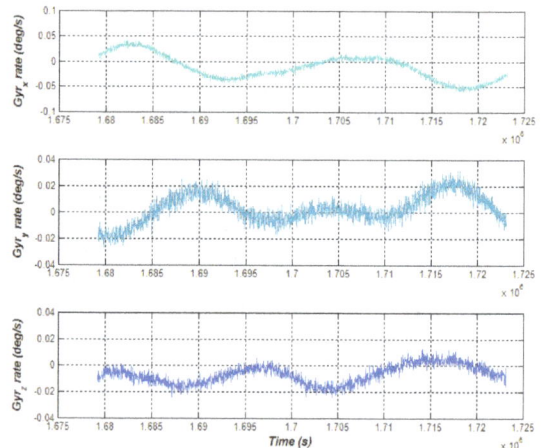

Fig. 6: tri-axis gyroscope data

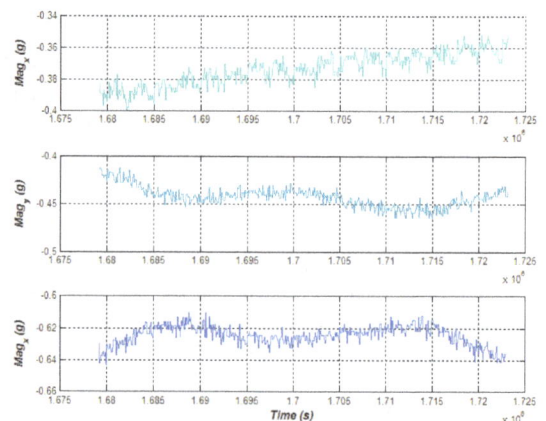

Fig. 7: tri-axis magnetometer data

## 2.5 Evaluation result

A glimpse into upper figures, it can be deduced that in addition to the noise in observation, there are drift and bias. In the following, Mahoney, Complementary and Xsense Kalman Filter are used for attitude determination by means of raw data of the sensor, shown in (fig. 8). By looking at (fig. 8), each of three attitude plots by the nearest approximation pursues each other.

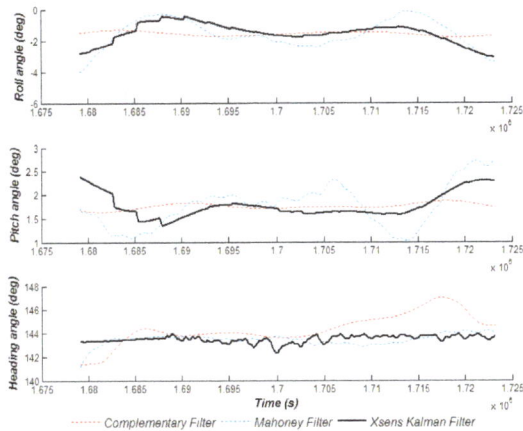

Fig. 8: Attitude determination by Mahoney, Complementary, XKF3i

It should be noted, we used Xsens Kalman Filter algorithm as the reference algorithm without any drift and bias, so for evaluation of the accuracy and precision of Mahoney and Complementary Filter, as it can be seen in (fig. 9) & (fig. 10), we compared them with Xsens Kalman filter algorithm.

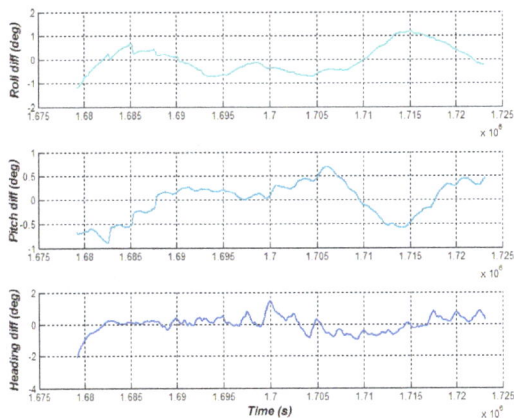

Fig. 9: Differences between Mahoney Filter and XKF3i

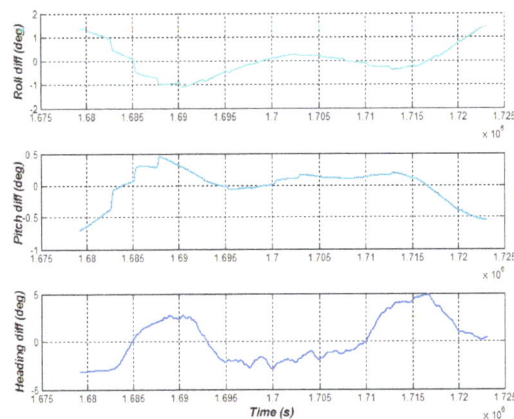

Fig. 10: Differences between Complementary Filter and XKF3i

Also, the standard deviation of this comparison brought in (tab. 1). Due to the (tab. 1) the mean differences between Mahoney filter and XKF3i for roll, pitch, and heading angles respectively

almost are $-1.45*10^{-15}$, $8.23*10^{-16}$, $-4.00*10^{-6}$ and the mean differences Complementary filter and XKF3i for roll, pitch, and heading angles respectively almost are $1.36*10^{-15}$, $1.73*10^{-15}$, 0.1855. On the other hand, the standard deviation of differences between Mahoney filter and XKF3i for roll, pitch, and heading angles respectively almost are 0.5644, 0.3872, 0.4990 and the standard deviation of the differences between the Complementary filter and XKF3i for roll, pitch, and heading angles respectively almost are 0.6349, 0.2621, and 2.3778.

| Statistics | Roll Ang | Pitch Ang | Heading Ang |
|---|---|---|---|
| Mahoney min | -1.1869 | -0.8922 | -2.0883 |
| Mahoney max | 1.2046 | 0.6971 | 1.4710 |
| Mahoney mean | -1.45E-15 | 8.23E-16 | -4.00E-06 |
| Mahoney Std | 0.5644 | 0.3872 | 0.4990 |
| Complementary min | -1.1286 | -0.7053 | -3.1761 |
| Complementary max | 1.4555 | 0.4540 | 4.8839 |
| Complementary mean | 1.36E-15 | 1.73E-15 | 0.1855 |
| Complementary Std | 0.6349 | 0.2621 | 2.3778 |

(Fig. 11, 12, 13) show the roll, pitch, and heading angles diagram of standard deviation between Mahoney and Complementary filters. As it's clear from (fig. 11), the standard deviation of Mahoney algorithm is lower than the Complementary algorithm, therefore, Mahoney algorithm for roll angle determination is more accurate.

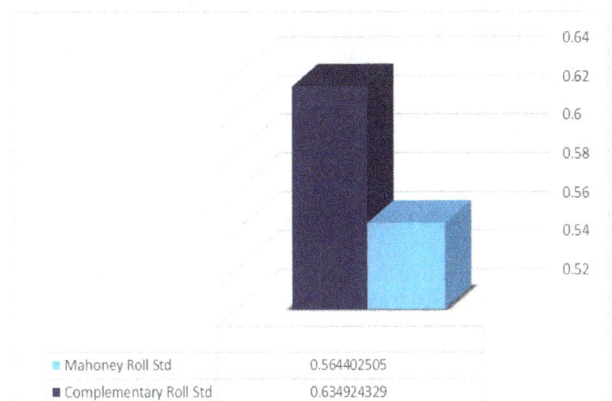

| ■ Mahoney Roll Std | 0.564402505 |
| ■ Complementary Roll Std | 0.634924329 |

Fig. 11: Mahoney and Complementary Roll Std

But this principle is not true for pitch angle determination (fig. 12). Because of the lower standard deviation of the Complementary algorithm, it is more accurate for pitch angle determination.

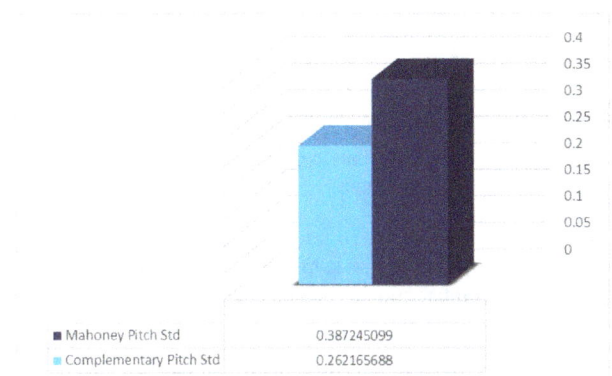

| ■ Mahoney Pitch Std | 0.387245099 |
| ■ Complementary Pitch Std | 0.262165688 |

Fig. 12: Mahoney and Complementary Pitch Std

Eventually, can be claimed that Complementary algorithm is not appropriate for heading angle determination, due to greater standard deviation with respect to Mahoney algorithm.

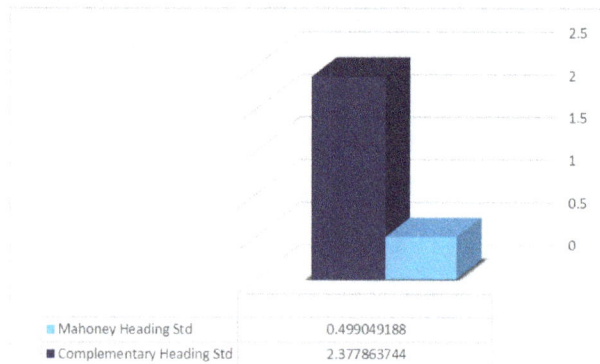

| | |
|---|---|
| ■ Mahoney Heading Std | 0.499049188 |
| ■ Complementary Heading Std | 2.377863744 |

Fig. 13: Mahoney and Complementary Heading Std

## 3. CONCLUSION

In this research, we used Mahoney, Complementary, and XKF3i algorithms for attitude determination from raw data of accelerometer, gyroscope, and magnetometer. In order to collect data, a test field by means of a lightweight buoy with onboard Xsens IMU is done in Kish Island. Each of algorithms for accuracy evaluation is compared with XKF3i, so, due to presented results, it is proved that Complementary algorithm is only sufficient for pitch angle determination, while, Mahoney algorithm is more accurate for roll and heading angles determination. Accordingly, it is suggested that presented algorithm be used for different uses such as Marine Engineering Sciences, Hydrography, and Oceanography.

### ACKNOWLEDGEMENTS (OPTIONAL)

The authors would like to thank hydrographic office of national cartographic center of Iran for helping to build buoy and data acquisition in Kish Island.

### REFERENCES

Barshan, B. and Durrant-Whyte, H. F., 1995. Inertial navigation systems for mobile robots. 11(3), pp. 328-342.

Bortz, J. E., 1971. A new mathematical formulation for strapdown inertial navigation. (1), pp. 61-66.

Crossbow Technology Inc, 2007. AHRS400 Series User's Manual. 4145 N. First Street, San Jose, CA 95134, rev. c edition.

Euston, M., Coote, P., Mahony, R., Kim, J., and Hamel, T., 2007. A complementary filter for attitude estimation of a fixed-wing uav with a low-cost imu. In 6th International Conference on Field and Service Robotics.

Foxlin, E., 1996. Inertial head-tracker sensor fusion by a complementary separate-bias kalman filter. In Proc. Virtual Reality Annual International Symposium, the IEEE 1996, pp. 185-194.

Ignagni, M. B., 1990. Optimal strapdown attitude integration algorithms. In Guidance, Control, and Dynamics, volume 13, pp 363-369.

InterSense, Inc., 2008. InertiaCube2+ Manual. 36 Crosby Drive, Suite 150, Bedford, MA 01730, USA, 1.0 edition.

Jan, p., 2003. Reading a IMU Without Kalman: The complementary Filter http://www.pieter-jan.com/node/11 (26 Apr. 2013).

Kallapur, A., Petersen, I. and Anavatti, S., 2009. A robust gyroless attitude estimation scheme for a small fixed-wing unmanned aerial vehicle, pp. 666-671.

Kalman, R. E., 1960. A new approach to linear filtering and prediction problems. Journal of Basic Engineering, 82, pp. 35-45.

Luinge, H. J., Veltink, P. H. and Baten, C. T. M., 1999. Estimation of orientation with gyroscopes and accelerometers. In Proc. First Joint [Engineering in Medicine and Biology 21st Annual Conf. and the 1999 Annual Fall Meeting of the Biomedical Engineering Soc.] BMES/EMBS Conference, volume 2, pp. 844.

Luinge, H. J. and Veltink, P. H., 2004. Inclination measurement of human movement using a 3-d accelerometer with autocalibration. 12(1), pp. 112-121.

Luinge, H. J. and Veltink, P. H., 2006. Measuring orientation of human body segments using miniature gyroscopes and accelerometers. Medical and Biological Engineering and Computing, 43(2), pp. 273-282.

Mahony, R., Hamel, T. and Pimlin, J.-M., 2008. Nonlinear complementary filters on the special orthogonal group. Automatic Control, IEEE Transactions on, 53(5), pp. 1203-1218.

Marins, J. L., Xiaoping Yun, Bachmann, E. R., McGhee, R. B. and Zyda, M. J., 2001. An extended kalman filter for quaternion-based orientation estimation using marg sensors. In Proc. IEEE/RSJ International Conference on Intelligent Robots and Systems, volume 4, pp. 2003-2011.

MicroStrain Inc., 2009. 3DM-GX3 -25 Miniature Attitude Heading Reference Sensor. 459 Hurricane Lane, Suite 102, Williston, VT 05495 USA, 1.04 edition.

MTi User Manual, 2015. MTi 10-series and MTi 100-series. Document MT0605P, Revision F, 27 February 2015.

PNI sensor corporation., Space point Fusion. 133 Aviation Blvd, Suite 101, Santa Rosa, CA 95403-1084 USA.

Sabatini, A. M., 2006. Quaternion-based extended kalman filter for determining orientation by inertial and magnetic sensing. 53(7), pp. 1346-1356.

SegBot, 2003. Complementary Filter http://www.arxterra.com/segbot-complementary-filter/ (04 Dec. 2014).

VectorNav Technologies, 2009. LLC. VN -100 User Manual. College Station, TX 77840 USA, preliminary edition.

Xsens Technologies B.V., 2009. MTi and MTx User Manual and Technical Documentation. Pantheon 6a, 7521 PR Enschede, The Netherlands.

Yazdi, N., Ayazi, F. and Najafi, K., 1998. Micromachined inertial sensors. 86(8), pp. 1640-1659.

# 16

# RECURSIVE LEAST SQUARES WITH REAL TIME STOCHASTIC MODELING: APPLICATION TO GPS RELATIVE POSITIONING

F. Zangeneh-Nejad [a], A. R. Amiri-Simkooei [b], M. A. Sharifi [a,*], J. Asgari [b]

[a] School of Surveying and Geospatial Engineering, Research Institute of Geoinformation Technology (RIGT), College of Engineering, University of Tehran, Iran- (f.zangenehnejad, sharifi@ut.ac.ir)
[b] Department of Geomatics Engineering, Faculty of Civil Engineering and Transformation, University of Isfahan, 81746-73441 Isfahan, Iran- (amiri, asgari@eng.ui.ac.ir)

KEY WORDS: Recursive least squares, Stochastic modeling, Least-squares variance component estimation, GPS relative positioning

ABSTRACT:

Geodetic data processing is usually performed by the least squares (LS) adjustment method. There are two different forms for the LS adjustment, namely the batch form and recursive form. The former is not an appropriate method for real time applications in which new observations are added to the system over time. For such cases, the recursive solution is more suitable than the batch form. The LS method is also implemented in GPS data processing via two different forms. The mathematical model including both functional and stochastic models should be properly defined for both forms of the LS method. Proper choice of the stochastic model plays an important role to achieve high-precision GPS positioning. The noise characteristics of the GPS observables have been already investigated using the least squares variance component estimation (LS-VCE) in a batch form by the authors. In this contribution, we introduce a recursive procedure that provides a proper stochastic modeling for the GPS observables using the LS-VCE. It is referred to as the recursive LS-VCE (RLS-VCE) method, which is applied to the geometry-based observation model (GBOM). In this method, the (co)variances parameters can be estimated recursively when the new group of observations is added. Therefore, it can easily be implemented in real time GPS data processing. The efficacy of the method is evaluated using a real GPS data set collected by the Trimble R7 receiver over a zero baseline. The results show that the proposed method has an appropriate performance so that the estimated (co)variance parameters of the GPS observables are consistent with the batch estimates. However, using the RLS-VCE method, one can estimate the (co)variance parameters of the GPS observables when a new observation group is added. This method can thus be introduced as a reliable method for application to the real time GPS data processing.

## 1. INTRODUCTION

The least squares (LS) parameter estimation has been extensively employed in geodetic data processing. There are two different forms for the LS namely batch form and recursive LS (RLS) form. In the batch form, the whole measurements are simultaneously processed through the adjustment procedure while the RLS processes the observations sequentially in time. The RLS method is therefore suitable for real time applications in which observations are collected sequentially over time.

To obtain the best linear unbiased estimation (BLUE) using the least squares method, either in batch form or in recursive form, the realistic choice of the stochastic model of the observables is an essential issue. This describes the statistical properties of observables by means of a covariance matrix. The covariance matrix of observables is relatively known and expressed as an unknown linear combination of known cofactor matrices for most geodetic applications. The estimation of the unknown (co)variance parameters is referred to as variance component estimation (VCE). There are many different methods for VCE such as best invariant quadratic unbiased estimator (BIQUE) (Koch 1978, 1999), minimum norm quadratic unbiased estimator (MINQUE) (Rao 1971 and Junhuan et al 2011),

restricted maximum likelihood (REML) estimator (Koch, 1986) and the least squares variance component estimation (LS-VCE) (Teunissen 1988; Teunissen and Amiri-Simkooei 2006, 2008 and Amiri-Simkooei 2007).

GPS data processing is usually performed using the LS method in a recursive manner via a sequential filter, i.e., a least squares sequential filter or discrete Kalman filter. A realistic stochastic model for the GPS observables is therefore necessary for high-precision GPS positioning. The noise characteristics of GPS observables has been investigated by Amiri-Simkooei and Tiberius (2007), Amiri-Simkooei et al. (2006, 2007, 2009, 2013), Bischoff et al. (2006), Tiberius and Kenselaar (2000, 2003), Teunissen et al. (1998), Hartinger and Brunner (1999), Wang et al. (1998, 2002), and Satirapod et al. (2002). Amiri-Simkooei et al. (2009, 2013) have been applied the LS-VCE algorithm to GPS observables using the geometry-free observation model (GFOM) and the geometry-based observation model (GBOM), respectively. They applied the LS-VCE to the GPS observables in a batch form. This causes the high computational time needed for the batch solution. To overcome such a problem, the entire time span of the GPS observations was divided into a few groups, each consisting of a few consecutive epochs. The unknown (co)variance parameters

---

$(\sigma_k; k = 1 : p)$ can then be separately estimated for each group using the LS-VCE method. For a Kronecker and block structure of the functional and stochastic models, one can show that the final estimates are just the arithmetic mean of the individual estimates over the groups.

In the case of GPS data processing, the stochastic model can represent unmodeled systematic errors, multipath and noise. Therefore to adapt the stochastic model with the environment, the stochastic model must be properly chosen in real time so that the unknown parameters and the stochastic model are both updated as new data arrive over time. In this contribution, we look for a recursive procedure providing a proper stochastic modeling for GPS observables using the LS-VCE method.

This paper is organized as follows. The RLS method is first introduced and described in details in the next section. We then explain how the recursive LS-VCE is implemented in GPS relative positioning. In order to assess the noise characteristics of GPS observables, the implementation results of the proposed method are then provided and compared with those obtained by the batch solution. Finally we draw some conclusions in the last section.

## 2. RECURSIVE LEAST SQUARES (RLS)

The RLS is an appropriate method for sequential rather than batch processing. Consider the following partitioned linear model of observation equations:

$$E\left\{\begin{pmatrix} y_1 \\ y_2 \end{pmatrix}\right\} = \begin{pmatrix} A_1 \\ A_2 \end{pmatrix} x \quad ; \quad D\left\{\begin{pmatrix} y_1 \\ y_2 \end{pmatrix}\right\} = \begin{pmatrix} Q_{y_1} & 0 \\ 0 & Q_{y_2} \end{pmatrix} \quad (1)$$

where $E$ and $D$ denote the mathematical expectation and dispersion operators, respectively, $y_1$ and $y_2$ are the $m_1$-vector of the old observations and the $m_2$-vector of the new observations, respectively, $x$ is the $n$-vector of unknown parameters, $A_1$ and $A_2$ are the $m_1 \times n$ and $m_2 \times n$ corresponding design matrices, respectively, and $Q_{y_1}$ and $Q_{y_2}$ are the covariance matrices of the observables $y_1$ and $y_2$, respectively. The correlation between $y_1$ and $y_2$ is assumed to be absent (i.e. $Q_{y_1 y_2} = 0$).

The model of observation equations in Eq. (1) is also equivalent to the following model (Teunissen 2000)

$$E\left\{\begin{pmatrix} \hat{x}_1 \\ y_2 \end{pmatrix}\right\} = \begin{pmatrix} I \\ A_2 \end{pmatrix} x \quad ; \quad D\left\{\begin{pmatrix} \hat{x}_1 \\ y_2 \end{pmatrix}\right\} = \begin{pmatrix} Q_{\hat{x}_1} & 0 \\ 0 & Q_{y_2} \end{pmatrix} \quad (2)$$

where $\hat{x}_{(1)}$ and $Q_{\hat{x}_{(1)}}$, the least squares estimate of the unknowns and its covariance matrix obtained from the first group of observations, i.e., $E(y_1) = A_1 x$; $D(y_1) = Q_{y_1}$, are of the form

$$\hat{x}_{(1)} = (A_1^T Q_{y_1}^{-1} A_1)^{-1} A_1^T Q_{y_1}^{-1} y_1 \quad ; \quad Q_{\hat{x}_{(1)}} = (A_1^T Q_{y_1}^{-1} A_1)^{-1} \quad (3)$$

The recursion of the LS solution of Eq. (2) is then of the form (Teunissen 2000, 2001, 2005)

$$\hat{x}_{(2)} = \hat{x}_{(1)} + K_2(y_2 - A_2 \hat{x}_{(1)}) \quad (4)$$

where $K_2 = (Q_{\hat{x}_{(1)}}^{-1} + A_2^T Q_{y_2}^{-1} A_2)^{-1} A_2^T Q_{y_2}^{-1}$ is the gain matrix of the recursive least squares. Also the covariance matrix of the least squares estimate of $\hat{x}_{(2)}$ is as follows

$$Q_{\hat{x}_{(2)}} = (Q_{\hat{x}_{(1)}}^{-1} + A_2^T Q_{y_2}^{-1} A_2)^{-1} \quad (5)$$

The RLS method can be implemented in GPS data processing so that the estimated unknowns (relative receiver position and double difference (DD) integer ambiguities on the L1 or L2 in the case of GPS relative positioning) are updated when new observations are added in sequential epochs. In the next subsection, the GBOM model is briefly explained, which here is considered to be the functional model for assessing the noise characteristics of GPS observables in a recursive manner.

### 2.1 GPS Geometry-Based Observation Model (GBOM)

The GBOM model is a commonly used model for high-precision GPS positioning from code and phase observations using a relative GPS receiver setup. In the model, the observations are the DD pseudorange and phase observation on the L1 or L2 frequency. The unknown vector consists of the unknown baseline components between the reference and rover receivers and the DD integer ambiguities on the L1 or L2 (Teunissen, 1997, Odijk 2008 and Amiri-Simkooei et al. 2013). Ignoring the DD atmospheric (ionospheric and tropospheric) delays, the observation equations of the GBOM at epoch $t_k$ is of the form

$$E\begin{pmatrix} P_1^{DD}(t_k) \\ P_2^{DD}(t_k) \\ \phi_1^{DD}(t_k) \\ \phi_2^{DD}(t_k) \end{pmatrix} = A_k \begin{pmatrix} g \\ a_1^{DD} \\ a_2^{DD} \end{pmatrix} \quad (6)$$

In the preceding equation, $P^{DD}$ and $\phi^{DD}$ denote the DD pseudorange and phase observations between two receivers and two satellites on the L1 or L2 frequency, respectively, $A_k$ is the design matrix, the vector $g$ consists of the unknown baseline components between the reference and rover receivers and $a$ is the vector of DD integer ambiguities on the L1 or L2.

At first, the unknowns of Eq. (6) are obtained using some initial epochs. The unknown vector can then be updated using Eqs. (3-5) recursively when new observations related to the consecutive epochs are added.

### 2.2 A realistic stochastic model of GPS observables

The complete structure of stochastic model of the GPS observables considering different variance components for observables (C1, P1, L1 and L2), correlations among observations, satellites elevation dependence of GPS observables precision and temporal correlations of observables is of the form (Amiri-Simkooei et al. 2009; 2013)

$$Q_y^D = \Sigma_C \otimes \Sigma_T \otimes \Sigma_E \qquad (7)$$

where $\otimes$ is the Kronecker product of two matries. Table (1) represents the matrices $\Sigma_C$, $\Sigma_T$ and $\Sigma_E$ given in Eq. (7).

$$\Sigma_C = \begin{vmatrix} \sigma_{C_1}^2 & \sigma_{C_1,P_2} & \sigma_{C_1,\phi_1} & \sigma_{C_1,\phi_2} \\ \sigma_{C_1,P_2} & \sigma_{P_2}^2 & \sigma_{P_2,\phi_1} & \sigma_{P_2,\phi_2} \\ \sigma_{C_1,\phi_1} & \sigma_{P_2,\phi_1} & \sigma_{\phi_1}^2 & \sigma_{\phi_1,\phi_2} \\ \sigma_{C_1,\phi_2} & \sigma_{P_2,\phi_2} & \sigma_{\phi_1,\phi_2} & \sigma_{\phi_2}^2 \end{vmatrix}$$

consisting of 10 unknown components (4 variances and 6 covariances).

$$\Sigma_T = \begin{vmatrix} \sigma_{(0)} & \sigma_{(1)} & \cdots & \sigma_{(K-1)} \\ \sigma_{(1)} & \sigma_{(0)} & \cdots & \sigma_{(K-2)} \\ \vdots & \vdots & \ddots & \vdots \\ \sigma_{(K-1)} & \sigma_{(K-2)} & \cdots & \sigma_{(0)} \end{vmatrix}$$

representing time correlation of the observables ($K$ is the epochs of observations).

$$\Sigma_E = 2 \begin{vmatrix} \sigma_{[1]}^2 + \sigma_{[2]}^2 & \sigma_{[1]}^2 & \cdots & \sigma_{[1]}^2 \\ \sigma_{[1]}^2 & \sigma_{[1]}^2 + \sigma_{[3]}^2 & \cdots & \sigma_{[1]}^2 \\ \vdots & \vdots & \ddots & \vdots \\ \sigma_{[1]}^2 & \sigma_{[1]}^2 & \cdots & \sigma_{[1]}^2 + \sigma_{[k]}^2 \end{vmatrix}$$

describing the satellite elevation dependence of GPS observables precession ($k$ is the number of satellites, and the satellite number [1] is assumed to be the reference satellite).

Table 1 Matrices $\Sigma_C$, $\Sigma_T$ and $\Sigma_E$ given in Eq. (7)

In this contribution, the time correlation of observations is ignored, i.e., $\Sigma_T = I_K$ where $I_K$ is an identity matrix of size $K$. However, to consider the satellites elevation effect on the GPS observables precision, an elevation-angle based sine function model is employed as follows

$$\sigma_{[i]}^2 = \frac{1}{\sin^2(E_i)} \qquad (8)$$

where $\sigma_{[i]}^2$, the variance factors of $\Sigma_E$ and $E_i$, denotes the elevation angle of the satellites. The unknown variance and covariance components of $\Sigma_C$ are then estimated by the LS-VCE method. These parameters can be estimated in a batch form or in a recursive form. The first has been already investigated in Amiri-Simkooei et al. (2009; 2013). For more information the reader is referred to Amiri-Simkooei et al. (2009; 2013). In this contribution, we look for a recursive procedure providing a proper stochastic modeling for GPS observables using the LS-VCE method.

## 3. LEAST-SQUARES VARIANCE COMPONENT ESTIMATION (LS-VCE)

The structure of the covariance matrix $Q_y$ is generally expressed as an unknown linear combination of some known cofactor matrices as

$$Q_y = Q_0 + \sum_{k=1}^{p} \sigma_k Q_k \qquad (9)$$

where $Q_0$ is the known part of the covariance matrix and $Q_k; k = 1,...,p$ are the cofactor matrices such that the sum $Q_y = Q_0 + \sum_{k=1}^{p} \sigma_k Q_k$ is non-negative definite. The unknown (co)variance parameters $\sigma_k$, $k = 1,...,p$ can then be estimated as $\hat{\sigma} = N^{-1}l$, where the entries of matrix $N$ and vector $l$ are given as (Amiri-Simkooei 2007)

$$n_{ij} = \frac{1}{2}tr(Q_i Q_y^{-1} P_A^{\perp} Q_j Q_y^{-1} P_A^{\perp}) \qquad (10)$$

and

$$l_i = \frac{1}{2}\hat{e}^T Q_y^{-1} Q_i Q_y^{-1}\hat{e} - \frac{1}{2}tr(Q_i Q_y^{-1}P_A^{\perp}Q_0 Q_y^{-1}P_A^{\perp}) \qquad (11)$$

where $P_A^{\perp} = I_m - A(A^T Q_y^{-1} A)^{-1}A^T Q_y^{-1}$ is an orthogonal projector and $\hat{e} = P_A^{\perp}y$ denotes the $m$-vector of residuals. The variances of the estimates $\hat{\sigma}$ is also expressed by the diagonal entries of the covariance matrix of the estimated (co)variance components, which is obtained as $Q_{\hat{\sigma}} = N^{-1}$. For more information about LS-VCE we refer to Teunissen and Amiri-Simkooei (2008) and Amiri-Simkooei (2007).

### 3.1 Application of LS-VCE to GPS observables stochastic modelling

The covariance matrices of the observables should be properly defined for high-precision GPS positioning applications. We aim to assess the noise characteristics of the GPS observables using the GBOM in a recursive manner.

To obtain the realistic stochastic model of the GPS observables, one should properly determine the three matrices $\Sigma_C$, $\Sigma_T$ and $\Sigma_E$ given in Eq. (7). In this contribution, the time correlation of observations is ignored, i.e., $\Sigma_T = I_K$ and the satellites elevation effect on the GPS observables precision is modeled using an elevation-angle based sine function given in Eq. (8), i.e., the matrix $\Sigma_E$ is also known. However, the components of the matrix $\Sigma_C$ are estimated using the LS-VCE method. These parameters can be estimated in a batch form or in a recursive form as follows:

- Batch form: In this method, the entire time span of the GPS observations is divided into a few groups, each consisting of a few consecutive epochs. The unknown (co)variance parameters (here the components of the matrix $\Sigma_C$) can then be separately estimated for each group. The final estimates of the unknown (co)variance parameters are just the arithmetic mean of the individual estimates over the groups (Amiri-Simkooei et al. 2009; 2013).

- RLS-VCE: In this method, the components of the matrix $\Sigma_C$ are estimated recursively so that the (co)variance parameters are estimated when the new group of observations is added. Further descriptions are given hereinafter.

### 3.2  Application of recursive LS-VCE to GBOM model

In the first step, assume $E(y_1) = A_1 x$; $D(y_1) = Q_{y_1}$ where $y_1$ and its covariance matrix are of the form

$$y_1 = \begin{pmatrix} P_1^{DD}(t_1) \\ P_2^{DD}(t_1) \\ \phi_1^{DD}(t_1) \\ \phi_2^{DD}(t_1) \end{pmatrix} \quad ; \quad Q_{y_1} = \Sigma_C \otimes \Sigma_T \otimes \Sigma_E \qquad (12)$$

The least squares estimate of the unknowns along with its covariance matrix can then be obtained which are denoted by $\hat{x}_{(1)}$ and $Q_{\hat{x}_{(1)}}$, respectively. The components of the matrix $\Sigma_C$ (4 variances and 6 covariances) can also be obtained using the LS-VCE method.

Assuming the new group of observations is also added to the problem. Now again consider the following partitioned model of observation equations which is equivalent to the model of observation equations of Eq. (1)

$$E\left\{ \begin{pmatrix} \hat{x}_1 \\ y_2 \end{pmatrix} \right\} = \begin{pmatrix} I \\ A_2 \end{pmatrix} x \quad ; \quad D\left\{ \begin{pmatrix} \hat{x}_1 \\ y_2 \end{pmatrix} \right\} = \begin{pmatrix} Q_{\hat{x}_1} & 0 \\ 0 & Q_{y_2} \end{pmatrix} \quad (13)$$

where $Q_{y_2} = \Sigma_C \otimes \Sigma_T \otimes \Sigma_E$.

The unknown components of the matrix $\Sigma_C$ (4 variances and 6 covariances) can also be obtained using the LS-VCE as $\hat{\sigma} = N^{-1}l$ where $N$ and $l$ are given by Eqs. (10 and 11). Note that in the batch method, these parameters are obtained for each group of observation, separately, i.e., we used only $E(y_2) = A_2 x$; $D(y_2) = Q_{y_2} = \Sigma_C \otimes \Sigma_T \otimes \Sigma_E$. However, in the RLS-VCE algorithm, these parameters are estimated using the model given in Eq. (13) including both the LS estimates of the unknowns obtained from the previous groups and the new observations related to the next one. This is the main difference of the method with the batch method. Similarly, the final estimates of the unknown (co)variance parameters are the arithmetic mean of the individual estimates over the groups.

### 4.  RESULTS AND DISCUSSIONS

To apply the RLS-VCE method, we used one zero baseline GPS data set. The receiver used in this experiment is Trimble R7 corresponded to January 1[th], 2004, from 12:25:00 to 13:15:00 UTC. The total number of epochs is then 3000. The data set is collected at a meadow area close to Delft, the Netherlands with 1-sec intervals providing the code and phase observations on both the L1 and L2 frequencies (namely, C1–P2–L1–L2).

The noise characteristics of GPS observations are now investigated for both batch and recursive solution procedures. This section consists of two subsections. In the first part, the

variances of the GPS observables for C1, P2, L1 and L2 observables are provided. The covariances/correlations among the GPS observables are given in the second subsection.

### 4.1  Variances of GPS observables

In this contribution, the matrices $\Sigma_E$ and $\Sigma_T$ are assumed known. The components of the matrix $\Sigma_C$ are then estimated using the LS-VCE method in a batch form or in a recursive form. The elements on the principal diagonal of the matrix are the variances of the GPS observables ($\sigma_{C_1}^2, \sigma_{P_2}^2, \sigma_{\phi_1}^2$ and $\sigma_{\phi_2}^2$) whereas the off-diagonal elements are the covariances between the observables.

At first, the entire time span of the GPS observations for the data set we used is divided into 300 groups, each consisting of a 10 consecutive epochs. The (co)variance parameters can then be estimated using the LS-VCE method in a batch manner or in a recursive manner. Applying the LS-VCE in a batch form, one can obtain the (co)variance parameters for each group of observations (here 300 groups) separately. In the recursive form, the (co)variances parameters can be estimated recursively when the new group of observations is added. Use is made of the model given in Eq. (13). Figures 1 and 2 show the groupwise estimate of the standard deviations of C1 and P2 and L1 and L2 for the Trimble R7 receiver, respectively, for both batch and recursive solutions. The results confirm the consistency between the batch estimates and the recursive ones. However, applying the LS-VCE method in a recursive manner provides a real time stochastic modeling for GPS observables as new data arrive over time. This is a great advantage of the RLS-VCE compared to the batch one.

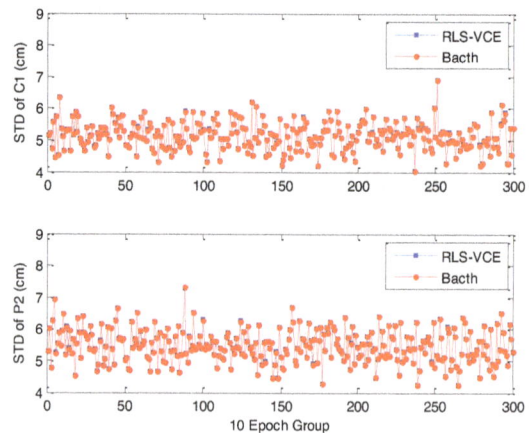

Fig. 1 Estimated standard deviations of GPS code observations (C1 and P2) for Trimble R7 receiver in cm for both batch and recursive methods

For a better view, the estimated standard deviations of four GPS observation types for both solutions are presented in Table 2 (in mm). As mentioned, the final estimates of the unknown (co)variance parameters are just the arithmetic mean of the individual estimates over the 300 groups for both batch and recursive methods.

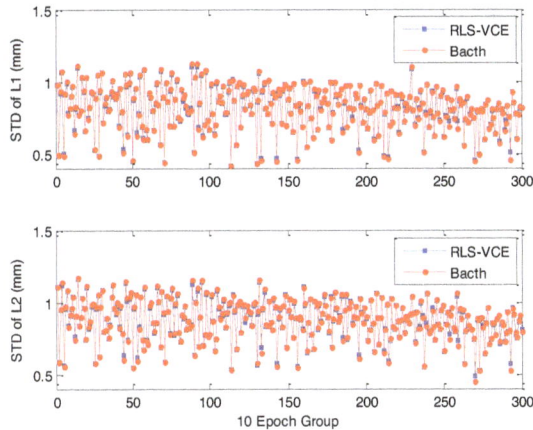

Fig. 2 Estimated standard deviations of GPS phase observations (L1 and L2) for Trimble R7 receiver in mm for both batch and recursive methods

Table 2 Estimated standard deviations of four GPS observation types for both solution forms in mm

| Method | C1 | P2 | L1 | L2 |
|--------|------|------|-----|-----|
| Batch | 51.7 | 55.1 | 0.8 | 0.9 |
| RLS-VCE | 51.8 | 55.1 | 0.8 | 0.9 |

#### 4.2 Correlations among GPS observables

According to the description provided in the previous section, the covariances/correlations among the GPS observables are estimated using the LS-VCE method in both batch and recursive forms. We estimated the six correlations among C1 and P2, C1 and L1, C1 and L2, P2 and L1, P2 and L2 and L1 and L2, among them , for example, figure 3 provides the estimated correlations among the phase observations L1 and L2ferquencies for Trimble R7 receiver using both batch and recursive methods. The results indicated a significant correlation between the phase observations L1 and L2 of about 0.9 for Trimble R7 receiver. Similarly again, the consistency between the batch estimates and the recursive ones is confirmed.

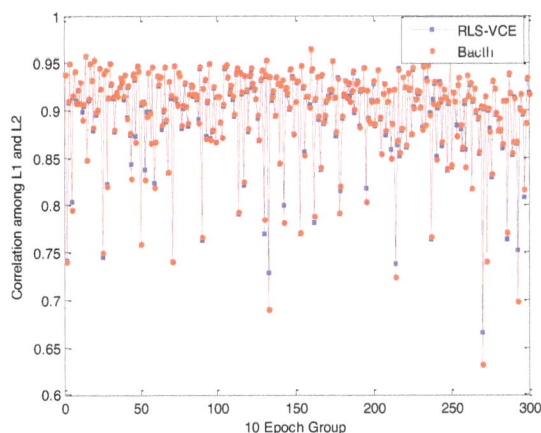

Fig. 3 Estimated correlations among phase observations on L1 and L2 frequencies for Trimble R7 receiver for both batch and recursive methods

## 5. CONCLUDING REMARKS

The least squares (LS) adjustment procedure can be extensively used in geodetic data processing in two different but equivalent forms: 1) batch form and 2) recursive least squares (RLS) form. For real time applications in which observations are collected sequentially over time, the RLS method is preferred compared to the batch one. Concerning the LS adjustment method either in batch form or in recursive form, both functional and stochastic models should be properly defined. The realistic choice of the stochastic model of the observables is required to obtain the best linear unbiased estimation (BLUE). This is important for GPS data processing as well. The model we used here is the geometry-based observation model (GBOM). The model consisting of two parts: the functional and the stochastic models. The former is completely established. Therefore, the choice of the functional model for the GPS observables is not the subject of discussion in this contribution. However, choosing a realistic stochastic model for the GPS observables is necessary for high-precision GPS positioning which was the subject of the research. At first, the covariance matrix of observables has been expressed as an unknown linear combination of known cofactor matrices. The unknown (co)variance parameters can then be estimated using the different variance component estimation (VCE) methods. Here, we investigated the noise characteristics of the GPS observables using the least squares variance component estimation (LS-VCE) method. The LS-VCE algorithm can be implemented to GPS observables in a batch manner or in a recursive manner. The former has already been investigated by Amiri-Simkooei et al. (2009, 2013). We proposed a recursive procedure providing a proper stochastic modeling for GPS observables using the LS-VCE method.

To consider the performance of the proposed method and to access the proper GPS observable covariance matrix, real data, collected by one GPS receiver Trimble R7 was used. We employed the LS-VCE algorithm on the GPS observations for the two different cases batch and recursive forms. Applying the LS-VCE in a batch form, one can obtain the (co)variance parameters of the GPS observables for each group of observations separately. In the recursive form, the (co)variances parameters can be estimated recursively when the new group of observations is added. This is a great advantage of the RLS-VCE rather than the batch one. The results confirmed that the proposed method has an appropriate performance so that the estimated (co)variance parameters of the GPS observables are completely consistent with the batch estimates. This method can thus be introduced as an efficient method to the GPS processing stage.

### REFERENCES

Amiri-Simkooei, A.R., 2007. Least-squares variance component estimation: Theory and GPS applications. Ph.D. thesis, Delft Univ. of Technology, Delft, The Netherlands.

Amiri-Simkooei, A.R., Tiberius, C.C.J.M., and Teunissen, P.J.G., (2006). Noise characteristics in high precision GPS positioning. P. Xu, J. Liu, A. Dermanis, eds., Proc., *6th Hotine-Marussi Symp. of Theoretical and Computational Geodesy*, Springer, Berlin, pp. 280–286.

Amiri-Simkooei, A.R., and Tiberius, C.C.J.M., 2007. Assessing receiver noise using GPS short baseline time series. *GPS Solutions*, 11(1), pp. 21–35.

Amiri-Simkooei, A.R, Teunissen, P.J.G., Tiberius, C.C.J.M., 2009. Application of least-squares varinace component estimation to GPS observables. *J Surv Eng.*, 135(4): pp. 149-160.

Amiri-Simkooei, A.R., Zangeneh-Nejad, F., and Asgari, J., 2013. Least-squares variance component estimation applied to GPS geometry-based observation model. *J Surv Eng.,* 139(4): pp. 176-187.

Bischoff, W., Heck, B., Howind, J., and Teusch, A., 2006. A procedure for estimating the variance function of linear models and for checking the appropriateness of estimated variances: A case study of GPS carrier-phase observations. *J. Geodesy, Berlin,* 79(12), pp. 694–704.

Hartinger, H., and Brunner, F. K., (1999). Variances of GPS phase observations: The SIGMA-ε model. *GPS Solutions*, 2(4), pp. 35–43.

Junhuan, P., Yun, S., Shuhui, L., and Honglei, Y., 2011. MINQUE of variance-covariance components in linear Gauss-Markov models. *J. Surv. Eng.,* 137(4), pp. 129–139.

Koch, K.R., 1978. *Schätzung von varianzkomponenten. Allgemeine Vermessungs-Nachrichten*, 85: pp. 264–269.

Koch, K.R., 1986. Maximum likelihood estimate of variance components. *Boll. Geod. Sci. Affini*, 60: pp. 329–338. (Ideas by A.J. Pope).

Koch, K.R., 1999. *Parameter estimation and hypothesis testing in linear models*. Springer Verlag, Berlin.

Odijk, D., 2008. GNSS Solutions: What does geometry-based and geometry-free mean in the context of GNSS? *Inside GNSS,* 3(2), pp. 22–24.

Rao, C.R., 1971. Estimation of variance and covariance components - MINQUE theory. *Journal of multivariate analysis*, 1(3): pp. 257–275.

Satirapod, C., Wang, J., and Rizos, C., 2002. A simplified MINQUE procedure for the estimation of variance-covariance components of GPS observables. *Survey Rev.,* 36(286), pp. 582–590.

Teunissen, P.J.G., 1997. The geometry-free GPS ambiguity search space with a weighted ionosphere. *J. Geodesy, Berlin*, 71(6), pp. 370-383.

Teunissen, P.J.G., 1988. Towards a least-squares framework for adjusting and testing of both functional and stochastic model. Internal research memo, Geodetic Computing Centre, Delft. A reprint of original 1988 report is also available in 2004, No. 26.

Teunissen, P.J.G., Jonkman, N.F., and Tiberius, C.C.J.M., 1998. Weighting GPS dual frequency observations: Bearing the cross of cross-correlation. *GPS Solutions*, 2(2), pp. 28–37.

Teunissen, P.J.G., 2000. *Adjustment theory: an introduction.* VSSD: Delft University Press. Series on Mathematical Geodesy and Positioning.

Teunissen, P.J.G., 2001. *Dynamic data processing: Recursive least-squares*. VSSD: Delft University Press. Series on Mathematical Geodesy and Positioning.

Teunissen, P.J.G., Simons, D.G., and Tiberius, C.C.J.M., 2005. *Probability and observation theory*. Delft: Delft University of Technology.

Teunissen, P.J.G., and Amiri-Simkooei, A.R., 2006. Variance component estimation by the method of least-squares. *Proc., 6th Hotine-Marussi Symp. of Theoretical and Computational Geodesy, IAG Symposia*, Vol. 132, P. Xu, J. Liu, and A. Dermanis, eds., Springer, Berlin, 273–279

Teunissen, P.J.G., and Amiri-Simkooei, A.R., 2008. Least-squares variance component estimation. *J. Geodesy, Berlin*, 82(2), pp. 65–82

Tiberius, C.C.J.M., and Kenselaar, F., 2000. Estimation of the stochastic model for GPS code and phase observables. *Survey Rev.,* 35(277), pp. 441–454.

Tiberius, C.C.J.M., and Kenselaar, F., 2003. Variance component estimation and precise GPS positioning: Case study. *J. Surv. Eng.,* 129(1), pp. 11–18.

Wang, J., Stewart, M.P., and Tsakiri, M., 1998. Stochastic modeling for static GPS baseline data processing. *J. Surv. Eng.,* 124(4), pp. 171–181.

Wang, J., Satirapod, C., and Rizos, C., 2002. Stochastic assessment of GPS carrier phase measurements for precise static relative positioning. *J. Geodesy, Berlin,* 76(2), pp. 95–104.

# 17

# HYDROTHEMAL ALTERATION MAPPING USING FEATURE-ORIENTED PRINCIPAL COMPONENT SELECTION (FPCS) METHOD TO ASTER DATA: WIKKI AND MAWULGO THERMAL SPRINGS, YANKARI PARK, NIGERIA

Aliyu Ja'afar Abubakar , Mazlan Hashim*, and Amin Beiranvand Pour

Geoscience and Digital Earth Centre (INSTeG)
Research Institute for Sustainability and Environment (RISE)
Universiti Teknologi Malaysia (UTM)
81310 UTM Skudai, Johor Bahru, Malaysia
*Corresponding author: mazlanhashim @utm.my

**Commission VI, WG VI/4**

**KEY WORDS:** FPSC; ASTER; Hydrothermal alteration; Yankari Park; Nigeria

**ABSTRACT:**

Geothermal systems are essentially associated with hydrothermal alteration mineral assemblages such as iron oxide/hydroxide, clay, sulfate, carbonate and silicate groups. Blind and fossilized geothermal systems are not characterized by obvious surface manifestations like hot springs, geysers and fumaroles, therefore, they could not be easily identifiable using conventional techniques. In this investigation, the applicability of Advanced Spaceborne Thermal Emission and Reflection Radiometer (ASTER) were evaluated in discriminating hydrothermal alteration minerals associated with geothermal systems as a proxy in identifying subtle Geothermal systems at Yankari Park in northeastern Nigeria. The area is characterized by a number of thermal springs such as Wikki and Mawulgo. Feature-oriented Principal Component selection (FPCS) was applied to ASTER data based on spectral characteristics of hydrothermal alteration minerals for a systematic and selective extraction of the information of interest. Application of FPCS analysis to bands 5, 6 and 8 and bands 1, 2, 3 and 4 datasets of ASTER was used for mapping clay and iron oxide/hydroxide minerals in the zones of Wikki and Mawulgo thermal springs in Yankari Park area. Field survey using GPS and laboratory analysis, including X-ray Diffractometer (XRD) and Analytical Spectral Devices (ASD) were carried out to verify the image processing results. The results indicate that ASTER dataset reliably and complementarily be used for reconnaissance stage of targeting subtle alteration mineral assemblages associated with geothermal systems.

## 1. INTRODUCTION

Remote sensing satellite sensors have been used for detection of prospective geothermal sites by capability of mapping hydrothermal alteration mineral assemblages and thermal anomaly in the visible and near-infrared (VNIR), shortwave infrared (SWIR) and thermal infrared (TIR) portions of the electromagnetic spectrum (Littlefield and Calvin, 2015). Availability and significant improvement in spectral resolution of satellite remote sensors such as the Advanced Spaceborne Thermal Emission and Reflection Radiometer (ASTER) in the VNIR and SWIR bands increases the capability of mapping subtle alteration mineral features associated with concealed and fossilized geothermal systems (Pour and Hashim, 2014).

The VNIR subsystem of ASTER has three recording channels between 0.52 and 0.86 μm with a spatial resolution of up to 15 m. The SWIR subsystem of ASTER has six recording channels from 1.6 to 2.43 μm, at a spatial resolution of 30 m. ASTER swath width is 60km (each individual scene is cut to a 60x60 km2 area) which makes it useful for regional mapping (Abrams, 2000; Yamaguchi and Naito, 2003; Pour and Hashim, 2015).

Concealed and fossilized geothermal systems are not characterized by obvious surface manifestations like hot springs, geysers and fumaroles, therefore, they could not be easily identifiable using conventional techniques (Norton, 1984). Geothermal systems are essentially associated with hydrothermal alteration mineral assemblages such as iron oxide/hydroxide, clay, sulfate, carbonate and silicate groups, which are produced by surface alteration of primary minerals as a result of metasomatism or contact of the hot fluids derived from magmatic intrusions (Pournamdari et al, 2014a,b; Calvin and Pace, 2016). Accordingly, depending on the spatial and spectral resolution of the satellite remote sensor, hydrothermal alteration mineral groups associated with the geothermal system could be identified using spectral bands in the VNIR and SWIR regions.

The main objective of this investigation is to evaluate the applicability of ASTE VNIR and SWIR bands for identifying subtle hydrothermal alteration mineral features associated with concealed and fossilized geothermal systems in Yankari Park, northeastern Nigeria.

## 2. MATERIALS AND METHODS

### 2.1 Geology of the study area

The Yankari Park is located within Latitude 09° 75′ 00″ N, and Longitude 10° 50′ 00″ E in the south-central part of Bauchi State, northeastern Nigeria (Fig. 1). It covers an area of about 2244 km². Yankari National Park lies in the southern part of the Sudan Savannah. It is composed of savannah grassland with well-developed patches of woodland. It is also a region of rolling hills, mostly between 200 m and 400 m. The Yankari Park is situated in the Benue Trough, bordered to the west by the basement complex crystalline rocks of the Jos Plateau, to the northeast by the Biu Plateau and to the southeast by the Adamawa highlands (Ajakaiye et al., 1988). Kerri formation characterized by Cenozoic to Mesozoic sedimentary rocks is dominated lithological sequence in the study area. However, there are many extinct volcanic features in the Jos Plateau and the northeastern and western part of the Yankari Park.

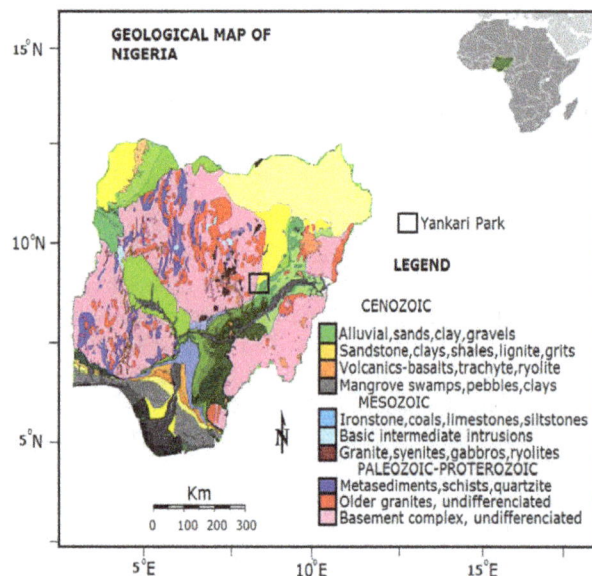

Figure 1. Geological Map of Nigeria showing the location of Yankari Park.

### 2.2 ASTER remote sensing data

Two cloud-free ASTER level 1T (Precision Terrain Corrected Registered At-Sensor Radiance) scene (AST_L1T_00301252006095330 and AST_L1T_00311142008095512 path/raw, 187/53) covering the Yankari Park were obtained from U.S. Geological EROS. The images have January 25th, 2006 and January 1st, 2003 acquisition dates, respectively. The ASTER Level 1 Precision Terrain Corrected Registered At-Sensor Radiance (AST_L1T) data contains calibrated at-sensor radiance, which corresponds with the ASTER Level 1B (AST_L1B), that has been geometrically corrected, and rotated to a north up UTM projection. The images were pre-georeferenced to UTM zone 32 North projection using the WGS-84 datum.

### 2.3 Image processing methods

Principal Component Analysis (PCA) (known as eigenvector-eigenvalue decomposition) is a statistical technique which is used to transform multidimensional data by reducing the variance and projecting the data along uncorrelated axes. It transforms a set of correlated input bands into uncorrelated spectral bands as principal components. The PCA is widely employed on multispectral data purposely to extract unique spectral responses such as hydrothermal alteration for geologic mapping and mineral exploration purposes (Gupta et al., 2013). In the present study, the feature-oriented principal component selection (FPCS) (Crosta and Moore, 1989), was employed in order to map specific hydrothermal alteration minerals of interest. Thus, the bands in which the targeted minerals, which are indicators of hydrothermal alteration related to geothermal systems were selected for the analysis. This was performed separately for iron oxide/hydroxide minerals (hematite, limonite, and goethite) that have diagnostic absorptions within the VNIR region, and for hydroxyl (OH) bearing minerals (clays, carbonates and sulfates) that contain absorption features in the SWIR portion.

The whole concept behind FPCS is that, by reducing the number of input bands for PCA in such a way as to ensure that certain minerals (or materials) will not be mapped while increasing the likelihood that other interesting targets will be unambiguously mapped into only one of the PC images (Loughlin, 1991).

A feature oriented guided PCA involves selection of bands in which specifically targeted minerals (or materials) of interest are theoretically known to have diagnostic reflectance and or absorption characteristics, while including a band in which such targets do not manifest, thus, facilitating the manifestation of the targets in the main input bands. The results can then be interpreted by identifying the Eigenvector matrix to observe the weighing. The Eigenvectors could be positive or negative which indicates reflection or absorption respectively, however, if the loadings are positive in the reflective band of a mineral the image tone is shown by bright pixels, and if they are negative, the image tone is shown by bright pixels for the enhanced target mineral (Gupta et al., 2013).

## 3. RESULTS AND DISCUSSION

ASTER subset scene (1515 × 1515 pixels) covering the zones of Wikki and Mawulgo thermal springs in Yankari Park area was used for FPCS analysis. Band 1 of VNIR region and bands 5, 6 and 8 of SWIR subsystem were selected to accomplish FPCS analysis here. Band 1 was selected due to high reflectance features of clay alteration minerals in VNIR. Band 5 (2.14-2.18 μm) was selected to identify alunite absorption feature in 2.16μm; band 6 (2.18-2.22 μm) was selected to detect Kaolinite (2.20 μm) and band 8 (2.29-2.36 μm) for calcites (2.35 μm) (Pour and Hashim, 2011a; Rowan et al., 2003). Band 2 (0.63-0.69 μm), band 3 (0.78-0.86 μm) and band 4 (1.6-1.7 μm) of ASTER were not selected here to avoid the effects of high abundance soil materials in Savannah environments.

Band 4 represents the spectral region where all alteration minerals of hydroxyl-bearing (OH) manifest strong reflection. Bands 2 and 3 show the manifestation of both iron oxides and vegetation due to strong reflections in 0.63 to 0.86 μm region, respectively. The statistical results derived from FPCS transformation to selected bands of ASTER are shown in Table 1.

| Input Bands | Band 1 | Band 5 | Band 6 | Band 8 | |
|---|---|---|---|---|---|
| Eigenvector Matrix | | | | | Variance (%) |
| PC 1 | 0.173479 | 0.513621 | 0.538524 | 0.645051 | 87.5 |
| PC 2 | -0.899214 | -0.056769 | -0.145648 | 0.408631 | 9.6 |
| PC 3 | -0.401646 | 0.349319 | 0.558380 | **-0.636292** | 2.5 |
| PC 4 | -0.000195 | **-0.781631** | **0.613995** | 0.109829 | 0.4 |

Table 1. Covariance eigenvector values of the FPCS for the selected bands (1, 5, 6, and 8 of ASTER) for Yankari selected subset scene.

As shown by the eigenvalues, the resulting PC1 eigenvector loadings and eigenvalues show that that PC1 contains 87.5% of the total data variance (Table 1). Overall scene brightness albedo and the topographic effect are responsible for the strong correlation between multispectral image bands (Loughlin, 1991). From the results in Table 1, it is observed that hydroxyl-bearing minerals will best be mapped in PC3 and PC4 because of the large negative eigenvector loadings at PC3 band 8 (-0.63) and large negative and positive loadings at PC4 bands 5 (-0.78) and 6 (0.61), which appear as opposite signs (Table 1). This means areas of possible carbonates alteration will appear as dark pixels in PC3 due to their strong absorption in bands 8, while sulfates and clays may appear as dark and bright pixels in PC4 because of the higher opposite signs in bands 5 and 6 in which these alteration minerals are known to have diagnostic absorption features in these bands.

The results of PC3 and PC4 are shown in Figures 2 and 3. This identified alteration zones served as a guide for understanding areas to focus subsequent analysis and field validation particularly in relation to their proximity to the thermal springs. The above results are observed to conform well to exposed alteration areas identified in the field especially around the Mawulgo thermal spring, which is less vegetated. The results also corroborate well especially with exposed alteration areas designated; W1, W2, M1 and M2 as can be seen in both PC3 and PC4 image maps (Figs. 2 and 3).

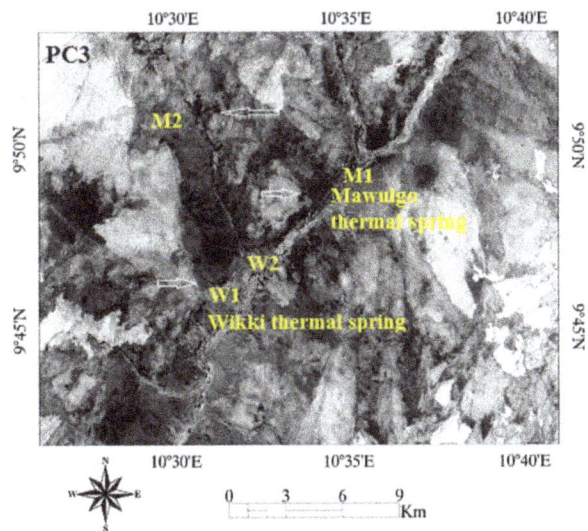

Figure 2. PC3 showing dark pixels as possible alteration zones due to hydroxyl-bearing minerals.

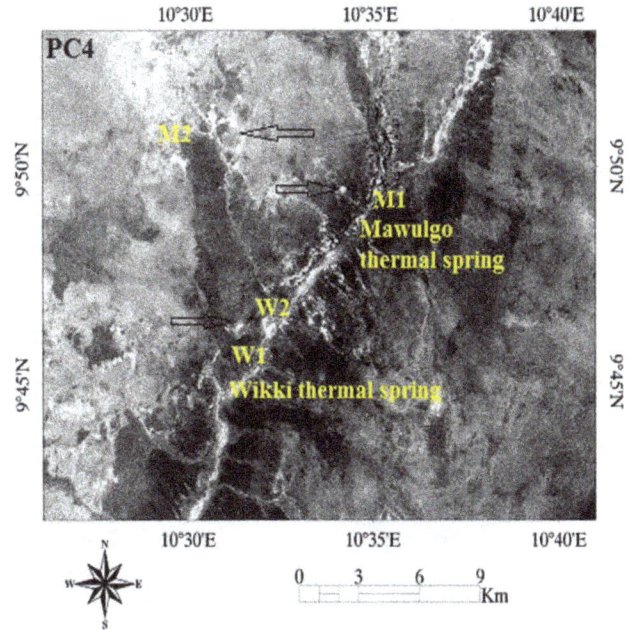

Figure 3. PC4 showing bright pixels as possible alteration zones due to hydroxyl-bearing minerals.

In this study, another FPCS was implemented to identify iron oxide alteration zones associated with geothermal systems using bands 1, 2 and 3 of the VNIR subsystem and band 4 of the SWIR subsystem of ASTER. Iron oxide/hydroxide minerals such as hematite, jarosite and limonite tend to have low reflectance in visible and higher reflectance in near infrared, coinciding with bands 1, 2 and 4 of ASTER data.

By observing the result of FPCS in Table 2, PC1 contains 80.32% of the total variance among the input bands, which implies that overall brightness and albedo is effectively mapped in PC1 as such may not contain spectra relevant for this analysis. The remaining PCs, however, shows a decreasing variance resulting from the PC 1, 2, 3 and 4. The PC2 appears to be appropriate to map iron oxide rich areas because of the large magnitude eigenvalue at PC2 band 2 (0.64) and moderate value at band 1 (0.40), which manifest possible areas of iron oxides as bright pixels in the PC2 image map (Fig. 4). It accounts for about 13% of the variance (Table 2). The negated low eigenvector value of PC2 in band 3 (-0.64) implies vegetated areas may appear as dark pixels because of vegetation manifest in the near infrared coincides with band 3 of ASTER. Consequently, along the Gaji river which is covered with vegetation appears as dark pixels (Fig. 4).

Generally, the results of the PCA effectively mapped and identified especially argillic alteration areas. However, there is some confusion in the iron oxide mapping in that area identified in the field as limonitic also appear as vegetated as observed in Figure 4. This may be attributable to the fact that most of the areas are covered by sparse vegetation and the alterations were territorially small not to be detected by the ASTER data, this was evident in the field especially around Wikki thermal spring.

| Input Bands | Band 1 | Band 2 | Band 3 | Band 4 | |
|---|---|---|---|---|---|
| Eigenvector matrix | | | | | Variance (%) |
| PC 1 | 0.232547 | 0.425563 | 0.630823 | 0.605707 | 80.32 |
| PC 2 | **0.406700** | **0.640638** | -0.647698 | **0.068307** | 13.09 |
| PC 3 | -0.461241 | -0.207888 | -0.415524 | 0.755896 | 5.29 |
| PC 4 | 0.753508 | -0.604370 | -0.099447 | 0.238902 | 1.29 |
| | | | | | |

Table 2. Covariance eigenvector values of the FPCS for the selected bands (1, 2, 3, and 4 of ASTER) for Yankari selected subset scene.

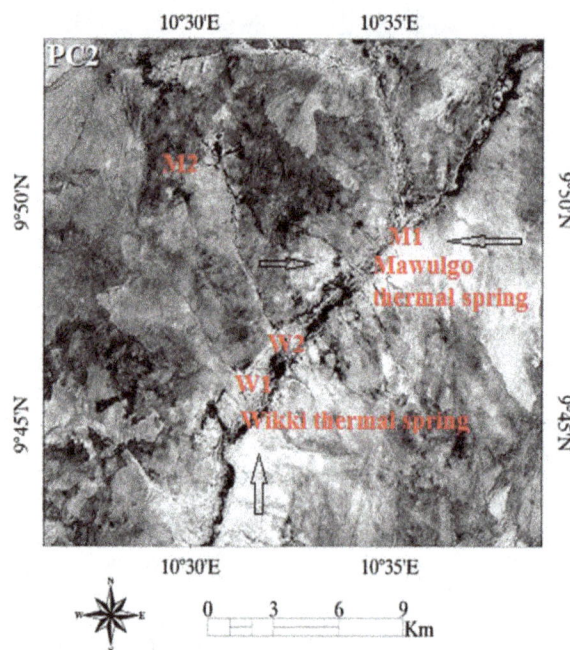

Figure 4. PC2 showing bright pixels as possible alteration zones due to iron oxide/hydroxide minerals.

It was observed that the alteration areas significantly conform to the identified areas during field work especially around Mawulgo thermal spring, where clay alterations were observed and the M designated samples are obtained. These are shown especially in PC3 dark pixels and PC4 dark and bright pixels for hydroxyl-bearing minerals in Figures 2 and 3, however, pixels appear unconformable when mapping iron oxides.

The result of the laboratory ASD and XRD analysis of rock samples acquired from alteration zones in the field were used to verify the image processing analysis. The results of representative collected rock samples from alteration zone analyzed in the laboratory conditions using XRD analysis are show the presence of kaolinite, illite, alunite, calcite, limonite, hematite, quartz and dickite as dominated alteration minerals in the collected rock samples.

During field survey was observed that the alteration types were predominantly limonitic around the Wikki thermal spring indicated by brownish to reddish iron oxides to hydroxide rock outcrops and argillic alterations indicated by bleached to gray-

whitish rocks around areas of the Mawulgo thermal spring (see Fig. 5 a-d).

Figure 5. Field photographs of hydrothermal alteration mineral zones associated with the geothermal systems; (a) Exposed clay (argillic) alteration zone around Mawulgo thermal spring;(b) Carbonate alteration at Wikki thermal springs; (c) Exposed argillic altered rocks around Mawulgo thermal spring; (d) Vegetated areas of altered rocks around Wikki thermal spring.

## 4.  CONCLUSIONS

In this study, we investigated the applicability of mapping hydrothermal alteration minerals associated with subtle geothermal systems using ASTER dataset as a proxy for identifying prospective sites of fossilized and concealed geothermal systems in the Yankari Park, northeastern Nigeria. Application of FPCS analysis to bands 5, 6 and 8 and bands 1, 2, 3 and 4 datasets of ASTER was used for mapping clay and iron oxide/hydroxide minerals in the zones of Wikki and Mawulgo thermal springs in Yankari Park area at the regional scale. The result of the laboratory, including ASD and XRD analysis of rock samples acquired from alteration zones in the field shows the presence of kaolinite, illite, alunite, calcite, limonite, hematite, quartz and dickite as dominated alteration minerals in the study zones. Therefore, the advanced argillic alteration was recorded for the Wikki and Mawulgo thermal springsin Yankari Park based on alteration mineral assemblages detected by image processing and fieldwork.

## ACKNOWLEDGEMENTS

This study was conducted as a part of Tier 1 (vote no: Q.J130000.2527.13H13), research university grant category. We are thankful to the Universiti Teknologi Malaysia for providing the facilities for this investigation.

## REFERENCES

Abrams, M. (2000). The Advanced Spaceborne Thermal Emission and Reflection Radiometer (ASTER): data products for the high spatial resolution imager on NASA's Terra platform. *international Journal of Remote sensing.* 21(5), 847-859.

Ajakaiye, D., Olatinwo, M. and Scheidegger, A. (1988). Another possible earthquake near Gombe in Nigeria on the 18-19 June 1985. *Bulletin of the Seismological Society of America.* 78(2), 1006-1010.

Calvin, W. M., Littlefield, E. F. and Kratt, C. (2015). Remote sensing of geothermal-related minerals for resource exploration in Nevada. *Geothermics.* 53, 517-526.

Calvin, W. M. and Pace, E. L. (2016). Mapping alteration in geothermal drill core using a field portable spectroradiometer. *Geothermics.* 61, 12-23.

Crosta, A. P. and Moore, J. M. (1989). Geological mapping using landsat thematic mapper imagery in Almeria province, south-east Spain. *International Journal of Remote Sensing.* 10(3), 505-514.

Gupta, R. P., Tiwari, R. K., Saini, V. and Srivastava, N. (2013). A simplified approach for interpreting principal component images.

Loughlin, W. (1991). Principal component analysis for alteration mapping. *Photogrammetric Engineering and Remote Sensing.* 57(9), 1163-1169.

Norton, D. L. (1984). Theory of hydrothermal systems. *Annual Review of Earth and Planetary Sciences.* 12, 155.

Pour, A. B. and Hashim, M. (2014). ASTER, ALI and Hyperion sensors data for lithological mapping and ore minerals exploration. *SpringerPlus.* 3(1), 1.

Pour, B.A., Hashim, M., 2015. Integrating PALSAR and ASTER data for mineral deposits exploration in tropical environments: a case study from Central Belt, Peninsular Malaysia. International Journal of Image and Data Fusion, 6(2), 170-188.

Pournamdary, M., Hashim, M., Pour, B.A., 2014a. Spectral transformation of ASTER and Landsat TM bands for lithological mapping of Soghan ophiolite complex, south Iran. Advances in Space Research 54 (4), 694-709.

Pournamdary, M., Hashim, M., Pour, B.A., 2014a. Application of ASTER and Landsat TM data for geological mapping of Esfandagheh ophiolite complex, southern Iran. Resource Geology 64 (3), 233-246.

U.S. Geological EROS, http://glovis.usgs.gov/; https://lpdaac.usgs.gov/dataset_discovery/aster/aster_products_table/ast_l1t.

Yamaguchi, Y. and Naito, C. (2003). Spectral indices for lithologic discrimination and mapping by using the ASTER SWIR bands. *International Journal of Remote Sensing.* 24(22), 4311-4323.

# 18

# SPATIO TEMPORAL DETECTION AND VIRTUAL MAPPING OF LANDSLIDE USING HIGH-RESOLUTION AIRBORNE LASER ALTIMETRY (LIDAR) IN DENSELY VEGETATED AREAS OF TROPICS

T. Bibi [a] *, K. Azahari Razak A. Abdul Rahman, A. Latif

[a] Department of Geoinformation, Faculty of Geoinformation and Real Estate, Universiti Teknologi Malaysia, Skudai 81310, Johor, Malaysia - tehmina_khan79pk@hotmail.com, alias.fksg@gmail.com,
UTM RAZAK School of Engineering and Advanced Technology, Universiti Teknologi Malaysia, 54100 Jalan Sultan Yahya Petra, Kuala Lumpur, Malaysia; khamarrul.kl@utm.my
[b] Dept. of Education, Azad Jammu and Kashmir, Muzaffarabad, adnan_ajk@hotmail.com

**KEY WORDS:** Landslide, Virtual mapping, Airborne LiDAR, DTM, Kundasang, Inventory, Spatio temporal

**ABSTRACT:**

Landslides are an inescapable natural disaster, resulting in massive social, environmental and economic impacts all over the world. The tropical, mountainous landscape in generally all over Malaysia especially in eastern peninsula (Borneo) is highly susceptible to landslides because of heavy rainfall and tectonic disturbances. The purpose of the Landslide hazard mapping is to identify the hazardous regions for the execution of mitigation plans which can reduce the loss of life and property from future landslide incidences. Currently, the Malaysian research bodies e.g. academic institutions and government agencies are trying to develop a landslide hazard and risk database for susceptible areas to backing the prevention, mitigation, and evacuation plan. However, there is a lack of devotion towards landslide inventory mapping as an elementary input of landslide susceptibility, hazard and risk mapping. The developing techniques based on remote sensing technologies (satellite, terrestrial and airborne) are promising techniques to accelerate the production of landslide maps, shrinking the time and resources essential for their compilation and orderly updates. The aim of the study is to provide a better perception regarding the use of virtual mapping of landslides with the help of LiDAR technology. The focus of the study is spatio temporal detection and virtual mapping of landslide inventory via visualization and interpretation of very high-resolution data (VHR) in forested terrain of Mesilau river, Kundasang. However, to cope with the challenges of virtual inventory mapping on in forested terrain high resolution LiDAR derivatives are used. This study specifies that the airborne LiDAR technology can be an effective tool for mapping landslide inventories in a complex climatic and geological conditions, and a quick way of mapping regional hazards in the tropics.

## 1. INTRODUCTION

The landslide hazard has been a topic for many research studies like Carrara et al. (1995), Klimeš (2007), Havlín et al. (2011), Bibi et al. (2016), Razak and Mohamad (2015), Freeborough et al. (2016) etc. It is a form of mass movement in which the rock, debris, or earth moves down along slope, under the influence of gravity (Cruden and Varnes, 1996). While the terms "mass movement", "slope failure", and "landslide" are commonly used words as synonyms. Although this natural phenomenon can cause serious hazard but it is quite difficult to predict it. Mapping, monitoring and modelling of such events are rather challenging tasks. The intensity of hazard increases many folds, when it came across with anthropogenic activities. The anthropogenic activities play vital role to slope failure in lose morphology, especially on tropical terrain where excessive rainfall and humidity exist. In this situation landslides convert from natural hazard to natural disasters.

The importance of mapping of new landslides with in fleeting time cannot be denied. It is equally important for updating of landslide inventories, susceptibility and hazard maps as well as for efficient post-disaster response (Van Den Eeckhaut., 2012). The availability of high resolution remote sensing data has been facilitating such efforts. Due to instant advances in Remote sensing techniques the research about landslides are becoming easier as compare to past (Jaboyedoff et al. 2012., Jing et al.2016). It is easy to map fresh landslide with the help of their scars and strong spectral contrast to their surrounds because of the absence of vegetation cover (Martha et al. 2010, Roback et al. 2017). But Various inactive or reactivated landslides are also a key threat especially in tropical areas where vegetation cover grow very fast because of excessive rain and high humidity. Generally, it is assumed that the probability of reoccurrence of a landslide in wet and moist climate increases many folds because of lithology, slopes, weathering rate, climate, etc., which are not changeable. In this situation, the significant mitigation and prevention measures should be implemented. For the mitigation and prevention, it is very important to have a historical record for old or dormant landslides. In other words, a landslide inventory could not be completed without historical records of landslides.

The use of Remote sensing techniques for landslides examination is experiencing rapid developments. The probability of obtaining 3D information of the terrain with high precision and spatial resolution is opening innovative ways of exploring the landslide phenomena. Recent advances in sensor electronics and data treatment make these techniques affordable. The two major remote sensing techniques that are rapidly developing in landslides investigation are interferometric synthetic aperture radar (InSAR) (Fruneau et al. 1996; Colesanti et al. 2003; Squarzoni et al. 2003; Mantovani et al. (2016), Ciampalini et al. (2015), Del Ventisette et al. (2014), Bianchini et al. (2013), Greif and Vlcko (2012), Bateson et al. 2015), and light detection and ranging (LIDAR) (Carter et al. 2001; Haugerud et al. 2003; Slob and Hack 2004; Chigira et al. 2004; Schulz, 2007; Booth et al. 2009; Guzzetti et al. 2012; Hölbling et al. 2012; Jaboyedoff, 2012; Pradhan et al. 2012; Wang et al. 2013; Lin et al. 2014; Scaioni et al. 2014; Chen et al. 2015; Li et al. 2015; Mahalingam and Olsen 2015; ).

InSAR techniques are usually ground-based (Stow 1996; Tarchi et al. 2003) or satellite-based (Carnec et al. 1996; Singhroy 2009), and only rarely airborne. Although InSAR is not dependent on clear skies for data collection as compare to other optical sensors but even then, the application of SAR data for landslide inventory mapping is inadequate due to foreshortening, layover effects, atmosphere propagation effects, and vegetation decorrelation in forested terrain (Rott, 2009). LIDAR (or laser scanning) provide high-resolution point clouds of the topography and has several applications that range from mapping (Ardizzone et al. 2007; Jaboyedoff et al. 2008a) to monitoring deformation (Gordon et al. 2001), landslides or rockfall displacements (Teza et al. 2007; Oppikofer et al. 2008; Abellan et al. 2010) to landslide in soils (Jaboyedoff et al. 2009a).

The availability of airborne laser scanning over the last few years (Jaboyedoff et al., 2010) has been used to identify and map landslide morphology in areas that are partially or completely covered by dense vegetation (Razak et al., 2011). LiDAR and its wide range of derivatives has become a powerful tool in landslide research, particularly for landslide morphology analysis (Glenn et al. 2006) and landslide identification and inventory mapping (Razak et al. 2011).

LIDAR is an advanced technology to investigate landslide in tropical region where vegetation is dense because of hot and moist climate. The LIDAR digital terrain models (DTM) is a very high resolution (VHR) accurate and precise DTM in raster grids or triangulated irregular networks (TINs) , known as 2.5D representations of the Surface. Furthermore, the true 3D point clouds of LIDAR have a high density of information. This density is dominantly based on the position of the sensor: centimetric to millimetric resolution for terrestrial laser scanning (TLS) and metric to decimetric resolution for airborne laser scanning (ALS) (Shan and Toth 2008). However, the Helicopter-based ALS can give a higher resolution as compare to aircraft-based ALS because it allows the scanner to orientate in all directions (Vallet and Skaloud 2004).

## 2. LANDSLIDE INVENTORY MAPPING ISSUES AND CHALLENGES

Landslides are classified as third in natural disasters in terms of death rate amongst the top ten natural disasters (UNISDR, CRED, EM-DAT, 2011). Petley (2011) mentioned in his study that more than 80,000 people around the world have been killed in the last 10 years by landslides. According to National slope master plan (2009-2023) the estimated economic losses for past 34 years is RM 3 billion and it is expected to increase up to RM 17 billion over the next 25 years without a comprehensive mitigation plan, (PWD, 2009). The precise landslide inventory mapping needs an extra effort of compiling and updating landslide maps at regional, national and global scales as the

number of events is often misjudged, and inaccurate hazard and risk maps are produced (Guzetti *et al.*, 2012)

Landslide occurs nearly every year in Malaysia which causes incredible damages to life and properties. The purpose of the Landslide hazard map is to identify the hazard regions for the execution of mitigation plans which can reduce the loss of life and property from future landslide incidences. Currently, the Malaysian research entities, academic institutions and Government agencies are trying to develop a landslide hazard and risk database for prone areas to back the prevention program, mitigation action, and evacuation plan. Though, there is a lack of attention regarding landslide inventory mapping as a basic input of landslide susceptibility and hazard mapping. The landslide inventory mapping provides detailed spatiotemporal information about the distribution of landslides. It contains the information regarding date, types, area/volume, depth, and so on of each particular landslide. The landslide inventory can be prepared by several ways e.g. image interpretation, virtual mapping, field mapping, by utilizing historical archive, local knowledge (interview), and with the help of combination of two or more methods.

Despite of incredible advances and extensive use of remote sensing data and its derivative products, it is notable that preparation and updating landslide inventories in rough and wooded terrains is difficult especially in tropical environment. The accuracy of inventory maps in the tropical region is still vague (Razak and Mohamed, 2015). Although the importance of landslide inventory maps for mitigation and planning cannot be denied but even then, they are rarely created (Guzzetti *et al.*, 2012). The preparation of landslide inventory map is a tedious process due to the detailed mapping of each individual landslide (Van Westen *et al.*, 2006).

Landslides covered a large portion of peninsula Malaysia. There limited government agencies which have the accountability for sustaining a landslide database, e.g., the Public Work Department (PWD). However, it only deals with already activated landslides which have some impact on life or property e.g. such as buildings or road. Other than that, the landslides which are for the time being not disturbing the life or value able property are ignored in database. Most of the time, after the completion of research projects the landslide inventory maps are not constantly updated. Hence the scarcity of reliable information on landslides effects the quality landslide susceptibility, hazard and risk assessment process at the regional, national and continental scales. To produce a precise landslide inventory map with respect to both, the area of interest and within the time investigated is rather a difficult task to complete (Ibsen and Brunsden, 1996; Glade, 2001).

There are different approaches for landslide inventory mapping which have some pros and cons e.g. geomorphology based landslide mapping via field can be highly accurate if supported by Global Navigation Satellite System (GNSS) and other instruments like laser rangefinder (Santangelo *et al.*, 2010), but it is an expensive and time-consuming process and, and therefore is not applicable widely (Santangelo *et al.*, 2010). The DTM generated by topographic maps are usually less precise because general the altitude is extracted from aerial photographs for topo maps. The accuracy of DTMs in thickly forested and built-up areas are normally depends on photogrammetry e, g, sunlight, flight height, camera and sensor type, and at least a relevant point on surface should be visible from at least two imaging locations (Kraus, 2007). However, the aerial photos based topo maps did not complete the criteria therefore, the representation of surface in photogrammetric DTMs of rough and wooded terrain leads to inadequate and capricious landslide inventories.

The Virtual examination based on an enormously high-resolution image by remote sensing, proved to be effective to map fresh landslides in a large forested terrain as the landslides have left very clear signs of their existence (Guzzetti *et al.*, 2012; Razak *et al.*, 2011a). The multispectral information is very useful for Semi-automatic detection of landslides for fresh or reactivated landslides (e.g., Mondini *et al.* 2011). Though, it is not very effective to characterize the old and dormant landslides under thick vegetation cover. Thus, the resulted inventory maps poorly represent the landslides in such places (Wills & McCrink, 2002; Brardinoni *et al.*, 2003).

The precision of airborne laser scanning (ALS) to map vegetated dormant landslides in temperate regions is varified e.g. the Flemish Ardennes, Belgium (Van den Eeckhaut *et al.* 2007). Though, only a few researches assessed its suitability for landslide mapping in the tropics. In the tropical environment, the landslides are always rapidly covered by vegetation cover because of which it becomes difficult to prepare a precise landslide inventory map. Figure 1 is showing the common most scenario of landslides in Malaysia. Therefore, it is difficult to detect landslide morphological features with the help of optical satellite images and aerial photographs in tropical regions as it leads to unsatisfactory landslide inventory, which can disturb the quality of landslide susceptibility, hazard and risk analysis. Furthermore, the unsuccessfulness of field mapping method is due to limited access to data coverage and poor synoptic view, which is making this method inappropriate for regional landslide assessment.

The existing mapping techniques show considerable limitations, while a precise virtual deforested LIDAR image could provide a better intimation of morphology and drainage pattern underneath thick vegetation cover. Additionally, landslide mapping is a challenging task in a tropical mountainous environments due to rapid growth of vegetation cover within months or seasons.

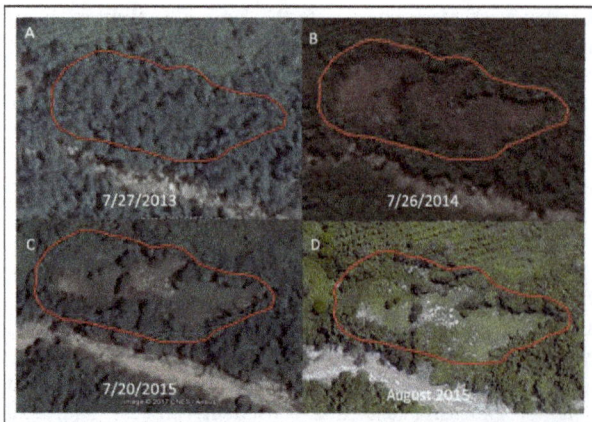

Figure 1: Landslides in a tropical conditions: A) Google image of kundassang captured in 2013 (B) Google image of kundassang for the same area captured in 2014 showing slope failer, (C) Google image of kundassang 2015 is showing the growth of vegetation cover on the slope failer. , (C) High resolution LiDAR Orthophoto of kundassang 2015 is showing the rapid growth of vegetation cover on the slope failer.

## 3. STUDY AREA

The Kundasang area is located in the Sabah highlands on the southeast side of Gunung Kinabalu. Kundasang township straddles the trunk road between Tuaran and Ranau. The study area is situated in Ranau district and is an attractive tourism spot. It is located in the foot hills of the world first Heritage site Mount Kinabalu in Malaysia. There are numerous tourism sites near Kg. Mesilau e.g. Kinabalu Golf Club, Mesilou Nature Resort, and Desa Dairy Farm. The study area is located on the southern flank of Mount Kinabalu and approximate height is in between 500 to 2,000 m from mean sea level. Geomorphologically the Mesilau river valley, Kundasang is consist on river terraces and plateau. Moreover, the major activity in the area is agriculture (Sarman and Komoo, 2000). An attractive portion of the land is under plantation other than recreational and tourism (Sarman and Komoo, 2000)

The geology of the upper and lower part of the study area consists of igneous rock and sedimentary rock, respectively. The oldest igneous rock of the study area is gabbro and ultramafic, followed by granodiorite. The oldest sedimentary rock is Trusmadi Formation, Crocker Formation, and Pinosuk

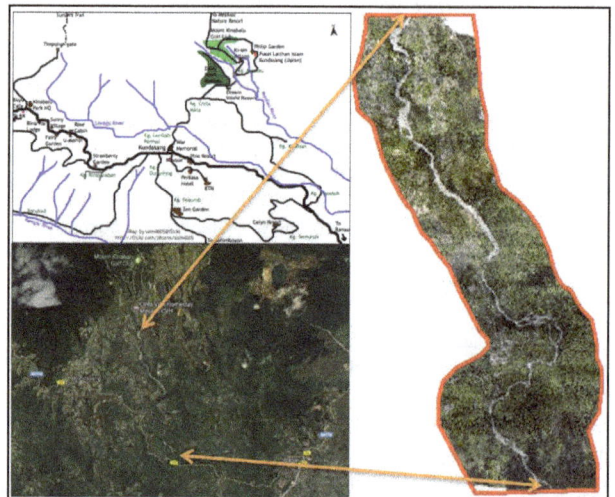

Figure 2: Location map of Sg mesilau, Kundasang, Saba state of Peninsular Malaysia (Source: http://sam4605.blogspot.my/)

Gravel. Trusmadi Formation is described as strongly folded and faulted grey and dark grey argilite, slate, siltstone and sandstone with volcanics, whereas Crocker Formation is referred as strongly folded and faulted sandstone, silstone, red and grey shale, mudstone and argilite. Pinosuk Gravel is poorly consolidated unsorted gravel up to boulder size in a sandy to clayey matrix (JMG, 2003).

The dominant lithology in the study area is sedimentary rock, specifically Pinosuk Gravel as shown in Figure 1. There are two major active faults exist in the study area known as Mensaban fault and Lobo-Lobo fault segments. Mensaban fault is a normal fault trending northwest-southeast and west-east. Lobo-Lobo Fault is left-lateral strike slip fault, which trends N20°E (2007).

The topography of the research area is low-lying terrain with meanders and small flood plain. The straight channels are indication active faults. The main morphology features in study area are terraces.

Figure 3: Landslide caused by heavy rain in Kampung mesilau (Source: http//: thesundaily.my)

Figure 4: Slope fauiler at the bank of river mesilau(Source: http//:thestar.com.my).

## 4. MATERIAL AND METHODS

### 4.1 LiDAR Data Capturing And Processing

The research focused on a debris flow channel of Mesilau river valley in Kudasang located in the foothills of mount Kinabalu, Saba, Malaysia. Kundasang is recognized as one of the main geological hazardous area in Malaysia because of the abundant landslides incidences at various places (Omar S et al 2016). For inventory mapping the airborne LiDAR data was used using helicopter based RIEGL system. In this process millions of light rays are sent through each pulse, among which some reflect after hitting the top most cover of the earth (vegetation and buildings) known as 'first return' or 'full feature' while some reached to the ground surface known as 'last return' or 'bare-earth' . Furthermore, with the help of precise Global Positioning Satellite (GPS) tracking technology, the 3D (XYZ) positions of each returned light point reflected by earth can be calculated. These raw points are denoted as LiDAR 'point cloud'. After collecting 3D raw point cloud with the help of software and manual manipulation the orthophoto and Digital elevation model (DEM) or Digital Terrain Models (DTM) can be generated using selected points from the LiDAR point cloud.

This 'bare-earth' DTM is the most interesting output of LiDAR point cloud for landslide identification. Most of the time the end users (geotechnical engineers and geologists) use the processed information in form of DTM as compare to directly working with the raw point cloud data of normally the raw LiDAR point clouds pre-processed and interpreted by the firms who collect the data because of specializing in this task.

The data captured (August 2015) and processed (raw point cloud) by BUMITOUCH plmc Sdn Bhd. The 3D point clouds recorded up to 120 points per square meter which is the highest point density reported over the disaster areas in Malaysia. The processed LiDAR data comprises of classified 3D point clouds, 0.25 m digital terrain model (DTM), 0.25 digital surface model (DSM), and 0.07 m orthophoto of the debris flow area. The procedural flow of the inventory preparation is illustrated in Figure 6. The process of preparing a comprehensive inventory map is also supported by multi temporal and historical records of landslide occurrences in the study areas. This inventory has been cross checked by spatio temporal images from google pro and historical records. After the processing of LiDAR dataset by image visualization in 2D and 3D the visual interpretation in GIS environment is done.

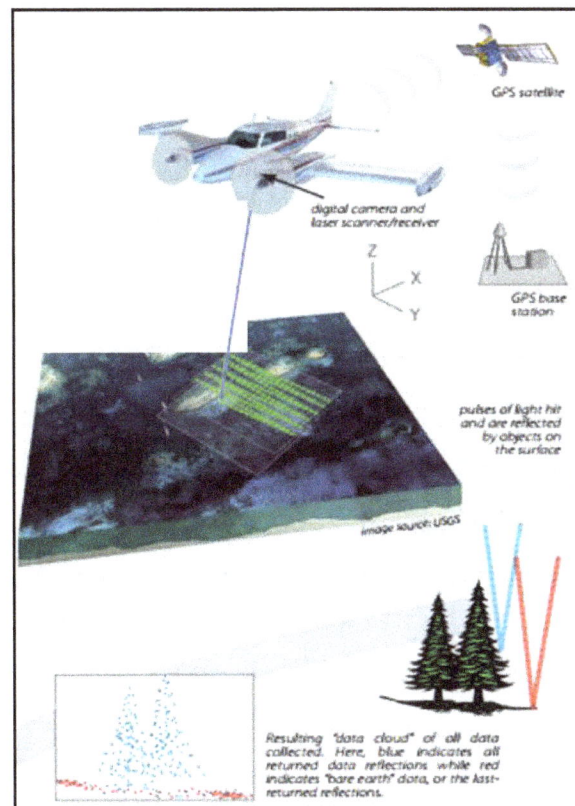

Figure 5: Data collection procedure of LiDAR

### 4.2 Virtual Mapping of Landslide Inventory

Landslide inventory is the most straightforward method to assess the landslide susceptibility, highlighting the spatial distribution of landslides, symbolized either as points (small scale) or as polygons (large scale) having attribute information about type and activity. Usually the inventory maps are considered as base for the landslide susceptibility and hazard assessment process. According to Guzzetti *et al*. (2012) until now the landslide inventories are produced and assembled either continuous in time or based on each particular triggering event e.g, rainfall, flood and earthquake. There are four possible stages of development of a landslide is described in the literature as: i) pre-failure (strained but intact slope); ii) failure (formation of a continuous surface of rupture); iii) post-failure (after failure until stop the movement); and iv) reactivation, (movement of slope along pre-existing surfaces of rupture) (Van Den Eeckhaut *et al*., 2007; Hungr *et al*., 2012; Razak *et al*., 2015).

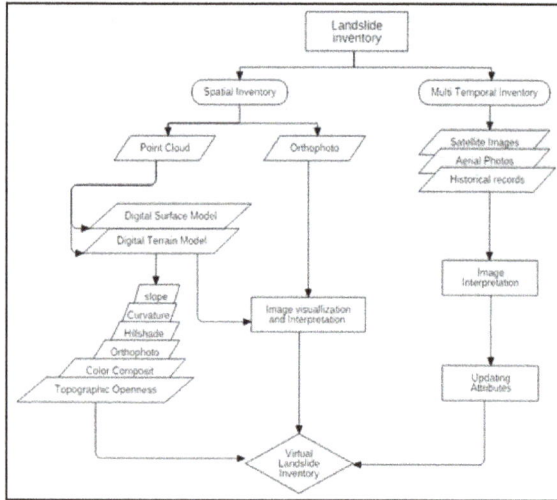

Figure 6: Procedural Flow for landslide inventory

shape of which normally looks like hummocky and bowl-shaped feature. Along the river banks, non-forested and developed areas, the accumulation area cannot be clearly seen as the displaced material is frequently removed after the event, either by river water or manually. That's why the delineation of accumulation area is quite challenging task.

Several material and methods have been used to identify landslide in natural terrain. The high-quality dataset includes, (0.07m) orthophoto, (0.25m) DTM which is further used to generate the LiDAR derivatives such as hillshade, color composite, 3D anaglyph, Slope, contour etc. to visualize landslide in multiple modelling view by using standard tools in the ArcGIS, ILWIS, SAGA-GIS, and QGIS Software. Landslide identification is performed by visual analysis and interpretation of the representation of the topographic surface on all the above-mentioned DTM derivatives. Each landslide is characterized based on type, activity and process in attribute table and systematically stored in geodatabase. The historical landslides are interpreted and mapped from historical images of Google pro for five years (2008, 2012, 2013, 2014, 2016).

Most of the time the dormant or inactive landslides are ignored while preparing landslide inventory database, which ultimately resultant as incomplete inventory database. This issue gains much more importance while dealing with landslides under thick forest cover and reactivated landslides (Razak *et al.*, 2014a; Mohamad *et al.*, 2015).

### 4.3 LiDAR Processing for Visualization

The importance of high resolution LiDAR image cannot be denied during identification and classification of landslide through Image visualization. The LiDAR-derived datasets are re-processed to improve the visualization in 2D and 3D models using open-source and commercial software's. The important most LiDAR-derivative for visualization is a default hillshade. Hillshade is a 3D representation of the surface in grayscale, with the relative position of sun, considered for shading the image. This function uses the altitude and azimuth properties (azimuth: 315, altitude: 45) to specify the sun's position. Hillshade image along with slope and curvature images are utilized for a better visualization and interpretation. In slope image, steep slope indicates the scarp or depletion zone and flatter slope can be indicated as body or accumulation zone. Moreover, in curvature image, concave slope denotes as denudation and scarp area and convex slope specifies debris or accumulation zone.

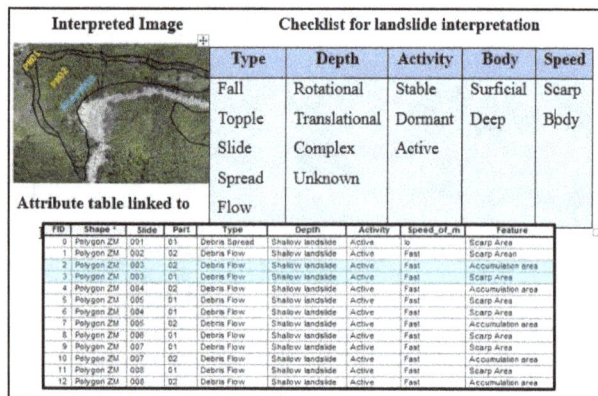

Figure 7: An example of an interpreted landslide map and its storage in GIS environment

An advanced mode of visualization is utilized known as topographic openness, generated in SAGA-GIS software from digital terrain models (DTM). The Positive values of openness signifies convex topography and negative values of openness shows concave topography (Yokoyama et al., 2002). Another important visualizing technique based on LiDAR-derived DTM is color composite image (Razak et al., 2011). This technique converts the one color DTM into a combination of three shades of color (red, green, and blue), indicating the solar illumination of the image in three directions compared to the conventional hillshade.

The virtual mapping of landslide inventories with the help of very high resolution (LiDAR) images can allow to map the insight detailed characteristics of each landslide, e.g., zone of accumulation and depletion (Figure 7). Whereas zone of depletion represents the scarp refers as erosional part while the zone of accumulation represents the body of the slide where the material is deposited. Hence it is very important to identify the topography, geomorphology, geology, LULC and past landslide events of the study area to have an understand with the regional landslide process, causal and conditioning factors to interpret a landslide. The identification of landslide body types is another essential element. Usually the shape of the body either it is accumulation part of scarp of the slide varies depending on the type and activity of the landslide. An active landslide generally has semi-circular shape with visible tension crack at the scarp. While old or dormant landslide mostly represented by a perfect semi-circular shape. However, in some cases, for complex landslide, the scarp has incomplete semi-circular shape because due to overlapped and erased by another scarp. To identify the accumulation area of the landslide it is important to count the

A high-resolution 3D orthophoto (0.7m) from LiDAR is used to identify recent landslide and observe the morphological- and geological features along with geo-indicators in forested terrain. These visualization techniques generate 2-D outputs for landslide interpretation. The 3D anaglyph image generated in SAGA-GIS by creating stereo pair images from DTM is another possibility of visualization and interpretation of landslide. It is used to visualize landslide in 3D by highlighting the depth of landslide to differentiate between deep-seated and shallow landslides using 3-D glasses. These visualization techniques increased the accuracy of inventory map.

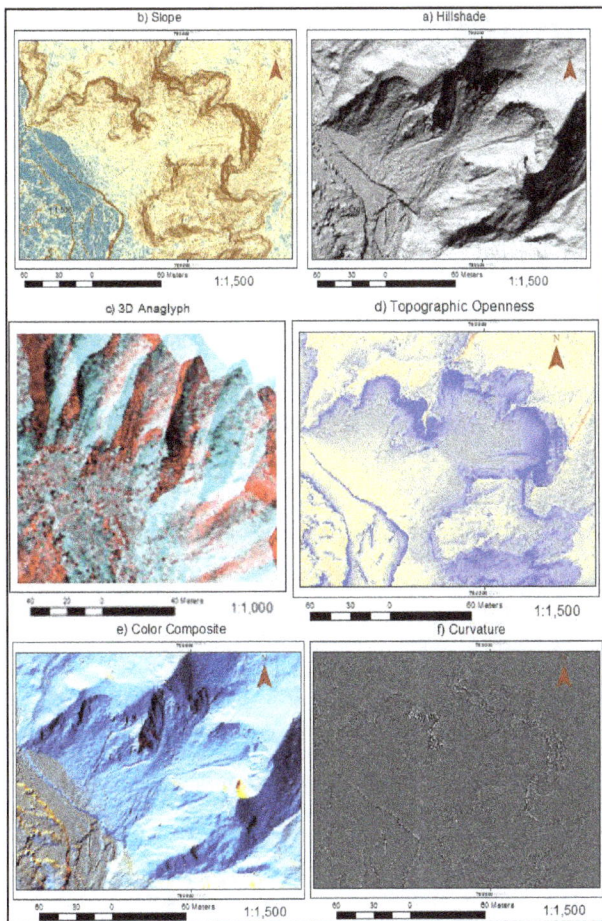

Figure 8: LiDAR derivatives used for interpretation and mapping of landslide inventory.

## 4.4 Landslide Interpretation through LiDAR Derivatives

The availability of high resolution data gains much importance in the visual interpretation of Landslides now days. All the LiDAR derivatives mentioned above are used to interpret the landslides precisely. Google pro images are used as data source to visualize the old or dormant landslides. In this study, a landslide geodatabase consists of landslide ID, body, type, activity and speed of movement, has been developed. Body attributes consist of scarp area, accumulation area (Figure: 7). Landslide types further divided in to shallow, flow, deep-seated, and rockfall. Under the Activity attribute the current state of landslide e.g. active, dormant, or stable is mentioned. The speed of movement of landslide divided in to two i.e. slow and fast. To interpret the landslides in the study area first geomorphology, geology, Landuse/ landcover (LULC), past landslide events, and topography of the study area were studied and mapped to understand regional landslide process in the area. The causal and conditioning factors were determined based on seismicity, climate, hydrology and topography. The parts of landslide body were interpreted and polygonised. Each landslide consists of two polygons as mentioned before (Figure: 7) though, the shape of polygon varies according to the type of the landslide. The Fresh, landslides generally have tension cracks at the scarp. Dormant or old landslide indicates a seamless semi-circular shape. The incomplete semi-circular shape of scarps indicates complex landslide because of overlapping and removal by the recent scarps. After delineating

the scarp of a landslide, the body or accumulation zone identified. The delineating accumulation area is quite challenging on steep slopes in the study area, because the accumulation or body is not distinctly seen. The displaced material frequently removed or washed out with rain or river etc.

## 5. RESULTS

The study area covers 5km middle channel of river Mesilou. The availability of high resolution LIDAR data, solved many issues and challenges faced by landslide inventory mapping in the past decades. The major finding of the study is that the "bare earth" images increased the visual interpretation of active as well as dormant landslides in dense forest many times. Indeed, the mapping and monitoring of tropical landslides with rapid growth of vegetation cover is not out of coverage now a day. The detailed mapping of forested terrain through ALS and TLS data improved the ability of mapping landslides inventories, which eventually effects the quality of susceptibility, hazard and risk maps in the country.

A detailed multi temporal landslide inventory is produces and stored in ArcGIS geodatabase by observing multi temporal satellite images and historical record as well as literature. It is concluded that the change in LULC is the main cause of landslides for the activation of landslides in the study area. The conversion of forest in agriculture is greatly contributing in the activation of landslide because of loose morphology of the area. The disconnected roads (Figure 9-A) along debris flow channel is very common impact of landuse change at Mesilau valley. The river Mesilau itself is contributing much in destabilizing the land by cutting the ground on meanders (Figure 9-B) and caused activation of landslides in most of the cases. (Figure 1) shows images of before, during and after the landslides event. It is desperately found that most of the houses are constructed either on the scarp or accumulation zone of the old landslides increasing the vulnerability to life and property. The damage in infrastructure e,g. roads and houses reveal the important of landslide inventory using VHR for land-use planning to avoid future life and property damages. From figure 10 It is very clear that the density and intensity of landslides is greater along the river bank and agriculture area as compare to the others. landslide inventory has been produced with the help of LiDAR-derivatives for the study area. The inventory is further upgraded by using multi temporal images from google pro, historical landslide records, news and literature etc. The visualization and interpretation techniques are based on plan curvature, hill shade image, color composite image, Slope image, topographic openness image, orthophoto and 3D anaglyph (Figure: 8).

These layers are used altogether to visualize and interpret the morphology of landslide e.g. scarp, body, concave-convex, type and topography. Though each individual image has some Pros and cons e.g. color composite image is brightening and its visual exposure is more as compare to the hillshade as it has shaded area in various perspectives. However, landslide scarp is sharply identifiable in topographic openness images (Figure 8). The visual interpretation of landslide is only possible through very high-resolution images as it is minor feature having significant impact on earth surface. LiDAR data is opening new horizons in the landslide investigations.

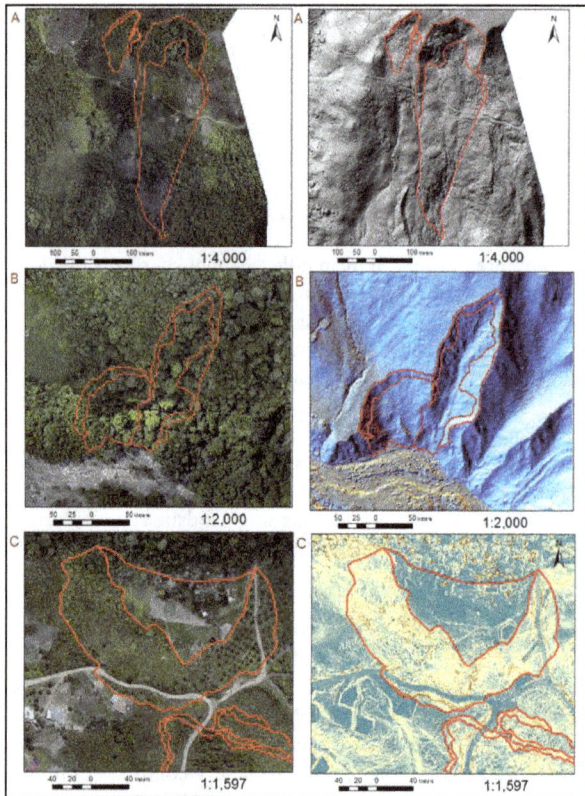

Figure 9: Diagnostic features of landslides in different morphological conditions.

Geomorphic features are playing key role in the interpretation of landslides e.g. from ortho photo fresh landslides can easily distinguished by observing vegetation pattern, river course, and anthropogenic features. Slope, color composite and topographic openness are good while interpreting the parts of landslide body. The color and bend of vegetation cover can also help in predicting landslides and for this purpose ortho photo play excellent role e, g. the tree density and pattern at the scarp and body have much different from each other. The vegetation color is darker in the scarp area as compare to body.

The tilt of the trees and the height of the vegetations are signs of fresh landslides. The difference of color in slope is also an

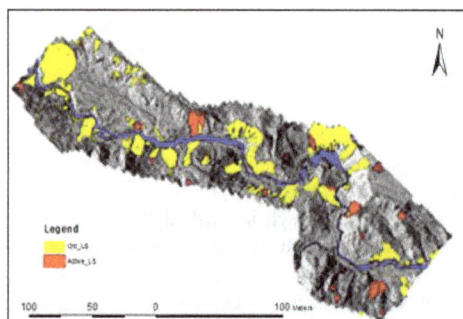

Figure 10: 3D illustration of old and new Landslides in part of Mesilau river valley.

indicator for recognizing scarp and body as the high and genital slope have distinct color tone. Another important predictor is curvature which is scale-dependent, and a suitable scale for

derivation of the geometry must be selected according to the scale of the features on ground to be detected. This method is very powerful to inevitably highlighting the location of shallow landslides and bank erosion, however by strong observation of interpreter it is possible to correctly recognize and outline the singularities. The brown patch is indicator of fresh slides on ground can easily be seen from very high resolution orthophotos. Consequently, there are numerous visualizing methods highlighting in the literature which can be helpful in the interpretation of landslides on a VHR images. The interpretation of LiDAR-derived datasets for landslide detection differs based on the accessibility and resolution of the dataset.

## 6. DISCUSSION AND CONCLUSION

From the observations and ground facts it is found that the slope failure in the research area is dominantly influenced by heavy rain followed by ground shaking incidences time to time. The major activity in the area is agriculture which is playing additional role in slope failure by making the material loose and pores. Landslide mapping via remote sensing techniques and are more efficient as compare to the field mapping in dense tropical forests and rugged mountains. It is an efficient way of producing and upgrading the event-based landslide inventory.
The interpretation for landslides utilizing various image visualization techniques is helpful in preparing a precise and update inventory as compare to rely on single interpretation technique e.g. hillshade image. The VHR LiDAR helps to visualize and interpret landslides precisely because of its high visibility for minor features. The virtual mapping of landslides is more efficient as compare to paper mapping in several ways e.g. i) it can have visualized on multi scales, ii) Maps can easily upgraded and printed any time, iii) rapid mapping of new events in case of emergency for disaster evacuation and relief etc.

Landslide interpretation via LiDAR-derivatives is not only efficient in interpretation of fresh and active event but also having the ability of identify the historical landslides e.g. the topographic openness is providing a very clear view of dormant morphological disturbances on even small scale. made possible to map and characterize recent landslide induced by earthquake, dormant, and historical landslides. It is also efficient in highlighting the disrupted drainages, small to medium crake's due to ground acceleration and cracked- and displaced roads. The addition of LiDAR dataset as primary tool, of field investigation is quite beneficial but still challenging due to its cost and sensitivity. It is proved from the study that the LiDAR dataset is highly beneficial in field investigations for interpreting and verifying landslides on various scales.
Based on landslide inventory it is very clear that the concentration of landslide is high in river incision areas, fault zones. The weak the lithology (Pinosuk Gravel) and presence of active faults (Mensaban Fault, Lobo-Lobo Fault) is a proof that the landslide activity in the Mesilau valley is dominantly controlled by structural and geological factors. The transport lines are more vulnerable in the area because of manual disturbance in to natural terrain for the construction of roads.

The paper presents the virtual mapping of landslide inventory in a loose morphology and tectonically active area in Mesilau, Kundasang (Borneo, Malaysia). It is an effort to contribute in disaster risk management activities facing by the community of Mesilau valley. The landslide inventory is mainly based on VHR airborne LiDAR coupled with temporal data and historical records in literature. LiDAR derivatives have transformed the assessment of hillslopes, morphology and structure in tropical conditions. Landslide interpretation and identification through LiDAR dataset along with geology, and topography revolutionized the landslide investigations. Field observation is getting easier and less time consuming by utilizing LiDAR data as field tool because of its wide-ranging characteristics in terms of scale, precision and exposure. There is still more room to enhance methods of preparing virtual landslide inventories by utilizing LiDAR datasets for disaster prone areas to cope with emergency situations. This study is an effort to capture the landslide activity in a part of Mesilau valley to contribute the inputs for planning and decision making at policy level for the mitigation action. As mentioned before that the change in LULC is destabilizing the geologically weak area of Mesilau valley that's why the landuse planning should be taken into consideration. The usefulness of landslide inventories for hazard, risk assessment and mitigation action cannot be denied.

## REFERENCES

Abellan A, Vilaplana JM, Calvet J, Blanchard J., 2010. Detection and spatial prediction of rockfalls by means of terrestrial laser scanning modelling. Geomorphology 119:162–171. doi:10.1016/j.geomorph. 2010.03.016

Ardizzone F, Cardinali M, Galli M, Guzzetti F, Reichenbach P., 2007. Identification and mapping of recent rainfall-induced landslides using elevation data collected by airborne Lidar. Nat Hazards Earth Syst Sci 7:637–650. doi:10.5194/nhess-7-637-2007

Bateson, L., Cigna, F., Boon, D., Sowter, A., 2015. The application of the Intermittent SBAS (ISBAS) InSAR method to the South Wales Coalfield, UK. Int. J. Appl. Earth Obs. Geoinf. 34, 249–257.

Bianchini, S., Cigna, F., Del Ventisette, C., Moretti, S., Casagli, N., 2013. Monitoring landslide-induced displacements with TerraSAR-X Persistent Scatterer Interferometry (PSI): Gimigliano Case Study in Calabria Region (Italy). Int. J. Geosci. 4 (10). http://dx.doi.org/10.4236/ijg.2013.410144

Bibi, T., Gul, Y., Rahman, A.A., Riaz, M., 2016. Landslide susceptibility assessment through fuzzy logic inference system (flis). Remote Sens. Spat. Inf. Sci. 42, 355e360. http://dx.doi.org/10.5194/isprs-archives-XLII-4-W1-355-2016. Booth AM, Roering JJ, Perron JT., 2009. Automated landslide mapping using spectral analysis and high-resolution topographic data: puget Sound lowlands, Washington, and Portland Hills, Oregon. Geomorphology. 109:132–147. DOI: 10.1016/j.geomorph.2009.02.027.

Brardinoni, F., Slaymaker, O., and Hassan, M. A.: Landslide inventory in a rugged forested watershed: A comparison between air-photo and field survey data, Geomorphology, 54, 3–4, 179– 196, 2003.

Carter W, Shrestha R, Tuell D, Bloomquist D, Sartori M., 2001. Airborne laser swath mapping shines new light on earth's topography. Eos, Trans, Am Geophys Union 82(46):549, 550, 555

Carrara, A., Cardinalli, M., Guzzetti, F. & Reichenbach P., 1995. GIS technology in mapping landslide hazard. Geographical Information Systems in Assessing Natural Hazards, 5, 135–175. DOI: 10.1007/978-94-015-8404-3_8.

Carnec C, Massonnet D, King C., 1996 Two examples of the use of SAR interferometry on displacement fields of small spatial extent. Geophys Res Lett 23(24):3579–3582. doi:10.1029/96GL03042

Chen RF, Lin CW, Chen YH, He TC, Fei LY., 2015. Detecting and characterizing active thrust fault and deep-seated landslides in dense forest areas of Southern Taiwan using airborne LiDAR DEM. Remote Sens. 7:15443–15466. DOI:10.3390/rs71115443.

Chigira M, Duan F, Yagi H, Furuya T., 2004. Using an airborne laser scanner for the identification of shallow landslides and susceptibility assessment in an area of ignimbrite overlain by permeable pyroclastics. Landslides. 1:203–209. DOI:10.1007/s10346-004-0029-x.

Ciampalini, A., Raspini, F., Frodella, W., Bardi, F., Bianchini, S., Moretti, S., 2015. The effectiveness of high-resolution LiDAR data combined with PSInSAR data in landslide study. Landslides, 1–12. http://dx.doi.org/10.1007/s10346-015-0663-5.

Colesanti C, Ferretti A, Prati C, Rocca F., 2003 Monitoring landslides and tectonic motions with the permanent scatterers technique. Eng Geol 68:3–14. doi:10.1016/S0013-7952(02)00195-3

Cruden,D.M., Varnes, D.J., 1996. Landslide Types and Processes, Special Report , Transportation Research Board, National Academy of Sciences, 247:36-75

Del Ventisette, C., Righini, G., Moretti, S., Casagli, N., 2014. Multitemporal landslides inventory map updating using spaceborne SAR analysis. Int. J. Appl. Earth Obs. Geoinf. 30 (1), 238–246.

EM-DAT., 2011. *The CRED International Disaster Database, Natural Disasters Trends.* http://www.emdat.be/naturaldisasters-trends.

Freeborough, K. A., Diaz Doce, D, Lethbridge, R, Jessamy, G, Dashwood, C, Pennington, C, and Reeves, H. J., 2016. Landslide Hazard Assessment for National Rail Network: The 3rd International Conference on Transportation Geotechnics (ICTG 2016), Volume 143, 2016, Pages 689–696.

Fruneau B, Achache J, Delacourt C., 1996. Observation and modelling of the Saint-E´tienne-de-Tine´e landslide using SAR interferometry. Tectonophysics 265(3–4):181–190. doi:10.1016/S0040-1951 (96)00047-9

Glade, T. & Crozier, M.J., 2005. The nature of landslide hazard and impact. *In*: Glade, T., Anderson, M. G. & Crozier, M.J. (Eds), *Landslide Hazard and Risk.* Wiley, London, pp. 43 - 74.

Gordon S, Lichti D, Stewart M., 2001 Application of a high-resolution, ground-based laser scanner for deformation measurements. In: Proceedings of the 10th international FIG symposium on deformation measurements, Orange, California, USA, 19–22 March 2001, pp 23–32

Greif, V., Vlcko, J., 2012. Monitoring of post-failure landslide deformation by the PSInSAR technique at Lubietova in Central Slovakia. Environ. Earth Sci. 66, 1585–1595. http://dx.doi.org/10.1007/s12665-011-0951-x.

Guzzetti, F., Mondini, A.C., Cardinali, M., Fiorucci, F., Santangelo, M., & Chang, K.-T., 2012. Landslide inventory maps: New tools for an old problem. *Earth-Sci. Rev.,* 112: 42-66. DOI: 10.1016/j.earscirev.2012.02.001.

Havlín, A., Bednarik, M., Magulová, B. & Vlčko J., 2011. Using logistic regression for assessment of susceptibility to landslides in the middle part of Chřib (Czech Republic) (in Czech). Acta Geologica Slovaca, 3(2), 153−161.

Haugerud R. A, Harding DJ, Johnson SY, Harless JL, Weaver CS, Sherrod BL., 2003. High-resolution lidar topography of the Puget Lowland, Washington—A Bonanza for earth science. GSA Today 13:4–10

Hölbling D, Füreder P, Antolini F, Cigna F, Casagli N, Lang S., 2012. A semi-automated object-based approach for landslide detection validated by persistent scatterer interferometry measures and landslide inventories. Remote Sens. 4:1310–1336. DOI:10.3390/rs4051310.

Hungr, O., Leroueil, S., & Picarelli, L., 2012. Varnes classification of landslide types, an update. In: Eberhardt, E., Froesse, C., Turner, A.K. & Leroueil, S. (eds) *Landslides and engineered slopes: protecting society through improved understanding,* vol 1. CRC Press, Boca Raton: 47–58

http://sam4605.blogspot.my/ (2009) map-kundasang-kundasang-war-memorial.html).

http//:thestar.com.my/news/nation/2015/06/26/things-go-sour-for-strawberry-farm-earthquaketriggered-landslide-brings-successful-business-crashing/ )

Ibsen, M. & Brunsden, D., 1996. The nature, use and problems of historical archives for the temporal occurrence of landslides, with specific reference to the south coast of Britain, Ventnor, Isle of Wight. *Geomorphology* 15: 241–258.

Jaboyedoff M, Oppikofer T, Abellán A, Derron MH, Loye A, M etzger R, Pedrazzini A., 2012. Useh of LIDAR in landslide investigations:    a    review.    Nat    Hazards.    61:5–28. DOI:10.1007/s11069-010-9634-2.

Jaboyedoff M, Oppikofer T, Locat A, Locat J, Turmel D, Robitaille D, Demers D, Locat P., 2009a. Use of ground-based LIDAR for the analysis of retrogressive landslides in sensitive clay and of rotational landslides in river banks. Can Geotech J 46:1379–1390. doi:10.1139/T09-073

Jaboyedoff M, Pedrazzini A, Horton P, Loye A, Surace I., 2008a., Preliminary slope mass movements susceptibility mapping using LIDAR DEM. In: Proceedings of 61th Canadian geotechnical conference, pp 419–426

JMG. *Geological Terrain Mapping of the Kundasang Area, Sabah.* 2003.

Jing, W., Yang, Y., Yue, X., and Zhao, X., 2016. A Comparison of Different Regression Algorithms for Downscaling Monthly Satellite-Based Precipitation over North China. Remote Sens. 2016, 8(10), 835; doi:10.3390/rs8100835.

K. Kraus, Photogrammetry: Geometry from Images and Laser Scans (de Gruyter, 2007).

Li X, Cheng X, Chen W, Chen G, Liu S., 2015. Identification of forested landslides using LiDar data, object-based image analysis, and machine learning algorithms. Remote Sens. 7:9705–9726. DOI:10.3390/rs70809705.

Lin ML, Chen TW, Lin CW, Ho DJ, Cheng KP, Yin HY, Chen MC., 2014. Detecting large-scale landslides using LiDar data and aerial photos in the Namasha-Liuoguey area, Taiwan. Remote Sens. 6:42–63. DOI:10.3390/rs6010042.

Mahalingam R, Olsen MJ., 2015. Evaluation of the influence of source and spatial resolution of DEMs on derivative products used in landslide mapping. Geomat Nat Haz Risk:1–21. DOI:10.1080/19475705.2015.1115431.

Mantovani, M., Devoto, S., Piacentini, D., Prampolini, M., Soldati, M., Pasuto, A., 2016. Advanced SAR interferometric analysis to support geomorphological interpretation of slow-moving coastal landslides (Malta, Mediterranean Sea). Remote Sens. 8 (6), 443.

Martha, T.R., Kerle, N., Jetten, V., van Westen, C., & Vinod Kumar, K., 2010. Characterising spectral, spatial and morphometric properties of landslides for semi-automatic detection using object-oriented methods. *Geomorphology* 116: 24–36.

Mohamad, Z., Razak, K.A., Ahmad, F., Abdul Manap, M., Ramli, Z., Ahmad, A., & Mohamed, Z., 2015. Slope hazard and risk assessment in the tropics: Malaysia's experience. *Geophys. Res. Abstr.,* 17: EGU2015- 7746.

Mondini, A.C., Chang, K.T., Yin, H.Y., 2011a. Combining multiple change detection indices for mapping landslides triggered by typhoons. Geomorphology 134 (3–4), 440–451. doi:10.1016/j.geomorph.2011.07.021

Omar S., Mohamed Z., Razak K.A., 2016. Landslide Mapping Using LiDAR in the Kundasang Area: A Review. In: Yusoff M., Hamid N., Arshad M., Arshad A., Ridzuan A., Awang H. (eds) InCIEC 2015. Springer, Singapore

Oppikofer T, Jaboyedoff M, Keusen HR., 2008. Collapse at the eastern Eiger flank in the Swiss Alps. Nat Geosci 1:531–535. doi:10.1038/ngeo258.

Pradhan B, Latif ZA, Aman SNA., 2012. Application          of airborne LiDAR-derived parameters and probabilistic-based frequency ratio model in landslide susceptibility mapping. AMM.                                         225:442–447. DOI:10.4028/www.scientific.net/AMM.225.442.

PWD (Public Works Department)., 2009. *National Slope Master Plan 2009 - 2023.* Public Works Department (PWD), Malaysia.

Petley, D.N., 2011. The landslide blog. http://blogs.agu. org/landslideblog/ .

Razak, K. A. and Mohamad, Z., 2015. Methodological Framework for Landslide Hazard and Risk Mapping using Advanced Geospatial Technologies: Malaysian journal of Remote Sensing and GIS. Vol. 4 Num. 2 Year 2015 ISSN: 1511-7049.

Razak, K.A., Mohamed, Z., Che Hasan, R., Aitin, A., Sheng, L.C., Abu Bakar, R., & Wan Mohd Akib, W.A.A., 2014a. Multi sensor lidar for hillslope geomorphology mapping: A step forward to geospatializing natural disaster in Malaysia. *7th AUN/SEED-Net Geological Engineering Conference (AGEC) and 2nd AUN/SEED- Net Natural Disaster Conference (ANDC)*, 29-30 September 2014. University of Yangon, Yangon, Myanmar.

Razak, K.A., Straatsma, M.W., van Westen, C.J. & de Jong, S.M., 2011a. Airborne laser scanning of forested landslides characterization: Terrain Model and visualization. *Geomorphology*, 126: 186-200.

Razak, K.A., van Westen, C.J., Straatsma, M.W., & de Jong, S.M., 2011b. Mapping of elements at risk for landslides in the tropics using airborne laser scanning. In: *FIG working week 2011: bridging the gap between cultures: Technical programme and proceedings*, Marrakech, Morocco, 18-22 May 2011. 16 p.

Roback, K., Clark, M. K., West, A. J., Zekkos, D., Li, G., Gallen, S. F., Chamlagain, D., Godt, J. W., 2017. The size, distribution, and mobility of landslides caused by the 2015 Mw7.8 Gorkha earthquake, Nepal, Geomorphology (2017), http://dx.doi.org/10.1016/j.geomorph.2017.01.030

Rott, H., 2009. Advances in interferometric synthetic aperture radar (InSAR) in earth system science. *Prog. Phys. Geogr.*, 33: 769-791.

Santangelo, M., Cardinali, M., Rossi, M., Mondini, A.C., Guzzetti, F., 2010. Remote landslide mapping using a laser rangefinder binocular and GPS. Natural Hazards and Earth System Sciences 10, 2539–2546. doi:10.5194/nhess-10-2539-2010.

Sarman M, Komoo I. *Kundasang-Ranau : Dataran Warisan Ais Gunung*. Geol. Soc. Malaysia Annu. Geol. Conf., 2000.

Scaioni M, Longoni L, Melillo V, Papini M., 2014. Remote sensing for landslide investigations: an overview of recent achievements and perspectives. Remote Sens. 6:9600–9652. DOI:10.3390/rs6109600.

Schulz WH., 2007. Landslide susceptibility revealed by LIDAR imagery and historical records, Seattle, Washington. Eng Geol. 89:67–87. DOI: 10.1016/j.enggeo.2006.09.019.

Singhroy V., 2009. Satellite remote sensing applications for landslde detection and monitoring. In: Sassa K, Canuti P (eds) Landslides—disaster risk reduction. Springer, Berlin/Heidelberg, pp 143–158

Slob S, Hack R., 2004. 3D terrestrial laser scanning as a new field measurement and monitoring technique. In: Engineering geology for infrastructure planning in Europe: a European perspective, Lectures Notes in Earth Sciences, Springer, Berlin/Heidelberg, 104:179–189

Squarzoni C, Delacourt C, Allemand P., 2003. Nine years of spatial and temporal evolution of the La Valette landslide observed by SAR interferometry. Eng Geol 68:53–66. doi:10.1016/S0013-7952(02)00198-9

Stow R., 1996 Application of SAR interferometry to the imaging and measurement of neotectonic movement applied to mining and other subsidence/downwarp modeling. In: Fringe 96 proceedings ESA workshop on applications of ERS SAR interferometry, Zurich, Switzerland.

Tarchi D, Casagli N, Fanti R, Leva DD, Luzi G, Pasuto A, Pieraccini M, Silvano S., 2003. Landslide monitoring by using ground-based SAR interferometry: an example of application to the Tessina landslide in Italy. Eng Geol 68:15–30. doi:10.1016/S0013-7952(02)00196-5

Teza G, Galgaro A, Zaltron N, Genevois R., 2007. Terrestrial laser scanner to detect landslide displacement fields: a new approach. Int J Remote Sens 28:3425–3446. doi:10.1080/01431160601024234.

Tjia HD. *Kundasang ( Sabah ) at the intersection of regional fault zones of Quaternary age*. Geol Soc Malaysia 2007;53:59–66.

Van Den Eeckhaut, M., Kerle, N., Poesen, J., & Hervas, J., 2012. Object-oriented identification of forested landslides with derivatives of single pulse LiDAR data. *Geomorphology*, 171-174: 30-42.

Van Den Eeckhaut, M., Muys, B., Verstraeten, G., Vanacker, V., Nyssen, J, Moeyersons, J., Van Beek, L.P.H., & Vandekerckhove, L., 2007. Use of LIDAR-derived images for mapping old landslides under forest. *Earth Surf. Proc. Land.*, 32: 754-769.

Vallet J, Skaloud J., 2004. Development and experiences with a fully-digital handheld mapping system operated from a helicopter. Int Archi Photogramm Remote Sens 35(B5):791–796

Van Westen CJ, Van Asch TWJ, Soeters R., 2006. Landslide hazard and risk zonation—why is it still so difficult? Bull Eng Geol Environ 65:67–184. doi:10.1007/s10064-005-0023-0

Wang G, Joyce J, Phillips D, Shrestha R, Carter W., 2013. Deli neating and defining the boundaries of an active landslide in the rainforest of Puerto Rico using a combination of airborne and terrestrial lidar data. Landslides. 10:503–513. DOI:10.1007/s10346-013-0400-x.

Wills, C. J. and McCrink, T. P.: Comparing landslide inventories: The map depends on the method, Environ. Eng. Geosci., 8, 4, 279–293, 2002.

# 19

# PERCEPTION MODELLING OF VISITORS IN VARGAS MUSEUM USING AGENT-BASED SIMULATION AND VISIBILITY ANALYSIS

Bienvenido G. Carcellar III [a]

Department of Geodetic Engineering, College of Engineering, University of the Philippines – Diliman
(bgcarcellar@up.edu.ph[a])

**KEY WORDS:** GIS, Perception Modelling, Agent-Based Simulation, Museum Management, GAMA

**ABSTRACT:**

Museum exhibit management is one of the usual undertakings of museum facilitators. Art works must be strategically placed to achieve maximum viewing from the visitors. The positioning of the artworks also highly influences the quality of experience of the visitors. One solution in such problems is to utilize GIS and Agent-Based Modelling (ABM). In ABM, persistent interacting objects are modelled as agents. These agents are given attributes and behaviors that describe their properties as well as their motion. In this study, ABM approach that incorporates GIS is utilized to perform analyticcal assessment on the placement of the artworks in the Vargas Museum. GIS serves as the backbone for the spatial aspect of the simulation such as the placement of the artwork exhibits, as well as possible obstructions to perception such as the columns, walls, and panel boards. Visibility Analysis is also done to the model in GIS to assess the overall visibility of the artworks. The ABM is done using the initial GIS outputs and GAMA, an open source ABM software. Visitors are modelled as agents, moving inside the museum following a specific decision tree. The simulation is done in three use cases: the 10%, 20%, and 30% chance of having a visitor in the next minute. For the case of the said museum, the 10% chance is determined to be the closest simulation case to the actual and the recommended minimum time to achieve a maximum artwork perception is 1 hour and 40 minutes. Initial assessment of the results shows that even after 3 hours of simulation, small parts of the exhibit show lack of viewers, due to its distance from the entrance. A more detailed decision tree for the visitor agents can be incorporated to have a more realistic simulation.

## 1. INTRODUCTION

### 1.1 Background

Visiting museums is one of the activities people do to learn more about a subject area such as history, art, or science. In most cases, museums have a number of exhibits and artwork. The bottom line in visiting museums is that the visitors want to maximize their viewing experience given some time constraint. The viewing experience of the visitors often depend on the time restrictions, their interests, the space of the museum, and mostly the tiredness that occurs when touring around the museum. (Lykourentzou, et al, 2013)

For the visitors, their goal is to maximize their viewing experience given their available time. For the museum managing team, they want to aid visitors in achieving this high quality of experience, as well as making sure that the artwork exhibits are viewed by the visitors. (Lykourentzou, et al, 2013)

To effectively analyze if the installed artworks are well arranged spatially, perception modelling of museum visitors can be done. In this concept, the visitors and the artworks are modelled inside a simulated environment. The visitors will decide which artwork to view next mainly based on their instantaneous perceived area.

### 1.2 Objectives and Expected Output

This paper aims to develop a model for assessment of visitor perceptions on the installed artworks in a museum. This model will be implemented using GAMA, an agent-based modelling software capable of incorporating GIS in BIM. This model will be tested to the chosen study area, which is the Vargas Museum.

Specifically, this study will create a 3-Dimensional GIS model of the second floor of the Vargas Museum. The model will also utilize an agent based simulation of the flow of the visitors, as well as to model the visitor count for each art work. Lastly, the model aims to generate visibility heat maps that can be used as an aid to the management of the Museum.

In this paper, Visibility Analysis and Agent Based Simulation are performed over the main exhibit of Vargas Museum in order to assess the perception of each of the artwork of the exhibits.

### 1.3 Scope and Limitations

This paper will be modelling the flow and perception of the visitors of the second floor of the Vargas Museum, which was the main exhibit of the museum before its renovation. The research aims to model the flow of visitors in utilizing GAMA platform version 1.7.

The 3D Model of the second floor, as well as the 2D model was generated from the given floor plan from the University Office of the Campus Architect. The overall CAD file, however only contains bare floor plan. It does not include the locations of the exhibits and artworks. The location of the art installations were ground validated and placed in the floor plan by the researcher.

The constraints and parameter values used in the simulation were gathered from the summary of the Art Perception Survey results, from a sample size of 31 respondents. The simulation parameter values were consulted to Ms. Veronica Fuentes (museum assistant) and Dr. Patrick D. Flores (museum curator) last May 20, 2016.

## 2. REVIEW OF LITERATURE

### 2.1 Visitor Movement in Museums

Numerous literatures have been done to further assess the experience of users in museum visits. The GPS has been one of the main technologies used in providing the museum managers visitor data (Roes, 2010). In most cases, GPS was utilized in indoor navigation as a way to gather visitor movement data in a museum to help managers in giving path recommendations to visitors. (Lykourentzou, et al, 2013)

Visitor movement was earlier categorized by Veron and Levasseur, around 1983. From this research, they identified the four main visiting styles namely the ant, fish, grasshopper, and the butterfly (Veron, Levasseur, 1983). The identified four visiting styles were mostly studied by different researches may it be in visualization (Chittaro, et al, 2004), or in utilizing the information for a museum guide development (Bianchi et al, 1999) (Gabrielli et al, 1999) (Zancanaro et al, 2007) (Hatala et al, 2005)

This four main visitor types were then modelled into an Agent-Based Simulation in the BLUE project. (Lykourentzou, et al, 2013). The identified models were the museum and the visitors. Each visitor type are modelled according to their movement, as well as quantifying the quality of experience of the visitors according to different social factors such as interests, crowd tolerance, and stamina.

In this research, the agent-based simulation is also utilized, but with a simpler approach in modelling the movement of the visitors.

### 2.2 Agent-based Modeling and GIS

One of the advantages of agent-based approach compared to a simple GIS spatial analysis is the integration of the behavioral aspect of the agents and the time element. Although GIS is very useful in representing and analyzing geospatial data, spatial data are mostly modelled in their basic form, mostly static in nature. Coupling GIS with ABM can maximize the ability of the model to fully abstract reality. But the linkage also depends on the modelling endeavor. (CASA UCL, 2006)

Agent-based relies on the principle that the environment is described by the actors or the agents that play within it (Bonabeau, 2002). These agents are autonomous in nature, can process information, and interact with other agents given a set of behaviors. The interaction also adapts to the environment and agents behave differently and independently. (CASA UCL, 2006)

One important concept of ABM is that mobility of agents are one of the useful features. The agents can have a behavior of moving at different speeds given different cases of environment factors. (CASA UCL, 2006). In this project, mobility and intellectual ability of the agents are the main factors considered in modelling the agents. Perception of visitors differ every change in their location. Thus, this must be incorporated in the ABM model to fully realize the perceived area of a visitor.

### 2.3 GAMA Simulation Software

GIS Agent-based Modeling Architecture, or simply GAMA is an agent based simulation platform which can easily incorporate spatial data to the environment to be modelled. The usual vector data in shapefile can be easily incorporated in the simulation, as well as its attributes. In this way, the model utilizes both the spatial strength of a GIS software and the complexity of the ABM software. (GitHub, 2015)

GAMA has its own scripting language which is GAML. A single model can contain multiple interacting agents that can be static or moving. Agents are modelled using the GAML to add agent behaviors and attributes. GAMA also contains support for data visualization in the form of creating different types of charts. (Taillandier, 2012)

## 3. METHODOLOGY

The main method has three stages: the data pre-processing stage, the visibility analysis, and the agent based modelling stage, as shown in Figure 1. The specific methods for each part will be thoroughly discussed in the following sections.

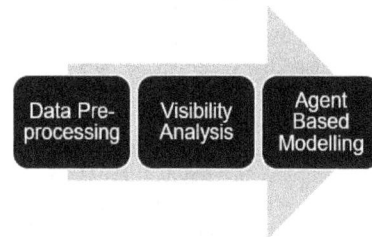

Figure 1. The main workflow

Figure 2 below shows a much detailed flowchart showing the interactions of the inputs and the outputs of each part to generate the visitor perception model. As seen in Figure 2, the floor plan and the on-site data are the main data needed to be inputted in the process. These information are to be preprocessed to be inputted to both the visibility analysis and the ABM. The output of the visualization model is then inputted in the simulation model to as one of the artwork attributes. After building the environment, the simulation will be generated over the environment to generate needed information for the perception model.

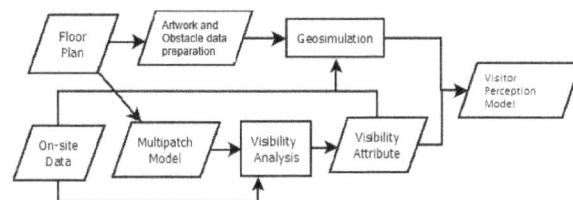

Figure 2. The general methodology

### 3.1 Study Area

Figure 3. Perspective view of Vargas Museum

Jorge B. Vargas Museum and Filipiniana Research Center, shown in Figure 3, is the in house Museum of the University of the Philippines – Diliman. The museum exhibits various art repositories even dating back to the early 1880s to the 1960s. It houses historical, as well as educational art works such as oil paintings, pastels, watercolors, drawings, and sculptures. One of its main agenda is to contribute to the Filipino heritage in its artistic aspect (Wordpress, 2015)

The building was erected in 1987, housing mostly the works of Mr Vargas. The building has a total of four floors. The basement is used mostly as an activity center for community arts programs and different workshops. The ground floor is the main access of the museum. Aside from the main reception, the lobby functions as a temporary exhibit, location for art launching and lectures, and other temporary activities. The second floor houses the main gallery of the museum wherein the artworks installed are mostly permanent. Aside from the normal art exhibit area, the third floor houses the historical archives and library are stored. The lobby area is also used as space for photography and contemporary art exhibits. (Wordpress, 2017)

Currently the second floor, which was used mainly for the permanent exhibits, is under renovation. The artworks initially installed in the main gallery are now located across the lobby of the third floor, as well as the first floor. (Wordpress, 2017) This study will implement an ABM geosimulation on the second floor of the said building to assess the previous configuration of its installations.

### 3.2    Environment Modelling

The building floor plan from the Office of the Campus Architect is used as a basemap of the generation of the 2D floorplan. Other information not included in the available CAD file such as the placement of the artworks and panel boards, are digitized manually. After adding the information from site visits, the 2D floor plan shapefile was generated. Figure 4 below shows the generated shapefile of the floorplan as well as the location of the artworks

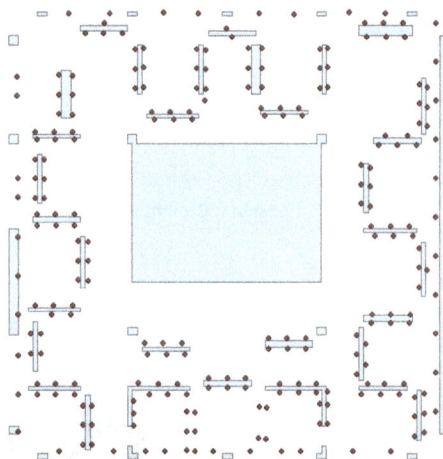

Figure 4. 2D Floor plan and artwork locations

The 3D model of the second floor was generated from the 2D model using the wall height attributes. The generated 3d layer was then converted to a multipatch, as shown in Figure 5. This multipatch model will be utilized in the visibility analysis.

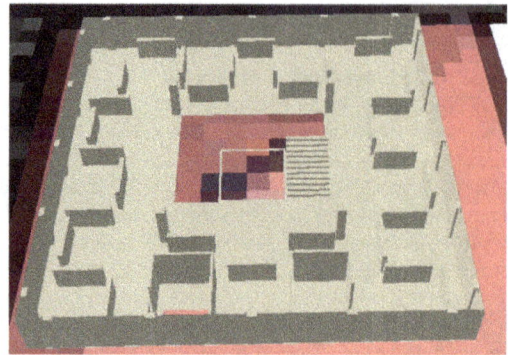

Figure 5. 3D model of the second floor.

### 3.3    Visibility Analysis

Here, the visibility analysis tests the overall visibility of the artworks from a generated 1 meter interval target points. Initially, the artwork locations identified earlier are set as source points while the target points are generated using fishnets. The *vispercent,* the value which determines the overall visibility was computed based on Equation 1 shown below.

$$vispercent = \frac{totvis}{targets} \qquad (1)$$

where         totvis = total number of visible targets
                    targets = total number of target points

Since the source points have height attributes in nature, the initial artwork locations were converted to a 3D point with height attribute. The assumed height of the artworks is 1.3 meters.

The target points serve as the location of the eye level of the visitors once they are standing in that location. The assumed height of the target points is 1.5 meters from the ground, as displayed in Figure 6. After generating these points, they are now subject to the line of sight analysis, utilizing the generated 3D model as the obstruction features. After which, equation 1 was used to solve for the overall visibility of each artwork.

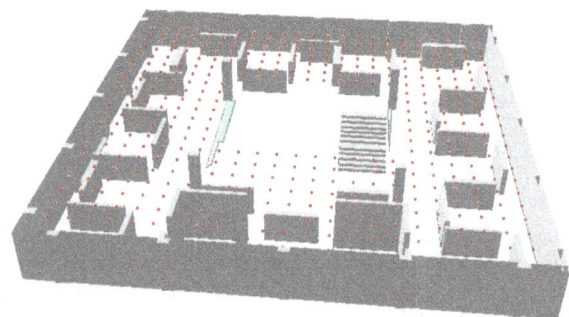

Figure 6. Source and Target points for the visibility analysis

### 3.4    Agent Based Modelling

The third part is the agent based modelling in GAMA. In this part, the 2D floorplan and the artwork locations with overall visibility attribute was used for the model.

Four main agents are identified that will take part in the simulation: Obstacles, artworks, people, and exits. Unlike the other three agents, the obstacle agent is an agent without an attribute. The other three agents will be discussed in the next sections. The Figure 7 shows the interrelationship of the four agents in a UML Diagram.

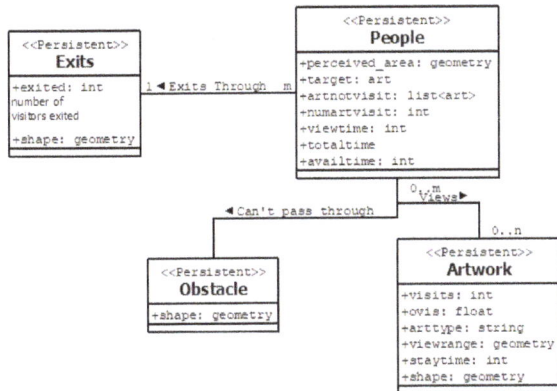

Figure 7. Class Diagram of the agents in the environment

### 3.4.1    Exit Agent

The exit agent is the designated location at which visitors will enter and leave the main exhibit hall. This agent's main goal is to count the total number of agents entering and leaving the scenario.

This agent has one attribute, which is the *exited*. This attribute counts the total number of visitors exited in that exit. The agent is also closely dependent on the global parameter *chance* which will be explained in section 3.4.4.

### 3.4.2    Artwork Agent

This agent represents the artworks in the exhibit area. The location of the artworks are based from the artwork location shapefile generated earlier in section 3.2. Each artwork agent has four attributes, which are *visits, ovis, viewrange, and staytime*. *Visits* attributes counts the number of visitors for each artwork. *Ovis* is the overall visibility attribute generated from the visibility analysis in section 3.3. *Viewrange* is a geometry type of attribute which corresponds to the minimum distance from the artwork a person can view that artwork. This is represented as a circle centered at the location of the artwork with 1 meter radius. Lastly, the *staytime* attribute is the viewing time of visitors on each artwork. This value is generated randomly from 50 to 90 seconds

### 3.4.3    People Agent

The people agent models the visitors as well as their perceived area as they move. Each people agent has a total of eight attributes which is updated for every cycle. *Perceived_area* is a geometry type of attribute representing the perceived area of a visitor. This area is default to be a cone with ±30 degrees view angle based from the heading direction of the visitor, and a radius of 5 meters. *Target* attribute is the art agent at which a visitor aims to view at that instant. *Artnotvisit* pertains to a list of artwork agents not yet visited by the people agent. The people agent will choose potential targets from this list. In case the visitor views a target artwork, it is removed to the list. *Numartvisit* counts the number of visited artworks. *Viewtime* tracks the number of seconds a visitor is viewing the target artwork. *Totaltime* tracks the total time of the visitor in the museum. Availtime is the total available time of the visitor in

the museum. This value is randomized from 4800 seconds to 6420 seconds. Lastly, *speed* is the built in attribute of moving agents. This value varies from 0.2 to 0.6 m/s depending on the state of the visitor agent.

The people agent is the only agent that has the ability to move. The movement, however, has a set of instructions the visitor will follow. Figure 8 shows the sequence of commands the visitor will follow when navigating inside the museum.

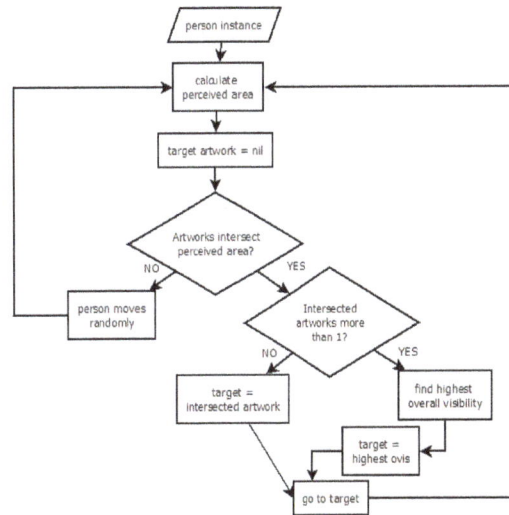

Figure 8. The flowchart showing the sequence of instructions for people agent mobility.

The visitor's perceived area attribute is updated every iteration to model the changing location of the people agent. The agent will end the visit once the available time attribute of the person is less than the total time attribute. By that time, no more targets will be set and the agent will go straight to the exit agent nearest to it.

### 3.4.4    Global Parameters

To help initialize the simulation, a museum art preference survey was performed to gather information on the parameter values such as the average length of stay and the average viewing time per artwork. Figure 9 shows a part of the online survey done.

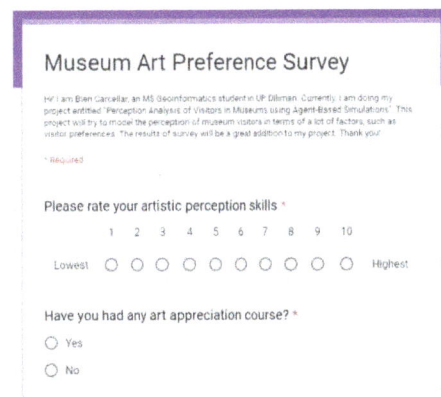

Figure 9. A screenshot of the online museum art preference survey

From the summary of the conducted survey, the parameter

values are generated. Perception distance parameter is set to 5 meters. The perceived area angle is set to 60 degrees. Speed of people agents vary from 0.2 to 0.6 m/s depending on their state. The tourist available time is also a random attribute from 1 hour 20 minutes up to 2 hours. The artwork viewing time attribute is also randomly selected from the range 50-90 seconds.

Another parameter, which is the chance, is included in the simulation. This chance parameter determines whether or not a new visitor will enter the museum in the next minute or not. It can also be interpreted as the probability that a new visitor will enter the museum in the next 60 seconds. Three values of chance parameter was used (10%, 20%, and 30% chance), corresponding to the three simulations performed as shown in the next section.

## 4    RESULTS AND DISCUSSIONS

### 4.4    Visibility Analysis

The following image shows the visibility analysis results. The green cells represent high value of overall visibility while the red cells represent the low visible artworks.

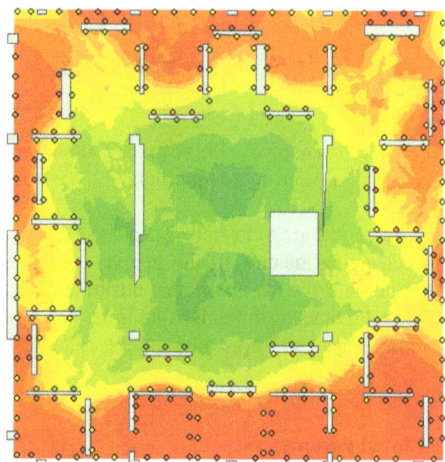

Figure 10. Overall visibility of artworks

Observing Figure 10, artworks near the center of the exhibit has the highest visibility from the target points. The placement of the panel boards significantly lowered the visibility of the artworks behind it. Although that fact is expected, it is observed that the areas mostly affected by these positioning of the panel boards are the southern portion of the exhibit. Even though this section is closer to the entrance as compared to the other artworks, the walls and the panel boards made the artworks in this location to have little to low visibility. Further checking the initial floor plan, it is found out that the southern area is a former comfort room and was just recently reconstructed as exhibit area for contemporary art. Walls and dividers from the previous configuration were retained, causing the artworks in this area not readily visible to the visitors entering the museum.

### 4.5    Simulation in GAMA Software

The dataset outputs from the previous processes were imported into the simulation software. The simulation is run for the three cases: 10%, 20% and 30% chance. Screenshots of the actual simulation is shown in Figure 11.

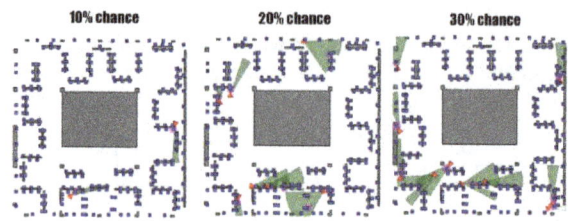

Figure 11. Screenshot of the simulation for 10%, 20%, and 30% chance respectively.

Two data charts were also displayed, along with the spatial display: the number of visitors graph, and the average visitors per artwork graph. Figure 12 shows the results of the number of visitors charts for all three chance use cases.

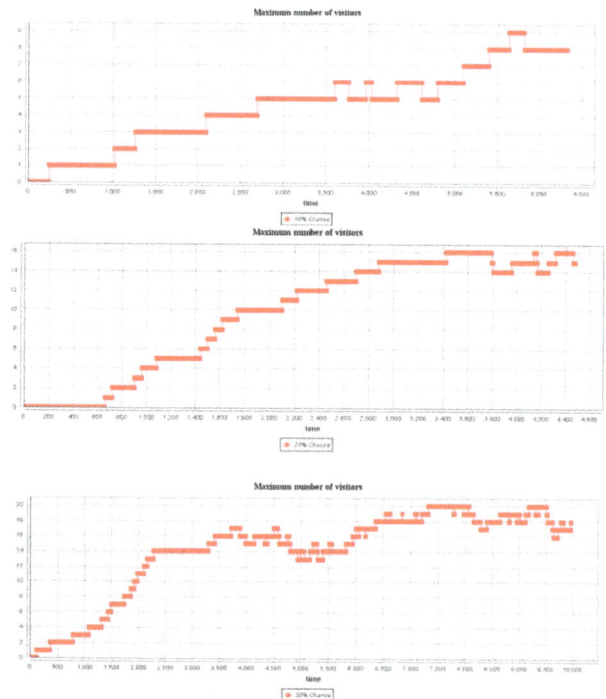

Figure 12. Chart of total visitors for the three chance cases

Summarizing the results from the graph in Figure 13, a maximum of 9 visitors for the 10% chance model was observed, 16 visitors for the 20% chance model, and 20 visitors for the 30% chance model. From these values, it is inferred that among the three implementation, the closest model to the actual is the 10% model with maximum of 9 people, as the Vargas Museum usually has around 5-10 visitors at a time on the average (V. Fuentes, personal interview, May 20, 2016).

Aside from the total visitors, a chart of the average visitors per artwork was also displayed. The following images show the charts for all three cases.

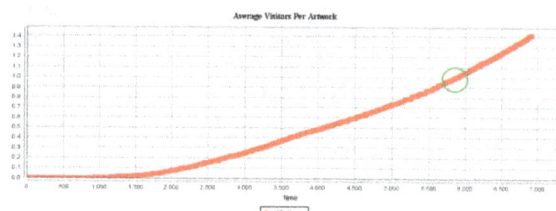

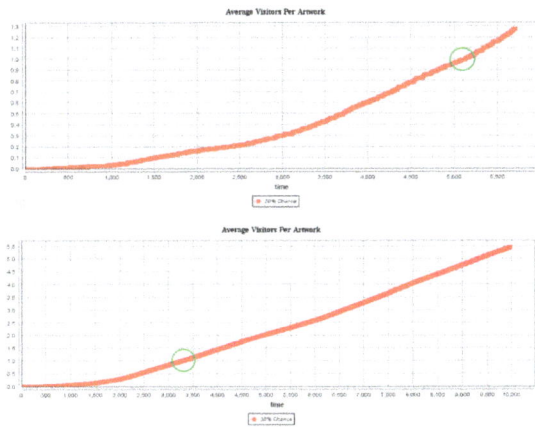

Figure 13. Average number of viewers per artwork for 10%, 20%, and 30% chance respectively.

From Figure 13, the green circles show the time for the average visitors per artwork to be 1. This means that after that identified time, on the average, the artworks will have a minimum exposure from the visitors. The identified time for the 10% chance case is 1 hour and 40 minutes. For the 20% case, the time is 1 hour and 20 minutes, while for the 30% chance, the time is 55 minutes. These values can be used as recommended minimum viewing time for the exhibit so that each artwork will have at least one viewer.

Aside from the two graphs, heatmaps were also generated to spatially examine the viewing patterns. Since the identified closest model to the actual is the 10% chance, the heatmaps were generated only for the 10% chance. Three heatmaps were generated, once every hour of simulation and for three hours. Figure 14 shows the third hour heatmap. Red cells show the least viewers while greener cells show high viewership.

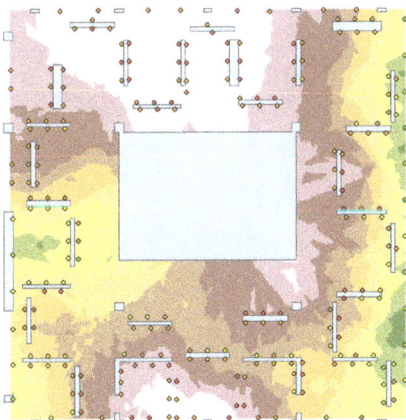

Figure 14. Heatmap of viewing patterns after three hours

Observing the three heat maps generated from the 10% chance models, it is spotted that after three hours, artworks from the eastern and western part of the exhibit area are viewed at least once. Areas such as the upper part of the exhibit shows the least number of visitors. This may be attributed to the fact that this part is the farthest from the entrance of the main exhibit. Southern part of the exhibit area has low perception results. This part has been a former comfort room, and has been consistently low in visibility value as well.

# 5  CONCLUSIONS AND RECOMMENDATIONS

## 5.4  Conclusions

This research project aims to create a visitor perception model using visibility analysis and ABM. From the three chance models generated from the model, it is observed that the 10% Chance simulation model is considered as the closest simulation model to the current Vargas Museum visitor profile of around 5-10 visitor count at a time.

From the 10% chance model, the average recommended minimum viewing time is around 1 hour and 20 minutes. For maximum viewing experience and maximum perception over all art exhibits, the recommended minimum time limit of museum viewing is 1 hour and 20 minutes.

As seen in the results, it is observed that GAMA is a potential tool in modelling social behavior scenarios particularly in art perception and preferences. Incorporating GIS and Agent Based Simulation is one important aspect in GAMA which can be utilized for spatial analysis and applications. This has been one of the advantages of GAMA, being able to easily incorporate GIS dataset to the simulation.

In general, Agent-Based Models such as these are helpful in scenarios which require to model actual movement of people. Such model can be used for management of visitors, as well as positions of artwork for maximizing viewing experience. One can get insights about the effectiveness of the current spatial distribution of artworks, and formulate possible museum guidelines for planning exhibit layouts.

## 5.5  Recommendations

This study models the social behavior of the visitors as a simple choice using the overall visibility as the sole parameter. This is however not the real case since a lot of factors may come into play into a visitor's mind when viewing an artwork. Factors such as visitor preferences, and art appreciation level of people can be used as one way of creating a more complex decision making visitors. Multi-Criteria Decision Making (MCDM) Analysis can be incorporated to this model for more realistic approach.

This study also modelled visitors as a single unit, meaning, each visitor enters the museum individually. The model can be more realistic if more complex visitor to visitor relationship (such as group movement, companionships) can be added in the model. This will be an added factor to the decisions of the visitors.

In this study, the agent-based model was utilized using a single set of constraint values. For future works, others may try to implement the model in different constraint values to explore more on the similarities and differences of each use case. Validating the model using methods other than the ones used in this research can be explored to further test the model generated.

## ACKNOWLEDGEMENTS

The researcher would like to thank Dr. Ariel C. Blanco and Asst. Prof. Edgardo Macatulad for their contribution to the conception, development, and completion of this research project. Also, the researcher would like to express gratitude to the people and agencies provided the needed datasets such as to the University of the Philippines Campus Architect and Vargas Museum management, and the respondents of the Art Perception Survey form.

## REFERENCES

Bianchi, A., and Zancanaro, M.: Tracking users' movements in an artistic physical space. In: The i³ Annual Conference, Siena, Italy, October 20-22, pp. 103–106 (1999). Web

Chen, X., and F. B. Zhan. "Agent-based Modelling and Simulation of Urban Evacuation: Relative Effectiveness of Simultaneous and Staged Evacuation Strategies." J Oper Res Soc Journal of the Operational Research Society 59.1 (2006): 25-33. Web.

Chittaro, L., and Ieronutti, L.: A visual tool for tracing users' behavior in virtual environments. In: AVI 2004: Proceedings of the working conference on Advanced visual interfaces, pp. 40–47. ACM, New York (2004). Web

Gabrielli, F., Marti, P., and Petroni, L.: The environment as interface. In: The i 3 Annual Conference, Siena, Italy, October 20-22, pp. 44–47 (1999). Web.

Hatala, M., and Wakkary, R.: Ontology-based user modeling in an augmented audio reality system for museums. User Modeling and User-Adapted Interaction (3-4), 339–380 (2005). Web.

I. Roes, Personalized Museum Tour with Real-Time Adaptation on a Mobile Device with Multi-Point Touch

Interface, Mater Thesis, Eindhoven University of Technology, 2010. Web.

K. Sookhanaphibarn and R. Thawonmas, A Movement Data Analysis and Synthesis Tool for Museum Visitors' Behaviors, in: Proceedings of the 10th Pacific Rim Conference on Multimedia: Advances in Multimedia Information Processing, Springer-Verlag, Bangkok, Thailand, 2009. Web.

Macatulad, Edgardo G. "3D GIS Based Multi-Agent Geosimulation Model for Building Evacuation." Thesis. National Graduate School of Engineering, University of the Philippines - Diliman, 2015. Print.

Macatulad, E. G., and A. C. Blanco. "3DGIS-Based Multi-Agent Geosimulation and Visualization of Building Evacuation Using GAMA Platform." Int. Arch. Photogramm. Remote Sens. Spatial Inf. Sci. ISPRS - International Archives of the Photogrammetry, Remote Sensing and Spatial Information Sciences XL-2 (2014): 87-91. Web.

"Principles and Concepts of Agent-Based Modelling for Developing Geospatial Simulations." Academia.edu. Centre for Advanced Spatial Analysis University College London, Sept. 2006. Web. 27 May 2016.

"The Building and Galleries." Wordpress. University of the Philippines, Jose B. Vargas Museum and Filipiniana Research Center, nd. Web. 29 May 2017.

Taillandier, Patrick, Duc-An Vo, Edouard Amouroux, and Alexis Drogoul. "GAMA: A Simulation Platform That Integrates Geographical Information Data, Agent-Based Modeling and Multi-scale Control." Principles and Practice of Multi-Agent Systems Lecture Notes in Computer Science (2012): 242-58. Web.

Veron, E., Levasseur, M.: Ethnographie de l'Exposition. Bibliothque publique d'Information, Centre Georges Pompidou, Paris (1983). Web.

Zancanaro, M., Kuflik, T., Boger, Z., Goren-Bar, D., Goldwasser, D.: Analyzing museum visitors' behavior patterns. In: Conati, C., McCoy, K., Paliouras, G. (eds.) UM 2007. LNCS (LNAI), vol. 4511, pp. 238–246. Springer, Heidelberg (2007). Web.

# 20

# 3D VISUALIZATION OF MANGROVE AND AQUACULTURE CONVERSION IN BANATE BAY, ILOILO

Gerard. A. Domingo, Mayann M. Mallillin, Anjillyn Mae C. Perez[a], Alexis Richard C. Claridades[a, b, c], Ayin M. Tamondong[a, b]

[a]Department of Geodetic Engineering, University of the Philippines, Diliman, Quezon City 1101
[b]Phil-LiDAR 2 Program Project 2: Aquatic Resources Extraction from LiDAR Surveys, University of the Philippines
[c]acclaridades@up.edu.ph

**KEY WORDS:** Mangroves, Aquaculture, Change Detection, Support Vector Machine Classification, 3D Visualization, Perception Survey

**ABSTRACT:**

Studies have shown that mangrove forests in the Philippines have been drastically reduced due to conversion to fishponds, salt ponds, reclamation, as well as other forms of industrial development and as of 2011, Iloilo's 95% mangrove forest was converted to fishponds. In this research, six (6) Landsat images acquired on the years 1973, 1976, 2000, 2006, 2010, and 2016, were classified using Support Vector Machine (SVM) Classification to determine land cover changes, particularly the area change of mangrove and aquaculture from 1976 to 2016. The results of the classification were used as layers for the generation of 3D visualization models using four (4) platforms namely Google Earth, ArcScene, Virtual Terrain Project, and Terragen. A perception survey was conducted among respondents with different levels of expertise in spatial analysis, 3D visualization, as well as in forestry, fisheries, and aquatic resources to assess the usability, effectiveness, and potential of the various platforms used. Change detection showed that largest negative change for mangrove areas happened from 1976 to 2000, with the mangrove area decreasing from 545.374 hectares to 286.935 hectares. Highest increase in fishpond area occurred from 1973 to 1976 rising from 2,930.67 hectares to 3,441.51 hectares. Results of the perception survey showed that ArcScene is preferred for spatial analysis while respondents favored Terragen for 3D visualization and for forestry, fishery and aquatic resources applications.

## 1. INTRODUCTION

### 1.1 Background of the Study

Mangrove forests in the Philippines, despite offering social, economic, and ecological benefits and functions, have been reduced drastically due to conversion to fishpond, salt ponds, reclamation, and other forms of industrial development (Melana, et. al., 2000). Aquaculture development, where ponds were built up into cultured ponds for production of shrimp, fish, and other aquatic resources, is known to be the leading cause of mangrove loss in the country (Garcia, et. al., 2013).

Banate Bay, the source of living of the people living in the coastal towns of Anilao, Banate, and Barotac Nuevo in Iloilo, is not only known for harboring tons of fishes but also a haven for mangroves (Overseas, 1998). Despite this fact, Iloilo has one of the largest percentages of mangrove areas being converted into aquaculture. In fact, as of 2011, it has 95% of mangrove to fishpond conversion percentage (Primavera, et. al., 2011). Although greater conservation and rehabilitation efforts have been in place (Samson & Rollon, 2008), it is expected that the mangrove ecosystem in the country will continue to degrade (Fortes, 2004). This is mainly due to planting of wrong species in the wrong areas (Primavera & Esteban, 2008).

Geovisualization is a helpful tool for the display of spatial information, and it provides a different perspective and insight into the datasets (Cartwright, 2004). 3D visualizations offer more realistic objects and users can comprehend patterns and relationships better than when presented in 2D or in still graphics (Laurini, 2017). Nowadays, it is widely used as a tool

for effective and efficient decision-making processes (Lange, 2005). This study uses four (4) various platforms to model land cover changes, particularly mangrove to aquaculture conversion of the study area, from 1973 to 2016 and to provide realistic 3D virtual environment allowing interaction and navigation of respondents. Also, results from the perception survey conducted can help future researchers on the proper selection of platform for different purposes and audiences of 3D visualization.

### 1.2 Study Area

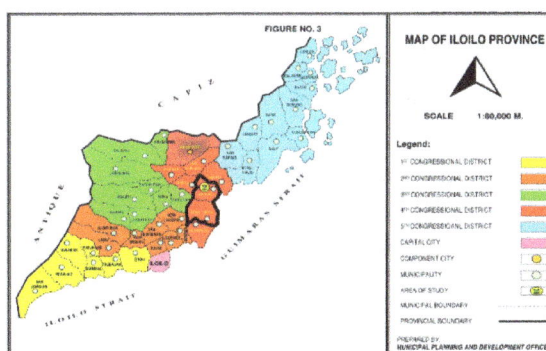

Figure 1. Study area (Source: Anilao Municipal Planning and Development Office)

The study area for this research involves portions of the Municipalities of Anilao and Barotac Nuevo, located in the province of Iloilo in Western Visayas (Region VI). Both municipalities belong to the fourth district of Iloilo. Anilao, a

4th class municipality, is composed of 21 barangays with a total area of 7,538 hectares and a population of 28,684 as of 2015 census. On the other hand, Barotac Nuevo, a 2nd class municipality, has 29 barangays, a total land area of 9,449 hectares with a population of 54,146 (Provincial Government of Iloilo, 2017). The province highly depends on agriculture and aquaculture production. Rice, corn, fruit vegetables, banana, and pineapple being their crop products while bangus, mudcrab, prawn, tilapia, catfish, oyster, and seaweeds their aquaculture harvests (Provincial Government of Iloilo, 2017).

### 1.3 Objectives

This research aims to generate 3D visualizations of mangrove to aquaculture conversion and vice versa using Google Earth, ArcScene, Virtual Terrain Project (VTP), and Terragen platforms; Specifically, this paper intends to determine mangrove and aquaculture conversion in Anilao and Barotac Nuevo, Iloilo from 1973 to 2016 using Landsat images and assess the usability, effectiveness, and potential of various platforms used in the visualization through the conduct of a perception survey.

### 1.4 Scope and Limitations

This study is limited only to using Landsat images acquired for the years 1973, 1976, 2000, 2006, 2010, and 2016 Landsat images for the classification. This inconsistent temporal dataset is due to the very persistent cloud cover in the area. The trial version for the Terragen platform is used for the model generation; hence, limitations in the software functionality hindered the production of visualization outputs. Also, the 3D objects used were built-in models from the different platforms resulting to distinct appearances of each scenario. Google Earth is considered as a visualization platform in 3D despite it having only 2.5D terrain information and 2D KML files, since it enables users to navigate through an area using 3D views and perspectives. Also, the visualization scenario provided by Google Earth provided a baseline for the users in their assessment of the platforms, since it is the most familiar and widely used. Also, no pre-assessment test was performed for the respondents of the perception survey. Their level of expertise therefore was classified based on their own discernments of their abilities.

## 2. METHODOLOGY

### 2.1 General Methodology

The following figure shows the steps undertaken for this research.

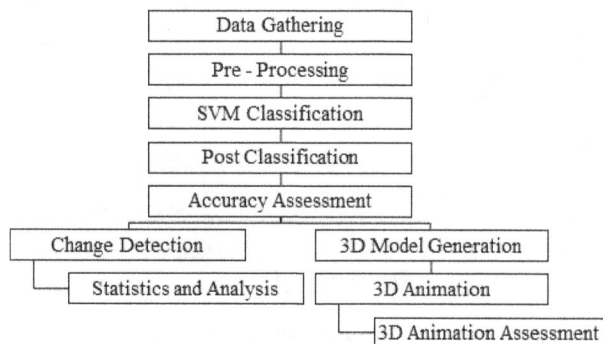

Figure 2. General workflow

### 2.2 Data Gathering

The summary of the Landsat scenes used along with their dates of acquisition and the images prior to processing is shown in Table 3. The 1973, 1976, and 2000 images were downloaded from Earth Explorer while the 2006, 2010, and 2016 images from the USGS Global Visualization Viewer.

| Date | Description |
|------|-------------|
| March 3, 1973 | Taken by Landsat 1 MSS with 4 bands (R, G, and 2 NIR) |
| May 7, 1976 | Taken by Landsat 2 MSS with 4 bands (R, G, and 2 NIR) |
| September 22, 2000 | Taken by Landsat 7 ETM with 8 bands (R, G, B, NIR, SWIR 1, SWIR 2, Thermal, and Panchromatic) |
| April 8, 2006 | Taken by Landsat 5 TM with 7 bands (R, G, B, NIR, SWIR 1, SWIR 2, and Thermal) |
| February 14, 2010 | Taken by Landsat 5 TM with 7 bands (R, G, B, NIR, SWIR 1, SWIR 2, and Thermal) |
| March 18, 2016 | Taken by Landsat 8 OLI with bands 11 bands (Ultra Blue, R, G, B, NIR, SWIR 1, SWIR 2, Panchromatic, Cirrus, Thermal Infrared 1, and Thermal Infrared 2) |

Table 1. Summary of Landsat images acquired

The Digital Elevation Model (DEM) used for the 3D models were acquired from the National Mapping and Resource Information Authority (NAMRIA).

### 2.3 Image Processing

The detailed methodology used for the image processing is shown in Figure 3. The Support Vector Machine (SVM) algorithm is used for its ability to handle small training data sets, often producing higher accuracy than traditional methods (Mantero et. al., 2005).

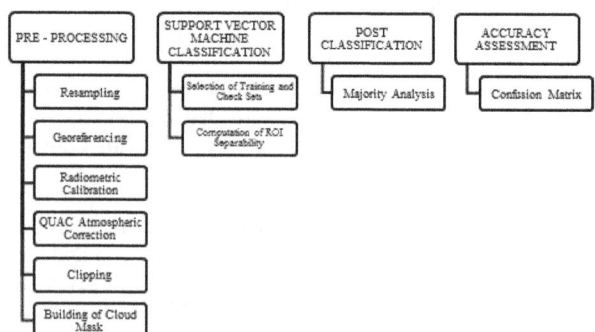

Figure 3. Detailed methodology for image processing

### 2.4 3D Visualization

The output classified images were first post-processed and the different feature classes from each year were extracted and exported as individual shapefiles using ArcMap. Trees were then divided further into two (2) types: mangrove and non-mangrove classes. These shapefiles served as the input files for the different visualization platforms. To generate 3D scenarios, the following platforms were utilized: (1) Google Earth; (2) ArcScene; (3) Virtual Terrain Project (VTP); and (4) Terragen.

### 2.4.1 Google Earth

The researchers selected Google Earth as one of the platforms for visualization since it is commonly used from geomatics practitioners to end-product users, due to its accessibility and affordability. This kind of visualization works by draping the classified images over the surface in Google Earth. It can be done through following steps described in Figure 4.

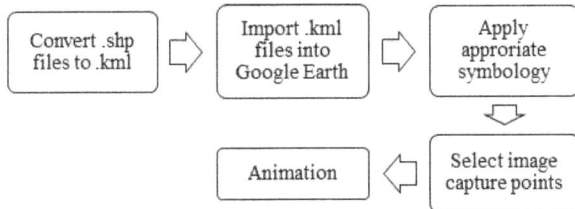

Figure 4. Methodology for Google Earth

### 2.4.2 ArcScene

ArcScene was also chosen as one of the platforms for visualization because of its spatial analysis capabilities. It is one of the platforms imbedded in a GIS software, ArcGIS. With this, handling of data sets became very easy and just needed a few adjustments to create a 3D scene. However, ArcScene is a proprietary software which requires a license to operate.

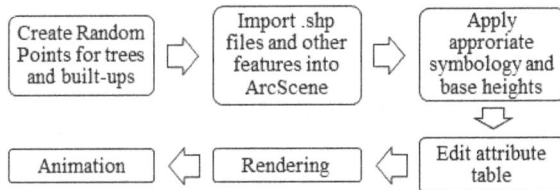

Figure 5. Methodology for ArcScene

To create a 3D visualization using ArcScene, mangrove, non-mangrove, and built-up shapefiles were first converted from polygon into points. Create Random Points tool was used multiple times to perform the conversion. These points were now imported to ArcScene together with the other features such as digital elevation model (DEM), fishpond boundaries, roads, and river. For each feature, appropriate symbology and base heights were applied.

After assigning these symbologies and base heights, attribute tables were edited. Start time and end time attributes were added for all features. This was utilized for the animation of the scenes. Then, scenes were rendered to come up with 30-second videos showing the conversions happened from 1973 to 2016. Four (4) 30-second videos were rendered and compiled into Windows Movie Maker for the final animation output.

### 2.4.3 Virtual Terrain Project (VTP)

Unlike ArcScene, VTP has no direct spatial analysis capabilities and is a free software. However, vector files such as .shp can be read by VTBuilder which is the pre-processing unit of VTP. Then conversions of these .shp files were first needed in order to view them into the Enviro, the viewer of VTP, in 3D. These files, in compatible formats, are needed to be saved in the specific folders generated by VTP upon installing the software.

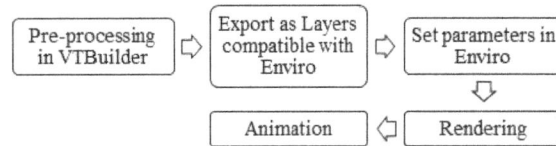

Figure 6. Methodology for VTP

For the pre-processing stage, the DEM in .tiff format was converted to .bt format through VTBuilder and saved in Elevation folder. Shapefiles of trees and vegetation were imported to VTBuilder. Using Generate Vegetation tool, trees and grass were created as points within the bounds of the shapefiles. Podocarpus sp. model was used for mangrove trees, Araucaria heterophylla model for non-mangrove trees, and Pennisetum setaceum model for vegetation. Then, they were exported as .vf files and saved into the PlantData folder. Shapefiles of built-up (the ones used in ArcScene) were imported as structures. These were exported in .vtst format and saved in BuildingData folder. Lastly, the road was converted to .rmf format and saved into RoadData folder.

After all conversion, the parameters were now set into the Enviro. In addition, scenes were created to portray features that are visible per year. In order to render the specific scenes, camera angles were first selected. Images were captured and animated using Windows Movie Maker.

### 2.4.4 Terragen

The last platform that was used is Terragen. It has totally no support for spatial analysis and no database access since it does not support vector files. Also, it is not a free software however, there is a free trial version available. Terragen is commonly used in movies and is a great animator. Comparing all four (4) platforms, Terragen provides the most realistic and most aesthetically pleasing objects and scenes.

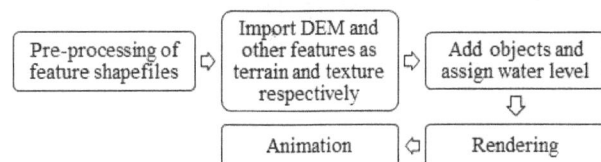

Figure 7. Methodology for Terragen

Since the researchers only used the free trial version, only three (3) objects at a time was allowed. Thus, it was decided to concentrate on the dynamics of mangroves and camera position were initially selected. Also, the fact that Terragen has no support for spatial analysis, importing shapefiles is not possible. The researchers needed to rasterize the mangrove shapefiles first. Then, rasterized files were then reclassified to have a 100 value for the specific target feature and 0 value for all other features.

Also, the DEM was converted into .ter format using Global Mapper. The DEM in .ter format was first imported as terrain. Next, the reclassified maps were imported as texture. By estimating the coordinates of the center of a cluster of mangroves and its boundaries, the researchers were able to add mangrove trees as objects. Also, water level was adjusted for it to become visible. Using the pre-set camera positions, the yearly scene were rendered and were compiled again using Windows Movie Maker to create an output video.

**2.5 3D Animation Assessment**

In order to assess the visualizations made, a perception survey was done. After the design of survey instrument, the researchers selected respondents with different levels of expertise in spatial analysis, 3D visualization, and forestry, fisheries, and aquatic resources. Content of the questionnaire include rating of the different platforms used divided into three categories – technology and purpose, interaction and navigation, and information content.

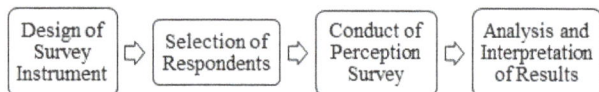

Figure 8. Methodology for 3D animation assessment

During the actual perception survey, the final video outputs generated from the 4 platforms were viewed by the respondents. Also, respondents were able to explore the platforms used on their own. A survey form was prepared to investigate the effectiveness of the platforms using various indicators. Inputs from the development of the visualizations were tabulated, together with the effectiveness ratings, to assess usability and potential.

# 3. RESULTS AND DISCUSSION

## 3.1 Image Processing

### 3.1.1 Land Cover Maps

For the 1973 and 1976 images, there are only five (5) classes created because of the difficulty of differentiating built up and bare soil classes given the spatial resolution of the early Landsat images. Using ArcMap, land cover maps from the SVMs classification were created and are presented in Figure 9.

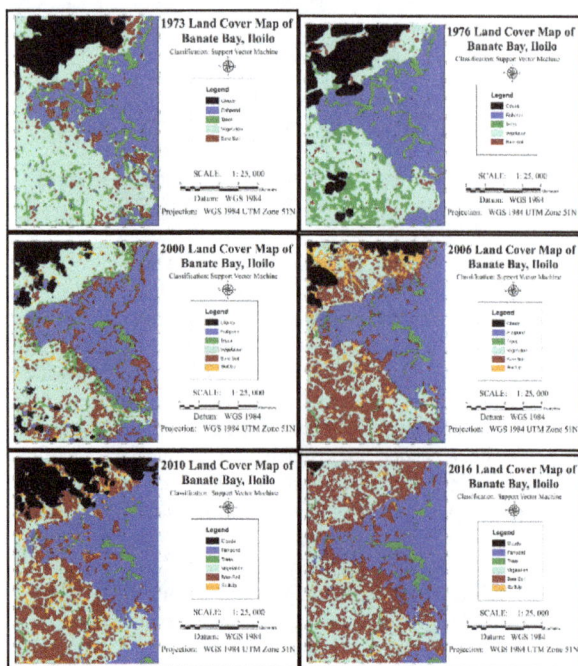

Figure 9. Land classification maps produced using SVMs
(L-R: 1973, 1976, 2000, 2006, 2010, 2016)

### 3.1.2 Mangrove and Aquaculture Cover Change

Confusion matrices for the ground truth ROIs were created for a pixel-based accuracy assessment. It should be noted that the 100% overall accuracy for the 1973 image was due to the limitations of the data. Classification was relatively simpler than the other images mainly because there were only four (4) classes in the image and they are spectrally separable even though there were only four (4) bands in the image. In addition, the vegetation and the bare soil classes exhibit a very obvious distinction.

All images have attained acceptable values for overall accuracy and kappa coefficient, and the values are summarized in Table 2.

| Year | Overall Accuracy (%) | Kappa Coeff. |
|------|----------------------|--------------|
| 1973 | 100.0000 | 1.0000 |
| 1976 | 81.6901 | 0.7550 |
| 2000 | 95.6989 | 0.9642 |
| 2006 | 91.5385 | 0.8942 |
| 2010 | 94.4079 | 0.9300 |
| 2016 | 96.2617 | 0.9531 |

Table 2. Overall accuracy and kappa coefficient for the classified images

Figure 10 summarizes the changes mangrove and aquaculture areas in the area from 1973 to 2016. Change detection analysis shows that the largest negative change in mangrove area happened from 2006 to 2010 with an annual rate of -8.56%, assuming a linear rate, parallel to a decrease of -1.21% annually in the same years. This is due to fishpond pixels classified as bare soil because of the draining of water for chemical removal. The highest increase in fishpond area occurred in 1973 to 1976 with an annual rate of 6.14%.

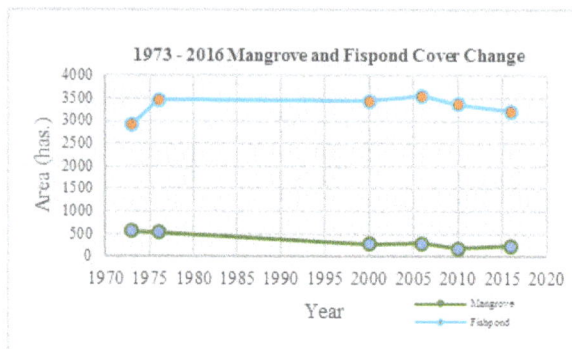

Figure 10. Change in mangrove and aquaculture area from 1973 to 2016

## 3.2 3D Visualization

The researchers were able to generate four (4) 3D scenarios portraying the mangrove and aquaculture conversions happened in a portion of Banate Bay, Iloilo from 1973 to 2016.

Figure 11. 3D visualizations generated from four (4) platforms: Google Earth (Upper left), ArcScene (Upper right), VTP (Lower left), Terragen (Lower right)

From these outputs, a perception survey was performed to assess the usability, effectiveness, and potential each platform in generating visualizations.

### 3.3 3D Animation Assessment

For the assessment, perception survey was done. There were 30 respondents which were categorized based on their perceived levels of expertise in spatial analysis, 3D visualization, and forestry/fishery/aquatic resources.

First is by categorizing the respondents based on their level of expertise in spatial analysis. The summary of the results of the survey is summarized in Figure 12. (TP – Technology and Purpose, IN – Interaction and Navigation, and IC – Information Content)

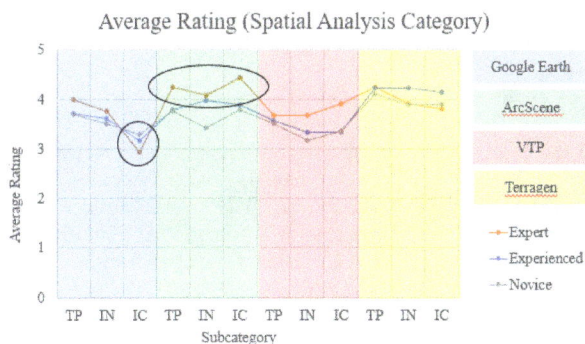

Figure 12. Average rating categorized by expertise in spatial analysis

As shown in the graph, Google Earth gave the least amount of information and the experts prefer ArcScene for visualizations focusing on spatial analysis. Re-categorizing the respondents, now based on their level of expertise in 3D visualization, the summary of the results of the survey is summarized in Figure 13.

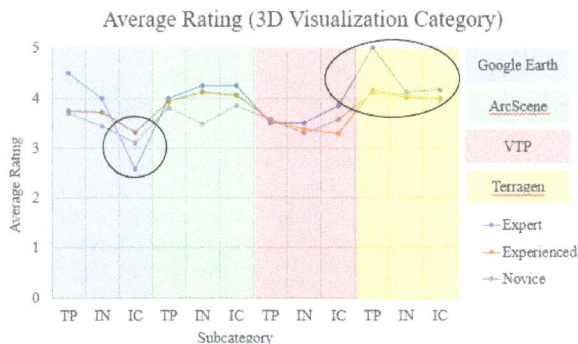

Figure 13. Average rating categorized by expertise in 3D visualization

The same trend can be observed in which the information relayed by Google Earth and Terragen is the preferred visualization of 3D visualization experts. Finally, re-grouping the respondents according to their level of expertise in forestry, fishery, and aquatic resources, the summary of the results of the survey is summarized in Figure 14.

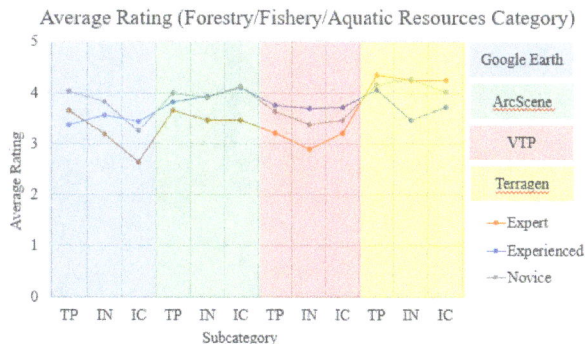

Figure 14. Average rating for forestry, fishery, and aquatic resources

For this application, Terragen is the preferred platform for 3D visualization involving forestry, fishery, and aquatic resources.

To assess the effectivity of the visualizations, the results of the perception survey were analyzed. The summary of the results of the perception survey for spatial analysis applications is presented in Figure 15.

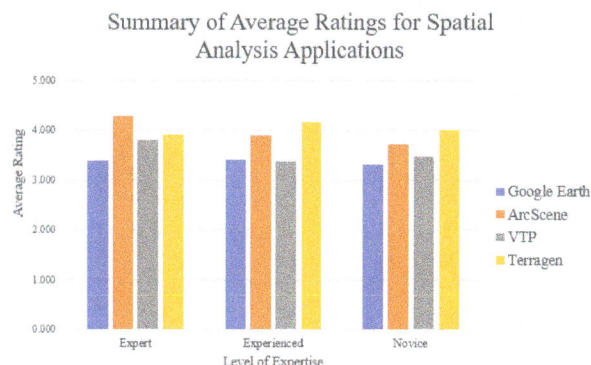

Figure 15. Effectiveness assessment for spatial analysis discipline

Based on the graph, the effective platform for the spatial analysis experts is ArcScene, Terragen for those with experience and the novices.

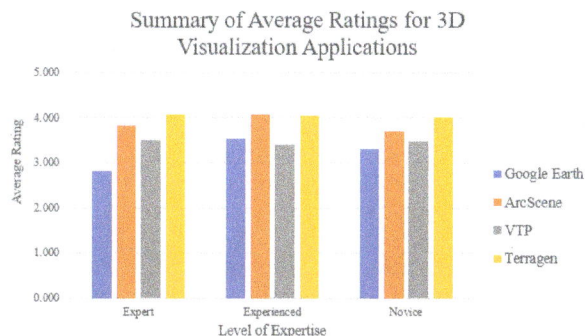

Figure 16. Effectiveness assessment for 3D visualization discipline

For the 3D visualization experts and novices, Terragen is the most preferred while those with experience rated almost the same for ArcScene and Terragen.

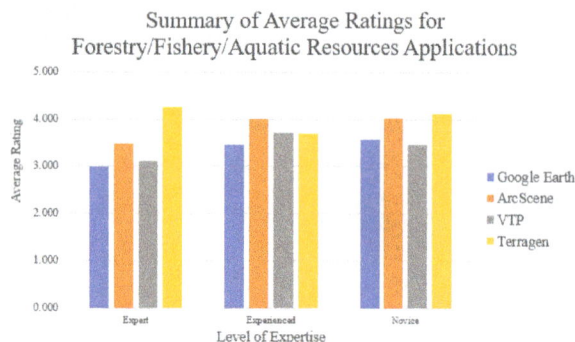

Figure 17. Effectiveness assessment for forestry, fishery, and aquatic resources discipline

From the forestry, fishery, and aquatic resources discipline, Terragen is the most effective platform for experts and the novices while ArcScene is the most effective platform for those with experience in this type of applications.

Aside from the actual perception survey, the researchers, as the developers of the visualization, also rated the different platforms to assess its usability.

To compare processing time among the platforms, a single machine was used by a single user in producing all visualization scenarios. A Windows 10 PC running on an Intel i5-4210U 1.70 GHz processor, with a 4GB memory and 2GB graphics card.

| Platform | Processing time (in hours) | Degree of Difficulty | | | |
|---|---|---|---|---|---|
| | | Data Input | 3D Object Creation | Scene Generation | Animation Creation |
| Google Earth | 2 | 4 | 1 | 5 | 4 |
| ArcScene | 25 | 5 | 4 | 3 | 3 |
| VTP | 20 | 2 | 3 | 2 | 4 |
| Terragen | 35 | 1 | 2 | 1 | 4 |

Table 3. Usability assessment of 3D visualization

From Table 3, Google Earth was identified as the easiest and fastest platform to use. However, it has very poor performance on generating 3D objects. ArcScene was rated 5 for pre-processing since almost no pre-processing step was needed in order for ArcScene to read the input files. On the other hand, it took 20 hours to generate the 3D visualization using VTP. Though generally, it is difficult to produce, handling of the software was made easy since not all data were processed into only one software. All files were prepared using VTBuilder and 3D scenes were viewed with Enviro.

Lastly, visualization from Terragen was the most difficult to produce. It took the researchers 35 hours to generate the final output. There were many conversions that happened and generating 3D objects was done manually. Also, rendering of scenes were slow averaging 10 minutes per image.

The researchers then combined the usability and effectiveness of the various platforms in order to get their individual potential for different applications.

| Application / Platform | Spatial Analysis | 3D Visualization | Forestry/ Fishery/ Aquatic Resources |
|---|---|---|---|
| Google Earth | Good | Poor | Good |
| ArcScene | Very Good | Good | Good |
| VTP | Good | Good | Good |
| Terragen | Good | Very Good | Very Good |

Table 4. Potential ratings for visualization of the four (4) platforms

Google Earth rated good for spatial analysis and forestry, fishery, and aquatic resources applications. However, it rated poor for 3D visualization. As expected, ArcScene rated very good for spatial analysis but good only for 3D visualization and forestry, fishery, and aquatic resources applications. Virtual Terrain Project rated good for all kinds of applications. Lastly, Terragen rated very good for 3D visualization and for forestry, fishery and aquatic resources applications. However, since Terragen has no spatial analysis support, it only rated good for spatial analysis applications.

## 4. CONCLUSIONS AND RECOMMENDATIONS

### 4.1 Conclusions

At the end of this study, the researchers were able to generate four (4) 3D visualization models using Google Earth, ArcScene, Virtual Terrain Project (VTP), and Terragen from the results of the Support Vector Machine (SVM) Classification of Landsat images. From the results of the classification, it was observed that the largest negative change of mangrove areas happened from 2006 to 2010, with an annual rate of -8.56% On the other hand, the largest positive change for aquaculture area happened from 1973 to 1976 with a rate of 6.14% annually. For the 3D visualization, the results of the perception survey showed that ArcScene is preferred for spatial analysis while respondents favored Terragen for 3D visualization and for forestry, fishery, and aquatic resources applications, thereby concluding that people with different levels of expertise prefer different visualization platforms depending on their discipline.

### 4.2 Recommendations

For further studies, the following are recommended by the researchers: First is to use a higher processing computer for faster rendering of features and layers. Secondly, it would improve the 3D visualization if 3D models and orthophotos generated from UAV Images are incorporated to arrive with more virtually realistic and accurate environment allowing immersion of the viewers and users to the 3D scenarios. Also, it is deemed necessary to include respondents from the local community for assessment of the classification results and acquire inputs and insights for improvement of the models generated as they are one of the end – users of the products. Finally, a post test for the respondents could be added for a more accurate assessment of the effectiveness of the visualization and determine their capability to operate and navigate the platforms used.

## ACKNOWLEDGEMENTS

This study would not be possible without the funding of the Department of Science and Technology (DOST), through its monitoring agency, the DOST-Philippine Council for Industry, Energy, and Emerging Technology Research and Development (DOST-PCIEERD).

The researchers would also like to acknowledge the following people and institutions for their significant contributions and participation to make this research possible. Without them, this study cannot be fulfilled within the given timeframe.

Special thanks to Dr. Ariel Blanco, local government of Anilao, Iloilo, especially to Ms. Emee Grace Miatapal, National Mapping and Resource Information Authority, Bureau of Fisheries and Aquatic Resources, and Dr. Rene Rollon of UP Institute of Environmental Science and Meteorology for providing us with necessary information and data for this study.

## REFERENCES

ArcGIS for Desktop, 2016. *Working with ArcGlobe and ArcScene.* Retrieved from http://desktop.arcgis.com/en/arcmap/10.3/main/get-started/choosing-the-3d-display-environment.htm.

Bailey, J.E., Whitmeyer, S.J. and D.G. De Paor., 2012. Introduction: The application of Google Geo Tools to geoscience education and research. *The Geological Society of America.*

Brown, W.H. and A.F. Fischer, 1918. Philippine mangrove swamps. Department of Agriculture and Natural Resources. *Bureau of Fishery Bulletin No.7.*

Canada Centre for Remote Sensing., 2011. *Fundamentals of Remote Sensing.* pp.147 – 148.

Cartwright, W., S. Miller, and C. Pettit. "Geographical Visualization: Past, Present and Future Development." Journal of Spatial Science 49.1 (2004): 25-36. Web.

CivilFX, 2016. Top 10 Benefits of 3D Visualization. Retrieved from http://www.civilfx.com/benefits-3d-architectural-visualization/

Clasen, M. and H. C. Hege., 2005. Realistic Illumination of Vegetation in Real-Time Environments. *Trends in Real-time Visualization and Participation, New Technologies for Landscape Architecture and Environmental Planning.*

Daupan, S., 2016. Community Participation in Mangrove Forest Management in the Philippines: Management Strategies, Influences to Participation, and Socio-Economic and Environmental Impacts. *PhD Dissertation*, University of Michigan.

Ding, S. and Chen, L., 2010. Intelligent Optimization Methods for High - Dimensional Data Classification for Support Vector Machines. *Intelligent Information Management.* Wuhan, China

Duric U., Abolmasov, B., Pavlovic, R. and B. Trivic., 2011. Using ArcGIS for Landslide "Umka" 3D Visualization. *The Geology in Digital Age: 17ᵗʰ Meeting of the Association of European Geological Societies.*

Fortes, M., 2004. National report on seagrass in the South China Sea - Philippines. Reversing environmental degradation trends in the South China Sea and Gulf of Thailand. *UNEP/GEF South China Sea Project.*

Garcia, K., Gevana, D., & Malabrigo, P., 2013, November. Philippines' Mangrove Ecosystem: Status, Threats, and Conservation. *Mangrove Ecosystems of Asia.*

Geman, S., Bienenstock, E., Doursat, R., 1992. Neutral networks and the bias/variance dilemma. Neural Computation 4 (1), 1 - 58.

Green, J., 2016. What Is 3D Visualization, Who Does It & Why Do You Need It? Retrieved from https://www.upwork.com/hiring/design/what-is-3d-visualization-who-does-it-why-do-you-need-it.

Guyon, I., Vapnik, V., Boser, B., Solla, S.A., 1992. Capacity control in linear classifiers for pattern recognition. *In: First IAPR International Conference on Pattern Recognition.* IEEE Computer Society Press, pp. 385-388.

Iloilo Provincial Government, 2017. Quick Facts. Retrieved from www.iloilo.gov.ph: http://www.iloilo.gov.ph/quick-facts

Lange, E., 2005. Issues and Questions for Research in Communicating with the Public through Visualizations. *Distance Learning Administration Conference Proceedings.*

Laurini, R., 2017. Geographic Knowledge Infrastructure: Applications to Territorial Intelligence and Smart Cities. ISTE Press.

Liu, X. and S. Shiotani., 2015. A virtual globe-based visualization and interactive framework for a small craft navigation assistance system in the near sea. *Journal of Traffic and Transportation Engineering.*

Malinverni, E.S., 2007. 3D Geoexploration of the Prealpi Giulie Nature Park. *International Archives of Photogrammetry and Remote Sensing Volume XXXVI-4/W45, 2007.*

Mantero, P., Moser, G., Serpico, S.B., 2005. Partially supervised classification of remote sensing images through SVM - based probability density estimation. *IEEE Transactions on Geoscience and Remote Sensing* 43 (3), 559-570.

Mei, G., Tipper, J. and N. Xu., 2013) 3D Geological Modeling and Visualization of Rock Masses Based on Google Earth: A Case Study. Cornell University Library. Retrieved from https://arxiv.org/abs/1301.3455.

Melana, E.E. and H.I. Gonzales., 1996. Field guide to the identification of some mangrove plant species in the Philippines. Ecosystems Research and Development Service, *Department of Environment and Natural Resources Region 7,* Banilad, Mandaue, Cebu City, Philippines. 29 p. + 8 p. appendices.

Melana, D., Melana, E.E., & Mapalo, A.M., 2000. Mangrove Management and Development in the Philippines. Oral presentation at *Mangrove and aquaculture management.* Kasetsart University Campus, Bangkok, Thailand.

Montgomery, D.C., Peck, E.A., 1992. *Introduction to Linear Regression Analysis*, 2nd ed. Wiley, New York.

Mountrakis, G., Im, J., and Ogole, C., 2010. Support vector machines in remote sensing: A review. *ISPRS Journal of Photogrammetry and Remote Sensing.* New York, USA.

Planetside Software, 2016. Terragen in Film & TV. Retrieved from https://planetside.co.uk/terragen-in-film-and-tv/#.

Planetside Software, 2016. What is Terragen. Retrieved from http://planetside.co.uk/terragen-overview.

Primavera, J., & Esteban, J.M.A., 2008. A review of mangrove rehabilitation in the Philippines: successes, failures and future prospects. *Wetlands Ecology and Management 16* (3): 173-253.

Primavera, J.H., Rollon R.N. & Samson M.S., 2011) The Pressing Challenges of Mangrove Rehabilitation: Pond Reversion and Coastal Protection.

Samson, M. S., & Rollon, R. N., 2008. Retrieved from Growth Performance of planted mangrove in the Philippines: revisiting forest management strategies. Ambio 37(4):234-240.

Schroth, O., 2010. From Information to Participation: Interactive Landscape Visualization as a Tool for Collaborative Planning. *Bibliografische Information der Deutschen Nationalbibliothek.*

Tupas, M.E., 2012. Sedimentation and land cover change modeling of Magat dam watershed area using remote sensing, geographic information system and cellular Automata-Markov analysis. *MS Thesis, University of the Philippines – Diliman.*

Turner, A.K., 1989. Three-Dimensional Modeling with Geoscientific Information Systems. *Kluwer Academic Publishers.*

VTP, 2017. Virtual Terrain Project. Retrieved from http://vterrain.org.

Warren-Kretzschmar, B. and S. Tiedtke. What Role Does Visualization Play in Communication with Citizens? – A Field Study from the Interactive Landscape Plan. *In Trends in Real-Time Landscape Visualization and Participation, edited by E. Buhmann,* 156–167. Heidelberg: Herbert Wichmann Verlag.

Wisniewski, P.K., Pala, O., Lipford, H.R. and D. Wilson., 2009. Grounding Geovisualization Interface Design: A Study of Interactive Map Use. *Spotlight on Works in Progress.* Boston, Massachusetts, USA.

# 21

# PERFORMANCE ANALYSIS OF LOW-COST SINGLE-FREQUENCY GPS RECEIVERS IN HYDROGRAPHIC SURVEYING

Mohamed Elsobeiey

Department of Hydrographic Surveying, Faculty of Maritime Studies, King Abdulaziz University
P. O. Box 20807 - Jeddah 21589, Kingdom of Saudi Arabia
melsobeiey@kau.edu.sa

**KEY WORDS:** Hydrographic Surveying; Single-Frequency; IHO; Precise Point Positioning

**ABSTRACT:**

The International Hydrographic Organization (IHO) has issued standards that provide the minimum requirements for different types of hydrographic surveys execution to collect data to be used to compile navigational charts. Such standards are usually updated from time to time to reflect new survey techniques and practices and must be achieved to assure both surface navigation safety and marine environment protection. Hydrographic surveys can be classified to four orders namely, special order, order 1a, order 1b, and order2. The order of hydrographic surveys to use should be determined in accordance with the importance to the safety of navigation in the surveyed area. Typically, geodetic-grade dual-frequency GPS receivers are utilized for position determination during data collection in hydrographic surveys. However, with the evolution of high-sensitivity low-cost single-frequency receivers, it is very important to evaluate the performance of such receivers. This paper investigates the performance of low-cost single-frequency GPS receivers in hydrographic surveying applications. The main objective is to examine whether low-cost single-frequency receivers fulfil the IHO standards for hydrographic surveys. It is shown that the low-cost single-frequency receivers meet the IHO horizontal accuracy for all hydrographic surveys orders at any depth. However, the single-frequency receivers meet only order 2 requirements for vertical accuracy at depth more than or equal 100m.

## 1. INTRODUCTION

Typically, commercial GPS receivers vary according to their receiving capabilities. There are different types such as single-frequency code receivers, single-frequency carrier-smoothed code receivers, single-frequency code and carrier receivers, dual-frequency receivers, and triple-frequency receivers. Single-frequency receivers access the L1 frequency only. Dual-frequency receivers access both the L1 and the L2 frequencies, while triple-frequency receivers access L1, L2, and L5 frequencies.

The first type of single-frequency GPS receivers, the single-frequency code receiver, is the cheapest and the least accurate type of receivers. It measures the pseudoranges with the C/A-code only. The second receiver type, the single-frequency carrier-smoothed code receiver, uses the L1 carrier phase to produce high-precision C/A-code measurements. Single-frequency code and carrier receivers output the raw C/A-code pseudoranges, the L1 carrier phase measurements in addition to the navigation message. Dual-frequency receivers are the most sophisticated and most expensive receiver type. They measure GPS data on both L1 and L2 frequencies. Triple-frequency receivers, on the other hand, are able to produce measurements on the legacy frequencies L1 and L2 and the modernized frequency L5.

Unlike dual-frequency receivers, however, ionosphere delay represents a major challenge for single-frequency receivers.

There are two main approaches to correct single-frequency data from the ionosphere delay (Cai et al., 2013). The first technique is to use ionosphere models to correct for the ionosphere delay. These models may be empirical models such as Klobuchar model, which can account for up to 60% of the delay at mid-latitudes (El-Rabbany, 2006). Klobuchar model coefficients are transmitted as part of the navigation message and can be improved by extending the eight parameters original Klobuchar model to ten-parameters to account for the ionosphere variation during the night time (Wang et al., 2016). Corrections from regional or global network may be estimated and then applied to single-frequency receivers. An example for the global ionosphere corrections is the Global Ionosphere Maps (GIMs) produced by the International GNSS service (IGS). Another option to correct for the ionosphere delay for single-frequency receivers in real-time is to broadcast ionosphere corrections from Space Based Augmentation Systems (SBAS) (Arbesser-Rastburg, 2002; Hofmann-Wellenhof et al., 2008).

The second technique to account for ionosphere delay is to form ionosphere-free linear combination using both code and carrier phase observations on L1 from the single-frequency receiver. This technique is based on the Group and Phase Ionosphere Calibration (GRAPHIC) (Cai et al., 2013; Schüler et al., 2011; Shi et al., 2012; Sterle et al., 2015).

Since the availability of low-cost single-frequency receivers, several attempts have been carried out to reduce the cost and increase the accuracy which can be obtained from such receivers compared with geodetic grade receivers. High-sensitivity low-cost receivers have thousands of correlators to reduce the search space of each correlator and are able to acquire signals with low decibel watt (dBW) (Schwieger, 2007). Hedgecock et al. (2013) introduced the standalone relative localization system using low-cost single-frequency receivers. They found that tracking the relative motion of the neighboring nodes is an order of magnitude better than taking the difference between the absolute coordinates of each node. Ambiguity resolution, on the other hand, is not possible from single epoch data from single-frequency receivers(Odijk et al., 2012). However, ambiguity resolution can be achieved using less than 10 minutes of accumulated data, by which sub-cm and a few cm levels can be achieved for horizontal and vertical directions, respectively (Odijk et al., 2014). The performance of low-cost single-frequency receivers can be improved by using a geodetic grade antenna instead of the low-cost single-frequency antenna (Takasu and Yasuda, 2008).

Low-cost single-frequency receivers have been used in numerous applications. An unmanned aerial vehicle (UAV) can be occupied by low-cost single frequency receivers, which are controlled through radio communication control. The system works in RTK mode and gets corrections from at least one reference receiver (Stempfhuber and Buchholz, 2011). Low-cost single-frequency receivers can be used along with SBAS corrections for autonomous guidance of agricultural tractors. The pass-to-pass error in trajectories is within 1m (Alonso-Garcia et al., 2011). It can be used in Structure Health Monitoring (SHM) applications such as monitoring the displacement. Averaging measurements from several GPS receivers can significantly reduce the noise level and the dynamic displacement response can be captured at 0.25m amplitude (Jo et al., 2013). Low-cost single-frequency receivers can be used also to monitor the snow liquid water content and avalanche prediction by measuring the changes of GPS L1 carrier strength (signal-to-noise ratio) (Koch et al., 2014).

In this paper, u-blox NEO-7P low-cost single-frequency GPS receiver, which fulfills accuracy requirements for all RTK surveying applications (Sioulis et al., 2015), is used to collect GPS data using a hydrographic vessel. The main objective of the paper is to evaluate the performance of such single-frequency receiver in hydrographic surveying applications.

## 2. IHO POSITIONING STANDARDS

The International Hydrographic Organization (IHO) published several standards that provide the minimum standards for hydrographic surveys execution to collect data to be used to compile navigational charts. Such standards must be achieved to assure both surface navigation safety and marine environment protection (IHO, 2008). According to the water depth, the IHO classifies surveys into four orders, namely; special order, order 1a, order 1b, and order 2. The first order is the special order which is the most rigorous among all hydrographic surveys orders. It is used for areas where under-keel clearance is critical

and where bottom characteristics are potentially hazardous to vessels such as critical navigation channels, harbors, and berthing areas (generally depth less than 40 m). the second order is order 1a which is for areas of depth more than 40 m and less than 100 m where under-keel clearance is less critical but man-made or natural features on the seabed are of concern to surface shipping may exist. Such areas like harbors, harbors approaching channels, and recommended tracks. The third order is order 1b, which is for areas of depth less than 100 m where under-keel clearance is not considered to be an issue for surface shipping in such areas. The fourth order is order 2, which is for areas deeper than 100 m where the general description of the sea floor is considered adequate and the man-made or natural features on the seabed will not have any impact on the surface navigation.

### 2.1 Total Horizontal Uncertainty (THU)

Hydrographic surveys measurements are affected by different sources of uncertainties, including random and systematic errors. All sources of measurement uncertainties are propagated, and the uncertainties of the computed parameters are then calculated, which is known as Total Propagated Uncertainty (TPU). The component of TPU in the horizontal plane is known as the Total Horizontal Uncertainty (THU), which is a 2 Dimensional (2D) quantity expressing latitude and longitude errors.

### 2.2 Total Vertical Uncertainty (TVU)

It is a 1 Dimensional (1D) quantity that expresses the component of the TPU in the vertical dimension. The depth uncertainty is affected by two errors, depth dependent and depth independent errors. The TVU can be computed at 95% confidence level as follows (IHO, 2008):

$$TVU = \pm\sqrt{a^2 + (b \times d)^2} \qquad (1)$$

where   a = the depth independent portion of uncertainty
        b = the coefficient that represents the depth
        dependent portion of uncertainty
        d = the depth

Table 1 summarizes the minimum standards for hydro-graphic surveys orders.

## 3. FIELD TEST

To test the performance of low-cost single-frequency GPS receivers in hydrographic surveying, raw GPS data was collected using New-7P u-blox GPS receiver. The test area was Sharm Obhur, Jeddah, Saudi Arabia where Faculty of Maritime Studies (FMS) is located. The base station was setup on the rooftop of FMS building using Ashtech ProFlex 500 GNSS receiver and the corresponding measurements simultaneously collected and employed to estimate the Real-Time Kinematic (RTK) solution, which considered as the reference solution. The length of the collected data was about 2.0 hours using King Abdulaziz University (KAU) vessel Hydrography 1, Figure 1. Figure 2, on the other hand, shows the trajectory of the data collected in November, 2015.

| Table Head | Hydrographic Surveys Order | | | |
| --- | --- | --- | --- | --- |
| | Special Order | Order 1a | Order 1b | Order 2 |
| Depth (d) | < 40m | < 100m | < 100m | > 100m |
| Area Characteristics | Harbors, berthing areas, and associated critical channels with minimum under-keel clearances | Under-keel clearance is less critical but features of concern to surface shipping may exist | Under-keel clearance is not an issue for surface shipping expected to transit the area | general description of the sea floor is considered adequate |
| Total Horizontal Uncertainty (THU) | 2m | 5m + 0.05d | 5m + 0.05d | 20m + 0.1d |
| Total Vertical Uncertainty (TVU) | a = 0.25m b = 0.0075 | a = 0.5m b = 0.013 | a = 0.5m b = 0.013 | a = 1.0m b = 0.023 |

Table 1. IHO Minimum Standards for Hydeographic Surveys (IHO, 2008)

Figure 1. KAU Hydrography 1 Vessel

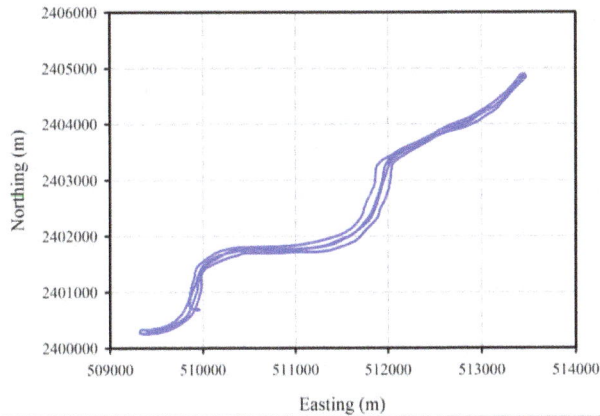

Figure 2. Test Trajectory

## 4. RESULTS AND DISCUSSION

The real-time solution from New-7P low-cost single-frequency GPS receiver is recorded during the surveying session at 1Hz sampling frequency. Moreover, the reference solution from Hydrography 1 vessel is recorded in real-time at the same sampling rate. The single-frequency solution is compared to the reference solution to investigate whether the low-cost single-frequency receivers satisfy IHO hydrographic surveys minimum standards according to Table 1. Both THU and TVU of the low-

cost single-frequency receiver are computed at 39% confidence level then transformed to 95% confidence level as follows (Mohamed El-Diasty and Elsobeiey, 2015):

$$THU_{39\%}^{2D} = \sqrt{\frac{\sum_{i=1}^{n}\left(\hat{N}_R - \hat{N}_{SFS}\right)_i^2 + \left(\hat{E}_R - \hat{E}_{SFS}\right)_i^2}{n}} \qquad (2)$$

$$THU_{95\%}^{2D} = 2.44\,THU_{39\%}^{2D} \qquad (3)$$

$$TVU_{39\%}^{1D} = \sqrt{\frac{\sum_{i=1}^{n}\left(\hat{U}_R - \hat{U}_{SFS}\right)_i^2}{n}} \qquad (4)$$

$$TVU_{95\%}^{1D} = 1.96\,TVU_{39\%}^{1D} \qquad (5)$$

Where $THU_{39\%}^{2D}$ = the total 2D horizontal uncertainty of Northing and Easting position error at 39% confidence level

$\hat{N}_R$ = the easting coordinate of the reference solution (the vessel integrated solution), $\hat{N}_{SFS}$ is the Northing single-frequency position

$\hat{E}_R$ = the Easting coordinate of the reference solution

$\hat{E}_{SFS}$ = the Easting single-frequency position

n is the total number of epochs

$THU_{95\%}^{2D}$ = the 2D total horizontal uncertainty of Northing and Easting position error at 95% confidence level

$TVU_{39\%}^{1D}$ = the total 1D vertical uncertainty of the Up component at 39% confidence level

$TVU_{95\%}^{1D}$ = the total 1D vertical uncertainty of the Up component at 95% confidence level

Figure 3 shows the Easting, Northing and the 2D horizontal error of the single-frequency solution compared with reference solution. In addition, Figure 4 shows the 2D horizontal error and the 1D vertical error. Figures 3 and 4 show that the maximum absolute 2D horizontal and 1D vertical errors are about 3.12m and 3.69, respectively. however, the mean error is about 1.26m and -0.62m for both 2D horizontal error and 1D vertical error, respectively. Moreover, Table2 summarizes the THU and TVU values at 39% and 95% confidence levels.

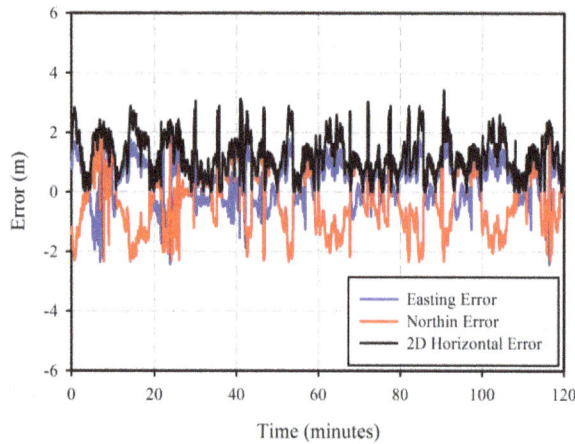

Figure 3. Northing, Easting, and the 2D horizontal Errors of Low-cost Single-frequency Receiver

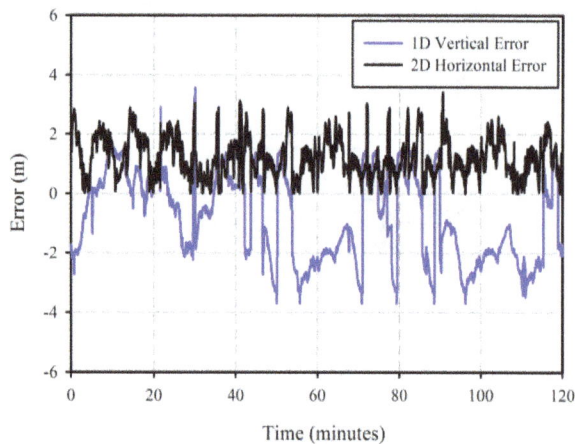

Figure 4. 2D horizontal Errors and 1D Vertical Errors of Low-cost Single-frequency Receiver

| Error Parameter | Low-cost Single-frequency Solution |
|---|---|
| $THU_{39\%}^{2D}$ | 0.692 |
| $THU_{95\%}^{2D}$ | $2.44 \times THU_{39\%}^{2D} = 1.690$ |
| $TVU_{39\%}^{1D}$ | 1.282 |
| $TVU_{95\%}^{1D}$ | $1.96 \times TVU_{39\%}^{1D} = 2.512$ |

Table 2. Summary Statistics of the Low-cost Single-frequency Solution

Our analysis showed that the THU is 1.69m at 95% confidence level which meets all hydrographic surveys types according to the IHO standards (Table 1) at any depth. However, the TVU is 2.512m at the same confidence level, which only meets hydrographic surveys order 2 at 100m depth.

## 5. CONCLUSIONS

This paper investigates the performance of real-time precise point positioning with low-cost single-frequency GPS receivers in hydrographic surveying applications. A session of two hours length is used to collect single-frequency data using King Abduaziz University vessel, Hydrography 1. The vessel solution, which consists of the RTK integrated solution is used as a reference for comparison. To check if the single-frequency receivers fulfil the IHO requirements, both THU and TVU are estimated at 95% confidence level. It is shown that the total horizontal uncertainty from the solution of low-cost single-frequency receivers is about 1.69m, which meet the IHO standards for all hydrographic surveys orders at any depth (special order, order 1a, order 1b, and order 2). However, the total vertical uncertainty from the solution of single-frequency receivers is found to be 2.512m, which meets only order 2 requirements for vertical accuracy at depth more than or equal 100m. in fact, ublox single-frequency GPS receivers accounting for troposphere and ionosphere errors by applying the corrections transmitted from SBAS. Such corrections are limited to an accuracy of about 0.2m, which affects the receiver solution, especially the height component. So, further investigations are required to improve the height component of single-frequency receivers in real-time solution, especially tropospheric delay, which highly correlated with ellipsoidal height.

## ACKNOWLEDGMENTS

The data set used in this research was collected using King Abdulaziz University vessel, Hydrography 1. The Author, therefore, acknowledges with thanks the technical stuff from hydrographic surveying department, Faculty of Maritime Studies.

## REFERENCES

Alonso-Garcia, S., Gomez-Gil, J., Arribas, J., 2011. Evaluation of the use of low-cost GPS receivers in the autonomous guidance of agricultural tractors. Spanish Journal of Agricultural Research 9, 377-388.

Arbesser-Rastburg, B., 2002. Ionospheric corrections for satellite navigation using EGNOS. Proc. of XXVII-th URSI General Assembly, Maastricht.

Cai, C., Liu, Z., Luo, X., 2013. Single-frequency Ionosphere-free Precise Point Positioning Using Combined GPS and GLONASS Observations. The Journal of Navigation 66, 417-434.

El-Rabbany, A., 2006. Introduction to GPS : the Global Positioning System, 2nd ed. Artech House, Boston, MA.

Hedgecock, W., Maroti, M., Sallai, J., Volgyesi, P., Ledeczi, A., 2013. High-accuracy differential tracking of low-cost GPS receivers, Proceeding of the 11th annual international conference on Mobile systems, applications, and services. ACM, pp. 221-234.

Hofmann-Wellenhof, B., Lichtenegger, H., Wasle, E., 2008. GNSS - global navigation satellite systems : GPS, GLONASS, Galileo, and more. Springer, Wien ; New York.

IHO, 2008. IHO STANDARDS FOR HYDROGRAPHIC SURVEYS, Special Publication No. 44, 5th Edition, February 2008, International Hydrographic Bureau, MONACO. http://www.iho.int/iho_pubs/standards/S-44_5E.pdf.

Jo, H., Sim, S.-H., Tatkowski, A., Spencer, B.F., Nelson, M.E., 2013. Feasibility of displacement monitoring using low-cost GPS receivers. Structural Control and Health Monitoring 20, 1240-1254.

Koch, F., Prasch, M., Schmid, L., Schweizer, J., Mauser, W., 2014. Measuring Snow Liquid Water Content with Low-Cost GPS Receivers. Sensors 14, 20975-20999.

Mohamed El-Diasty, Elsobeiey, M., 2015. Precise Point Positioning Technique with IGS Real-Time Service (RTS) for Maritime Applications. Positioning 6, 71-80.

Odijk, D., Teunissen, P., Zhang, B., 2012. Single-Frequency Integer Ambiguity Resolution Enabled GPS Precise Point Positioning. Journal of Surveying Engineering 138, 193-202.

Odijk, D., Teunissen, P.G., Khodabandeh, A., 2014. Single-Frequency PPP-RTK: Theory and Experimental Results, in: Rizos, C., Willis, P. (Eds.), Earth on the Edge: Science for a Sustainable Planet. Springer Berlin Heidelberg, pp. 571-578.

Schüler, T., Diessongo, H., Poku-Gyamfi, Y., 2011. Precise ionosphere-free single-frequency GNSS positioning. GPS Solut 15, 139-147.

Schwieger, V., 2007. High-Sensitivity GPS–the low cost future of GNSS. FIG Working Week, Hong Kong, SAR 13, 2007.

Shi, C., Gu, S., Lou, Y., Ge, M., 2012. An improved approach to model ionospheric delays for single-frequency Precise Point Positioning. Advances in Space Research 49, 1698-1708.

Sioulis, A., Tsakiri, M., Stathas, D., 2015. Evaluation of low cost, high sensitivity GNSS receivers based on the ISO RTK standards. INTERNATIONAL JOURNAL OF GEOMATICS AND GEOSCIENCES 6, pp 1597- 1606

Stempfhuber, W., Buchholz, M., 2011. A precise, low-cost RTK GNSS system for UAV applications. International Archives of Photogrammetry, Remote Sensing and Spatial Information Science 38, 1-C22.

Sterle, O., Stopar, B., Pavlovčič Prešeren, P., 2015. Single-frequency precise point positioning: an analytical approach. J Geod 89, 793-810.

Takasu, T., Yasuda, A., 2008. Evaluation of RTK-GPS performance with low-cost single-frequency GPS receivers, Proceedings of international symposium on GPS/GNSS 2008 Tokyo, Japan, pp 852–861, pp. 852-861.

Wang, N., Yuan, Y., Li, Z., Huo, X., 2016. Improvement of Klobuchar model for GNSS single-frequency ionospheric delay corrections. Advances in Space Research.

# 22

# IDENTIFYING THE ROLE OF NATIONAL DIGITAL CADASTRAL DATABASE (NDCDB) IN MALAYSIA AND FOR LAND-BASED ANALYSIS

N.Z.A. Halim [a], S.A. Sulaiman [a], K. Talib [a], O.M. Yusof [a], M.A.M. Wazir [a], M.K. Adimin [b]

[a] Center of Studies for Surveying Science and Geomatics, Faculty of Architecture, Planning and Survey, University Technology Mara, Shah Alam, Malaysia – nurzurairah@gmail.com
[b] Cadastral Division, Department of Survey and Mapping Malaysia, Kuala Lumpur, Malaysia – kamali@jupem.gov.my

KEY WORDS: National Digital Cadastral Database, Delphi technique, disaster management

**ABSTRACT:**

This paper explains the process carried out in identifying the significant role of NDCDB in Malaysia specifically in the land-based analysis. The research was initially a part of a larger research exercise to identify the significance of NDCDB from the legal, technical, role and land-based analysis perspectives. The research methodology of applying the Delphi technique is substantially discussed in this paper. A heterogeneous panel of 14 experts was created to determine the importance of NDCDB from the role standpoint. Seven statements pertaining the significant role of NDCDB in Malaysia and land-based analysis were established after three rounds of consensus building. The agreed statements provided a clear definition to describe the important role of NDCDB in Malaysia and for land-based analysis, which was limitedly studied that lead to unclear perception to the general public and even the geospatial community. The connection of the statements with disaster management is discussed concisely at the end of the research.

## 1. BACKGROUND

### 1.1 Role of land information for disaster management

Land registers and cadastre have the role to play in supporting governments and citizens in their efforts at mitigating climate change and trying to adapt to its impact (Van der Molen, 2009a). For example, a land acquisition process of land or property that are exposed to disaster risk can be done to reduce the effect of such unwanted catastrophe as the decision can be made priorly from a land-based analysis, from the long-term perspective and sustainable measure (Banba, 2017). Mitchell, Enemark, and van der Molen (2015) have argued that effective land administration and management be a necessary prerequisite for improving adaptive capacity and for climate resilient urban development. They have drawn on literature that articulates how improved tenure security reduces vulnerability to natural disasters and the importance of addressing land issues in risk reduction. They have also outlined the benefits of improved land-use planning to climate risk factors and vulnerability, and concluded the information on the people to the land relationship is crucial for post-disaster management to enable spatially accurate land planning and redevelopment.

The cadastre is the engine of land administration systems which is responsible for registering the land parcel's rights, restrictions and responsibilities (Bennett, Wallace, & Williamson, 2008). The combination of disaster risk information and cadastre information on land tenure, land value, and land use enables the required risk inhibition and mitigation measures to be identified and assessed in line with legal, economic, physical, and social consequences (Williamson, Enemark, Wallace, & Rajabifard, 2010).

### 1.2 Land Information System in Malaysia

The cadastral system in Peninsular Malaysia and Federal Territory of Labuan consists two main components which are the land registry and the cadastral survey. Except for Sabah and Sarawak, the rest of the states in Malaysia adopts the Torrens System and strictly ties to the National Land Code, 1965 or NLC, 1965 and the respective state's land law and legislation. The land parcel's information of states adopting the NLC,1965 can be retrieved from the Computerised Land Registration System (SPTB), at their respective state land offices (Ismail, 2011). The information provides textual attributes such as the land's ownership, rights, restrictions, and responsibilities. The land parcel's spatial information, on the other hand, is shown as cadastral maps (Certified Plan and Title Plan) can be retrieved either as a hardcopy or a digital map. Maps have long served as a fundamental purpose of understanding the geographical context of a disaster, and with the ability of GIS today, certain aspects of disaster situation can be represented (Tomaszewski, 2014) and spatially analysed. A Title Plan provides a 1:1 link relationship between land parcels and land registration. Alternatively, the spatial information of the cadastral maps can be obtained from the National Digital Cadastral Database or NDCDB.

### 1.3 National Digital Cadastral Database (NDCDB)

NDCDB was developed to replace the drawbacks of the former Digital Cadastral Database (PDUK) that was predominantly GIS technology unfriendly and not survey accurate (Kadir et al., 2003). It houses more than 7 million cadastral fabrics' in vector form, which is the spatial component of the cadastral system in Peninsular Malaysia and Federal Territory Labuan (to be mention as Malaysia in this paper from this point onwards).

The development of NDCDB included the 1st class and 2nd class survey information that was previously stored in PDUK, as well as the current 'free class' ones under the new eKadaster system (Yusoff & Halim, 2012). The least square adjustment method was opted for network readjustment to meet the acceptable horizontal tolerances of 5cm compared to the generic error method computed from loop closures of a closed traverse (JUPEM, 2012).

One of the main characteristics of NDCDB is the coordinate system adopts the Geocentric Cassini coordinate system and provides the positional spatial accuracy of 5cm for urban areas and 10cm for suburban areas. In other words, the spatial information is of survey accurate. Compared to PDUK, NDCDB was developed to be GIS friendly and available in either shapefiles or tab files. Each cadastral lots are also given a unique parcel identifier or UPI that responds as a foreign key for data relationship and attachments, thus enable data integration and data sharing. These characteristics permit NDCDB to be primarily used, but not exclusively, for spatial analysis. Even though the significance of the cadastral data has been discussed extensively internationally, the significant role of NDCDB in Malaysia specifically on land-based analysis is limitedly discussed. The situation has led to various perceptions and misinformation on NDCDB.

Generally, the societal impression of the cadastral data has been limitedly relevant to a single core statutory purpose of land registration that also included maintaining cadastral information consisting of rights, restrictions, and responsibilities (Van der Molen, 2009b). The same perception also applies in Malaysia where NDCDB is seen as a database that stores spatial information available in cadastral maps and nothing more. Nevertheless, a common stand was established among researchers (Mohamed, Chia, & Chan, 1998; Nordin, 2001; Omar, Kadir, & Sidek, 2006; Sim, 2012) that PDUK and now NDCDB, have impacted to the rise of GIS-based system development in Malaysia by multi-users.

## 1.4 Methods for building decisions and form consensus

In order to determine the significant role of NDCDB in Malaysia and evaluate its key themes, subjective insights and judgments on a collective basis from individuals with sundry expertise seemed appropriate. Among the best-known methodologies to achieve consensus are the Nominal Group Method and the Delphi technique (McMillan, King, & Tully, 2016). The Nominal Group method gathers all participants at the same time and location commonly in a workshop, being a cost-effective and time-efficient method. The method, however, requires an accurate pre-planning from the moderator and participants, and availability of participants since they are assembled, face to face in a single session.

The Delphi technique, on the other hand, minimises the influence of individuals and maximises the reliability of results by providing anonymity to the participants. Their participants are typically experts with a specific criterion to the subject discussed. The Delphi technique has application whenever policies, plans, or ideas have to be based on informed judgment (Yousuf, 2007), and because of that, the Delphi technique was chosen in this study. Moreover, the technique provides a platform where consensus by subject matter experts can be

achieved on a topic despite the limited evidence, lack of precise information and prior research (Avella, 2016; Yousuf, 2007).

Use of the web and conducting panel activities by e-mail is principally effective to meet the expert's busy timetable, logistics and provides Delphi participants with anonymity, privacy, and confidentiality. Equal status and equal opportunity to participate is permissible with anonymity. The disadvantages associated with face-to-face meetings such as personality influences or individual dominance can be avoided (Avella, 2016; McGeary, 2009) to elude influential result and inaccurate analysis.

## 2. LITERATURE ON THE DELPHI TECHNIQUE

### 2.1 Overview of the Delphi Technique

The Delphi technique is recognised as a consensus method, with the capability of providing insights for decision making (Förster & von der Gracht, 2014) and enhance consensus building as well as consistency from a group of experts regarding a topic (McGeary, 2009; Tottossy, 2005). Yousuf (2007) highlighted the technique is suitable for gathering current and historical data that is vaguely known or unavailable, evaluating possible budget allocations, exploring urban and regional planning options, university campus and curriculum planning development, putting together an educational model, delineating the pros and cons associated with potential policy choices, distinguishing and clarifying real and perceived information and motivation, and exploring priorities of personal values, social goals, etc.

The Delphi technique, however, does not have a specific standard for adoption (Skulmoski, Hartman, & Krahn, 2007). Instead, its method is flexible for modification to suit a research objective. Nevertheless, there are four distinct characteristics of the Delphi technique (Heiko, 2012; Rowe & Wright, 1999), which are anonymity of participants, iteration process to determine the level of consensus, controlled feedback to ensure stability in responses and statistical "group response" to measure agreement and stability. A type of method that adheres to these characteristics is distinguished as the Classical Delphi technique, and most researchers (Avella, 2016; Förster & von der Gracht, 2014; Kent & Saffer, 2014; Markmann, Darkow, & von der Gracht, 2013; McGeary, 2009; Skulmoski et al., 2007; Yang, 2003) agree that the method combines the qualitative and quantitative research design.

The steps commonly used in the Delphi technique from previous studies (Avella, 2016; Diamond et al., 2014; Holey, Feeley, Dixon, & Whittaker, 2007; Kermanshachi, Dao, Shane, & Anderson, 2016; McGeary, 2009; Ogbeifun, Agwa-Ejon, Mbohwa, & Pretorius, 2016b; Skulmoski et al., 2007; Yang, 2003) can be summarised as follows: i) identifying and selecting the panel of experts; ii) getting consent and approval from the selected experts as the research participant and the Delphi panel member; iii) data collection from the panel where different types of either closed or open-ended questions can be used. Otherwise, the questions can be more structured to guide the Delphi participants towards the research objective; iv) qualitative, quantitative, or mix mode analysing data from the panel; v) collating information on a new questionnaire and send back to panel to evaluate; vi) compile every panel's evaluation and send back to each participant for comments and revision; vii) the panel of experts have the opportunity to provide brief written arguments to support their evaluations; viii) descriptive statistical analysis method on the compiled and revised panel evaluation; ix) the same panel of members are involved during each iterative process survey rounds, and x) the survey round or iteration process is terminated the momment a full consensus or dissent are reached among panel experts.

The mode of interaction in the Delphi technique was initially paper and pen-based that consequently requires the panel to post relevant questionnaire or feedbacks either via snail mail or hand delivered. However, with the advent of ICT, panel activities and questionnaires can be communicated online and sent digitally to each participant through e-mail. Recent research was conducted to compare the result from both modes of interaction on the same issue, and the result indicates high similarity (Markmann et al., 2013; Ogbeifun, Agwa-Ejon, Mbohwa, & Pretorius). Nonetheless, rapid turnarounds reduced operational cost and eliminating participant's mortality are among the advantages of optimising the later mode of interaction (Hanafin, 2004; McGeary, 2009).

## 2.2 Participants

Selecting experts to participate as a panel member is the utmost importance in the Delphi technique to ensure research reliability. Even so, there are no standards that stipulate the panel selection criteria. The selection of panel, however, is rigorous, should not be random, and must have an explicit criterion. Warth, Heiko, and Darkow (2013) highlighted four important criteria for selecting the experts in their study, which required experts to have: i) extensive knowledge and experience of the research topic; ii) willingness and commitment to participate; iii) sufficient time for panel activities; and iv) communication skills. To define a participant as an expert, some researchers added the panel selection criteria for the Delphi technique to meet the specific number of years the expert has experience and practicing in the specific study area (Kent & Saffer, 2014; Ogbeifun et al., 2016a; Paul, 2014), while others impose professional body membership or certification, and senior designation or managerial levels in the specific study domain (Förster & von der Gracht, 2014; Markmann et al., 2013).

The appropriate panel composition has never been established for this technique, but a heterogeneous panel of experts are preferred as indicated by previous research (Avella, 2016; Förster & von der Gracht, 2014; Reefke & Sundaram, 2017; Warth et al., 2013) to avoid invalid consensus among similarly thinking experts or bias in the result and non-representative domain experts. Nonetheless, covers the wider spectrum, the point of views and diverse perspectives. Heterogeneity includes aspects related to the panel's professional experience and knowledge. There is also no specific number of groups to establish a heterogeneous panel of experts. However, the literature shows two or three groups are common, and participation should be based on those groups most directly affected by the topic of the study (Avella, 2016). The probability of participants to have similar opinions may create a biased result (Förster & von der Gracht, 2014) in a homogeneous panel since the panels have similar information base and desirability.

The standard size of the overall panel has also never been established. However, typical panels seem to fall in the 3 to more than 100-member range and consist of either two or three expert groups (Avella, 2016; Paul, 2014; Skulmoski et al., 2007) that should be balanced to eliminate group domination. Nevertheless the greater the heterogeneity of the group, the fewer number of experts are recommended where 5 to 10 members are considered ideal (Bueno & Salmeron, 2008).

## 2.3 Level of consensus

Delphi studies have used subjective analysis, descriptive statistics, and inferential statistics for the definition of a stopping criterion in each iteration process. Apart from consensus measurement, researchers (Diamond et al., 2014; Förster & von der Gracht, 2014; Heiko, 2012) recommended the stability or consistency be measured as well between panel responses in each iteration processes or rounds, to ensure accurate Delphi analysis and data interpretation. Reference (Heiko, 2012) and (Holey et al., 2007) added the elements of a reduced number of comments and evolution of statements to be a part of the stopping criterion as the indicators that consensus and stability have been reached in the Delphi study. The minimum round for consensus can be two but, three rounds are common in most studies (Holey et al., 2007). If group consensus is desirable and the sample is heterogeneous, then three or more rounds may be required (Skulmoski et al., 2007). Round one is generally the foundation for identifying issues of the research area, while the following rounds two and three are consensus building rounds (Reefke & Sundaram, 2017).

## 3.   RESEARCH METHODOLOGY & RESULT

Considering all the inputs from the literature, the research methodology for this study has adopted the common steps of a Classical Delphi technique, recommendations from previous research and modification to suit the research questions. The following are the methodology carried out in identifying the significant role of NDCDB in Malaysia and for land-based analysis:

### 3.1   Questionnaire design, validity, and reliability

Questionnaires are the research instrument to collect data in this study. In designing the questionnaires for Round 1, open-ended close structured questions were formed based on the overall research framework. The Round 2 questionnaire was the result of the Round 1 qualitative analysis result. Given the importance of this study on policies revolving the usage of NDCDB specifically for land-based analysis, the research instruments which are the questionnaires were assessed for validity and pretested for reliability.

The open-ended close structured and close-ended close structured Questionnaire's content for Round 1, and Round 2 used in this study were validated by four experts who are either practitioners or academia. Content validity index (I-CVI) for individual questions were used where the experts were asked to review the relevance of each question on a 4-point Likert scale of 1-Extremely irrelevant, 2-not relevant, 3-relevant and 4-very relevant. Polit and Beck (2006) highlighted all experts must agree on the content validity of an item (I-CVI of 1.00) when the panel consists of five or fewer experts. All experts in this study were in agreement that all of the twelve questions related to the significant role of NDCDB in Malaysia and land-based analysis be very relevant, which resulted in the calculated I-CVI value as 1.00. Experts were also instructed to give comments on the questionnaire's readability, feasibility, layout and style, and clarity of wording, if relevant.

Pretesting is not common in the Delphi technique (Avella, 2016; Hanafin, 2004), but if conducted the recommended minimum participants are 10% of the actual panel size (Waweru & Omwenga, 2015). Pretesting was performed of the Round 1 open-ended close structured questionnaire followed with the close-ended close structured questionnaire in Round 2 before the actual respective Delphi rounds. The pretesting was deemed necessary in this study to determine the reliability of the questionnaire in establishing whether the response to each question can be adequately interpreted in relation to the information required. Two rounds of test-retest pilot studies were carried out respectively for Round 1 and Round 2 questionnaires with three participants, which was more than the 10% numbers of participants recommended for pretesting. These participants provided comments and feedbacks, but their responses were not involved in the actual study. Participants in the pilot study have provided valuable feedback that contributed to the overall refinement of the study such as rephrasing questions, refine instructions and helps to identify unclear or ambiguous statements in the research protocol besides providing credibility to the entire research project.

### 3.2 Composition and panel size

Considering NDCDB has a direct association with the land, cadastre and GIS domain, a heterogeneous group of a panel of experts was set up in this study to diversify member's background and reduce bias result. The panel consists of experts involved directly in the decision-making process of their respective organisations. Initially, 18 potential members were carefully selected based on the 4 criterion used by previous Delphi researchers (Markmann et al., 2013; Warth et al., 2013), such as; i) number of years of job-relevant experience in their domain with 20 years as the threshold; ii) the expert's certification and accreditation to reflect their skills and knowledge; iii) current management level; and iv) labelled experts by social acclamation approach. Finally, the expert's responsibilities inside and outside of the organisation were also taken into consideration for panel selection (Avella, 2016; Markmann et al., 2013) in this study. The final totaled panel size throughout the Delphi process was 14 which can be summarized in three major groups: policy makers, interest group and stakeholders who are distinguished by individuals' professional experience as a specialist in Malaysia of either in the land, cadastre or GIS domain.

### 3.3 Round 1

After piloting, an email inviting participation, explaining the study, outlining the Delphi process and requesting responses to the question was sent to the potential participants. Simultaneously, participants were provided with an open-ended semi-structured questionnaire that has been pretested, to be completed by them. The questions were generally focused on the participants' expertise views based on themes of legal significance, technical significance, role significance and land-based analysis significance. However, the significance of the

NDCDB's role and land-based analysis theme are the focus of this paper.

Upon agreeing to participate the Delphi process, participants were initially given seven days to respond to the questionnaire questions, but some requested an extension of time due to the participant's busy schedule and commitments. A cut-off date was set to determine the timeline of Round 1. Out of 18 potential participants, 15 agreed to participate and returned the completed questionnaire through e-mail and hand-delivered, while 2 participants explained their busy schedule might hinder the study, and 1 participant did not return the questionnaire upon the cut-off date. Each response was of anonymity and was then compiled, collated and sent via e-mail to each participant, instead of a face-to-face group discussion. They were instructed to review the response from other anonymous respondents and were able to revise their responses within seven days. Only 2 participants had minor amendments by the cut-off date.

### 3.4 Round 1 Result

Responses from Round 1 were analysed by using the Thematic Analysis process. The qualitative data was firstly scrutinised, and emerging patterns or findings that were identified as nodes were processed using the Nvivo 11 Plus software. The nodes were listed under the Free Nodes theme. From the Free Nodes, a deductive approach was used to generate a Parent Node theme based on the research question and research objective, which focuses on the NDCDB's significant role in Malaysia and significant role for land-based analysis. Similar ideas related to the later theme were clustered together into emerging sub-themes. The emerging sub-themes were identified and named as shown in Table 1.

| No | Emerging sub-themes for NDCDB role in Malaysia | No | Emerging sub-themes for NDCDB for land-based analysis |
|----|------------------------------------------------|----|-------------------------------------------------------|
| 1 | Reference datasets | 1 | Sufficient knowledge on NDCDB |
| 2 | Aid sustainable development | 2 | Means for correct adoption of NDCDB |
| 3 | Support decision making | 3 | NDCDB as the fundamental layer |
| 4 | Planning activities | 4 | Geocentric coordinated system |
| 5 | Spatial enabler | | |
| 6 | Underpins national development program | | |

Table 1: Emerging sub-themes for Parent Node Theme NDCDB Role and NDCDB for land-based analysis

Statements that best described the essence of the majority of opinions within each sub-theme in Table 1 were generated and presented in Table 2 and 3. These statements on the NDCDB's significant role in Malaysia and significant role for land-based analysis have provided the basis for Round 2.

### 3.5 Round 2

Participants were presented with an online questionnaire that includes both themed statements shown in Table 2 and 3. A five-point Likert scale was used to gather participant's opinion and perception on the statements and linguistic scales from 5 points as being "strongly agree", 4 points for "slightly agree", 3

| No | Statements |
|----|------------|
| 1 | NDCDB is a spatial enabler that is also recognised as one of the key geospatial reference frames in Malaysia which facilitate the development of large-scale geospatial database and large-scale Spatial Data Infrastructure that is essential towards the conception of a spatially enabled platforms nationwide. |
| 2 | NDCDB underpins the national land-based programs by providing the homogeneity of a land parcel's legal status, its geospatial information, and other relevant data so a cohesive land-based spatial analysis can be performed, thus allowing accurate decision-making specifically for sustainable land governance. |
| 3 | NDCDB enables users to have cadastral fabrics that are a reliable reference for land administrative purposes when the land and people's relationship is the concern during planning or development stages. |

Table 2. Round 2 statements on NDCDB's significant role in Malaysia

| No | Statements |
|----|------------|
| 1 | NDCDB should be the fundamental dataset for any land-based spatial analysis because it provides meaningful insights on the land parcel's legal status and spatial information relationship with other geospatial or non-geospatial datasets. |
| 2 | NDCDB does improve the result of a land-based spatial analysis once users have good comprehension on its characteristics besides applying the correct adoption for spatial analysis. |
| 3 | NDCDB is most suitable as base maps for planning purposes specifically large-scale spatial analysis and long distance network studies. |
| 4 | NDCDB enables other overlayed datasets and information to be linked with Malaysia's geocentric datum and analyse spatial data with survey accurate results. |

Table 3. Round 2 statements on NDCDB's significant role in land-based analysis

| Statement | Round 2 | | | | | Round 3 | | | | | Mean Percent Change 100% |
|-----------|---------|-----|--------|------|----------------|----|-----|--------|------|----------------|---------------------------|
|           | N | IQR | Median | Mean | Std. Deviation | N | IQR | Median | Mean | Std. Deviation |    |
| Statement 1 | 14 | 1.00 | 5.00 | 4.64 | 0.497 | 14 | 1.00 | 5.00 | 4.71 | 0.469 | 1.40% |
| Statement 2 | 14 | 1.00 | 5.00 | 4.64 | 0.497 | 14 | 0.00 | 5.00 | 4.79 | 0.426 | 3.00% |
| Statement 3 | 14 | 1.00 | 5.00 | 4.57 | 0.514 | 14 | 1.00 | 5.00 | 4.86 | 0.363 | 5.80% |

Table 4: Statistical analysis of Round 2 and 3 – NDCDB's role in Malaysia

| Statement | Round 2 | | | | | Round 3 | | | | | Mean Percent Change 100% |
|-----------|---------|-----|--------|------|---------------|----|-----|--------|------|---------------|---------------------------|
|           | N | IQR | Median | Mean | Std Deviation | N | IQR | Median | Mean | Std Deviation |    |
| Statement 1 | 14 | 1.00 | 5.00 | 4.57 | 0.514 | 14 | 0.00 | 5.00 | 4.93 | 0.267 | 7.20% |
| Statement 2 | 14 | 1.00 | 4.00 | 4.43 | 0.514 | 14 | 0.00 | 5.00 | 4.79 | 0.426 | 7.20% |
| Statement 3 | 14 | 1.00 | 5.00 | 4.57 | 0.514 | 14 | 0.00 | 5.00 | 4.79 | 0.426 | 4.40% |
| Statement 4 | 14 | 1.00 | 4.50 | 4.50 | 0.519 | 14 | 1.00 | 5.00 | 4.71 | 0.469 | 4.20% |

Table 5: Statistical analysis of Round 2 and 3 – NDCDB's role in land-based analysis

points for "less agree", 2 points as "disagree" ad lastly 1 point as "strongly disagree". Neutral answer scale was not a choice because of it is equivalent to a no-judgement situation, whereas the Delphi technique should promote a discussion or otherwise to reduce bias result. Five-point or seven-point scales are generally preferred since smaller scales cannot transmit as much information and can stifle respondents whereas larger scales are less accurate (Reefke & Sundaram, 2017). Space was provided for optional comments justifying their scale allocation decisions. All 15 original participants were offered via e-mail to participate in Round 2 and were requested to respond within seven days. Similar to Round 1, each response was of anonymity and were then compiled, collated and sent via e-mail to each participant who was then instructed to review the response from other anonymous respondents and was able to revise their responses within seven days. Only comments were added by the cut-off date.

### 3.6  Round 2 Result

The number of participants allocating points to each statement in Round 2 was totaled as 14. 1 participant was transferred to a new agency and had to be withdrawn from the panel member. Likert scale points allocated were analysed with descriptive statistics using SPSS version 23 software, and statements with means lower than four are respectively removed. However, none were eliminated in this round as the means were higher than 4.00 as shown in Table 4 and 5. Additional comments received suggested minor amendments which required statements to be rephrased appropriately with the aim of moving towards group consensus and stability, as presented in Table 6 and 7.

### 3.7  Round 3

The Round 2 methodology was repeated, but with the original Statement 1 and rephrased Statement 2 and 3 for NDCDB's role in Malaysia while rephrased Statement 1 and 2, and original

Statement 3 and 4 for NDCDB's role in the land-based analysis was the basis for Round 3 questionnaire. 14 participants were offered to participate in Round 3 and were requested to respond within three days.

## 3.8  Round 3 Result

The number of respondents received in Round 3 was totaled as 14 or 100% feedback. Similar to Round 2, the Likert scale point's responses were analysed with descriptive statistics. The increase in mean values was found in the rephrased statements as stated in Table 5 and 6, with means value are more than 4.50. The Median and Interquartile Range (IQR) are best statistical choices to measure central tendency and dispersion to calculate data scored on an ordinal scale in Delphi processes (Heiko, 2012). The median values in Round 3 were all 5.0 while the IQR values were within 0.00 to 1.00, which is suitable as consensus indicator for 5-unit scales (Yang, 2003). The values showed consistency and high agreement, and therefore it was decided the Delphi process has come to a group consensus and stability which in result supports the decision that further Delphi rounds were unnecessary. Moreover, additional comments were not received in this round.

## 4.  ANALYSIS

### 4.1  Statistical result analysis

Analysis of the Delphi technique results showed a change in participants' views towards consensus (agreement) and stability in Round 3. Therefore, the Delphi is concluded to have reach panel consensus and stability in Round 3 as indicated by a trend towards the following:

i.   The median for the 5-Point Likert scale is the highest measure (5 – "strongly agree") to all statements;

ii.  The IQR values are within 0.00 to 1.00 which indicate high agreement (Peck & Devore, 2011);

iii. The increase of mean values in each round;

iv.  Percent change between each statement from both rounds are less than 15% and is considered stable (Diamond et al., 2014; Heiko, 2012);

v.   A decrease in comments in each round; and

vi.  The evolution of statements towards consensus

### 4.2  Agreed statements on the significant role of NDCDB in Malaysia

The statements agreed by the members of Delphi panel describing the significant role of NDCDB in Malaysia are established as follows:

*Statement 1: NDCDB is a spatial enabler that is also recognised as one of the key geospatial reference frames in Malaysia which facilitate the development of large-scale geospatial database and large-scale Spatial Data Infrastructure that is essential towards the conception of a spatially enabled platforms nationwide.*

*Statement 2: NDCDB underpins the national land-based programs by providing the homogeneity of a land parcel's legal status, its geospatial information, and other relevant data so a cohesive land-based spatial analysis can be performed to support sustainable land governance.*

*Statement 3: NDCDB provides reliable references of cadastral fabrics for land administrative purposes when the land and people's relationship becomes the concern for accurate decision making during planning or development stages.*

### 4.3  Agreed statement on the significant role of NDCDB for land-based analysis

The statements agreed by the members of Delphi panel describing the significant role of NDCDB for land-based analysis are established as follows:

*Statement 1: NDCDB should be recognised as the fundamental dataset for land-based spatial analysis because it provides meaningful insights on the land parcel's legal status and spatial information relationship with other geospatial or non-geospatial datasets.*

*Statement 2: NDCDB can help provide the required result of a land-based spatial analysis when users understand its characteristics and know how to correctly adopt it for spatial analysis.*

*Statement 3: NDCDB is most suitable as base maps for planning purposes specifically large-scale spatial analysis and long distance network studies.*

*Statement 4: NDCDB enables other overlayed datasets and information to be linked with Malaysia's geocentric datum and analyse spatial data with survey accurate results.*

## 5.  DISCUSSION AND CONCLUSION

The statements that best described the essence of the majority of opinions were critically analysed in Round 1 of this research. The majority of the Delphi panels highlighted that NDCDB would provide an impact in facilitating the concept of a spatially enabled society and government in Malaysia, and fundamental information for decision-makers when people and the land relationship is the concern in sustainable decision-making. The descriptions of the NDCDB's role in Malaysia and for land-based analysis were then transpired to the seven statements established in this research by using the Delphi technique. It can be concluded that NDCDB does have significant roles for land-based analysis, which is also essential for sustainable development in Malaysia.

Sustainable development has the direct linkage to disaster management (Ujang, 2017) and a spatially enabled society or government allows decision-making that can be associated with sustainability (Steudler, 2016). One can interpret and critically reflect on spatial information, interconnect with the assistance of maps and other spatial representations, and express location specific opinions using geoinformation and associated supports. NDCDB provides the spatial information and basic land information that is fundamental to spatial analysis. In other words, better judgement on land-based analysis which is crucial for disaster management is permissible by optimising NDCDB along with other related datasets. Accurate and timely information on land is the fundamental information to enable the concept (Bennett et al., 2008).

| No | Statements | Rephrased Statements |
|---|---|---|
| 1 | NDCDB is a spatial enabler that is also recognised as one of the key geospatial reference frames in Malaysia which facilitate the development of large-scale geospatial database and large-scale Spatial Data Infrastructure that is essential towards the conception of a spatially enabled platforms nationwide. | No amendments. |
| 2 | NDCDB underpins the national land-based programs by providing the homogeneity of a land parcel's legal status, its geospatial information, and other relevant data so a cohesive land-based spatial analysis can be performed, thus allowing accurate decision-making specifically for sustainable land governance. | NDCDB underpins the national land-based programs by providing the homogeneity of a land parcel's legal status, its geospatial information, and other relevant data so a cohesive land-based spatial analysis can be performed *to support* sustainable land governance. |
| 3 | NDCDB enables users to have cadastral fabrics that are a reliable reference for land administrative purposes when the land and people's relationship is the concern during planning or development stages. | NDCDB *provides reliable references of cadastral fabrics* for land administrative purposes when the land and people's relationship *becomes* the concern *for accurate decision making* during planning or development stages. |

Table 6: Round 2 statements on NDCDB's significant role in Malaysia

| No | Statements | Rephrased Statements |
|---|---|---|
| 1 | NDCDB should be the fundamental dataset for any land-based spatial analysis because it provides meaningful insights on the land parcel's legal status and spatial information relationship with other geospatial or non-geospatial datasets. | NDCDB should be *recognised as* the fundamental dataset for land-based spatial analysis because it provides meaningful insights on the land parcel's legal status and spatial information relationship with other geospatial or non-geospatial datasets. |
| 2 | NDCDB does improve the result of a land-based spatial analysis once users have good comprehension on its characteristics besides applying the correct adoption for spatial analysis. | NDCDB *can help provide the required* result of a land-based spatial analysis *when users understand* its characteristics and *know how to correctly adopt it* for spatial analysis. |
| 3 | NDCDB is most suitable as base maps for planning purposes specifically large-scale spatial analysis and long distance network studies. | No amendments. |
| 4 | NDCDB enables other overlayed datasets and information to be linked with Malaysia's geocentric datum and analyse spatial data with survey accurate results. | No amendments. |

Table 7: Round 2 statements on NDCDB's significant role in land-based analysis

In fact, according to Steudler (2016) spatial is everywhere and our ability to leverage and harness the ubiquity of spatial information will correlate to benefits in terms of wealth creation, social stability and environmental management.

Both disaster management and sustainable development require sound land governance to reduce the impacts of climate change and post-disaster effects (Ujang, 2017). A sound land governance is where issues pertaining land are administered, and managed with emphasis given to the relations between people, policies and places in support of sustainability and the global agendas.

Among the many global agenda goals listed in Agenda 2030 mostly relates to the integral role of land, cadastre, and people in adapting to climate change, disaster recovery, environmental degradation and rapid urbanisation (UN, 2015). With these statements on NDCDB are in place, the clear description of NDCDB's role could encourage more research to be done, and NDCDB can be optimised by decision makers to aid land-based decision-making that also includes for disaster management and post-disaster effects analysis.

## 6. CONTRIBUTION

This study has contributed to the existing body of knowledge in understanding the NDCDB. Previously, there were limited empirical data available on unfolding the significance of NDCDB, specifically concerning its role in Malaysia and for land-based analysis. This research has helped to describe the importance of NDCB based on the consensus of experts related to land, cadaster and GIS domain by optimising the Delphi technique. The outcome of this research provides a new paradigm to the existing perception of the NDCDB and statements that aids land surveyors, land administrators, and GIS users, as well as the general public, to recognize its role in the nation and for land-based analysis. With the statements in place, it is hoped the usage of NDCDB for land-based analysis increases to encourage spatially accurate analysis result, including in the disaster management domain.

## ACKNOWLEDGEMENTS

The authors would like to Universiti Teknologi MARA, DSMM, and all related agencies for providing valuable information for this study. Thanks are also due to all my colleagues who assisted during this study data and last, but not least the unknown reviewers are gratefully acknowledged.

## REFERENCES

Avella, J. R. (2016). Delphi Panels: Research Design, Procedures, Advantages, and Challenges. International Journal of Doctoral Studies, 11.

Banba, M. (2017). Land Use Management and Risk Communication. In M. Banba & R. Shaw (Eds.), Land Use Management in Disaster Risk Reduction: Practice and Cases from a Global Perspective (pp. 13-17). Tokyo: Springer Japan.

Bennett, R., Wallace, J., & Williamson, I. (2008). Organising land information for sustainable land administration. Land Use Policy, 25(1), 126-138. doi:http://dx.doi.org/10.1016/j.landusepol.2007.03.006

Bueno, S., & Salmeron, J. L. (2008). Fuzzy modeling Enterprise Resource Planning tool selection. Computer Standards & Interfaces, 30(3), 137-147. doi:https://doi.org/10.1016/j.csi.2007.08.001

Diamond, I. R., Grant, R. C., Feldman, B. M., Pencharz, P. B., Ling, S. C., Moore, A. M., & Wales, P. W. (2014). Defining consensus: a systematic review recommends methodologic criteria for reporting of Delphi studies. Journal of clinical epidemiology, 67(4), 401-409.

Förster, B., & von der Gracht, H. (2014). Assessing Delphi panel composition for strategic foresight—A comparison of panels based on company-internal and external participants. Technological forecasting and social change, 84, 215-229.

Hanafin, S. (2004). Review of literature on the Delphi Technique. Dublin: National Children's Office.

Heiko, A. (2012). Consensus measurement in Delphi studies: review and implications for future quality assurance. Technological forecasting and social change, 79(8), 1525-1536.

Holey, E. A., Feeley, J. L., Dixon, J., & Whittaker, V. J. (2007). An exploration of the use of simple statistics to measure consensus and stability in Delphi studies. BMC Medical Research Methodology, 7(1), 52.

Ismail, M. S. (2011). National Land Code 1965: Electronic Land Administration System In Land Registries. Jurnal Pentadbiran Tanah, Bil 1/2011, 20.

JUPEM. (2012). Final Report JUPEM T8/2011 Memperkasakan NDCDB. JUPEM: JUPEM.

Kadir, M., Ses, S., Omar, K., Desa, G., Omar, A. H., Taib, K., & Nordin, S. (2003). Geocentric datum GDM2000 for Malaysia: Implementation and implications. Paper presented at the Seminar on GDM2000, Department of Survey and Mapping Malaysia, Kuala Lumpur, Malaysia.

Kent, M. L., & Saffer, A. J. (2014). A Delphi study of the future of new technology research in public relations. Public Relations Review, 40(3), 568-576.

Kermanshachi, S., Dao, B., Shane, J., & Anderson, S. (2016). Project Complexity Indicators and Management Strategies – A Delphi Study. Procedia Engineering, 145, 587-594. doi:http://dx.doi.org/10.1016/j.proeng.2016.04.048

Markmann, C., Darkow, I.-L., & von der Gracht, H. (2013). A Delphi-based risk analysis—Identifying and assessing future challenges for supply chain security in a multi-stakeholder environment. Technological forecasting and social change, 80(9), 1815-1833.

McGeary, J. (2009). A critique of using the Delphi technique for assessing evaluation capability-building needs. Evaluation Journal of Australasia, 9(1), 31.

McMillan, S. S., King, M., & Tully, M. P. (2016). How to use the nominal group and Delphi techniques. International Journal of Clinical Pharmacy, 38(3), 655-662. doi:10.1007/s11096-016-0257-x

Mitchell, D., Enemark, S., & van der Molen, P. (2015). Climate resilient urban development: Why responsible land governance is important. Land Use Policy, 48, 190-198. doi:http://dx.doi.org/10.1016/j.landusepol.2015.05.026

Mohamed, A. M., Chia, W. T., & Chan, H. S. (1998). Cadastral Reforms in Malaysia. Paper presented at the XXI FIG Congrees, Brighton, United Kingdom.

Nordin, A. F. (2001). Institutional Issues in the Implementation of the Coordinated Cadastral System for Peninsular Malaysia: A Study on the Legal and Organizational Aspects. (Masters), Universiti Teknologi Malaysia.

Ogbeifun, E., Agwa-Ejon, J., Mbohwa, C., & Pretorius, J. (2016a). The Delphi technique: A credible research methodology. Paper presented at the International Conference on Industrial Engineering and Operations Management, Kuala Lumpur.

Ogbeifun, E., Agwa-Ejon, J., Mbohwa, C., & Pretorius, J. (2016b). The Delphi technique: A credible research methodology.

Omar, A. H., Kadir, A. M. A., & Sidek, R. M. S. (2006). Development of Automated Cadastral Database Selection And Visualization System To Support The Realization of Modern Cadastre In Malaysia. Retrieved from

Paul, S. A. (2014). Assessment of critical thinking: A Delphi study. Nurse Education Today, 34(11), 1357-1360. doi:https://doi.org/10.1016/j.nedt.2014.03.008

Peck, R., & Devore, J. L. (2011). Statistics: The exploration & analysis of data: Cengage Learning.

Polit, D. F., & Beck, C. T. (2006). The content validity index: are you sure you know what's being reported? Critique and recommendations. Research in nursing & health, 29(5), 489-497.

Reefke, H., & Sundaram, D. (2017). Key themes and research opportunities in sustainable supply chain management – identification and evaluation. Omega, 66, Part B, 195-211. doi:https://doi.org/10.1016/j.omega.2016.02.003

Rowe, G., & Wright, G. (1999). The Delphi technique as a forecasting tool: issues and analysis. International journal of forecasting, 15(4), 353-375.

Sim, C. Y. (2012). Investigation of data models and related requirements affecting the implementation of a multipurpose cadastre system in Malaysia. (Master Dissertation/Thesis), University of Glasgow.

Skulmoski, G. J., Hartman, F. T., & Krahn, J. (2007). The Delphi method for graduate research. Journal of information technology education, 6, 1.

Steudler, D. (2016). The Six Fundamental Elements for Spatially Enabled Societies From a Fit-For-Purpose Perspective. Paper presented at the World Bank Conference on Land and Poverty, Washington DC.

Tomaszewski, B. (2014). Geographic Information Systems (GIS) for Disaster Management. Florida: CRC Press.

Tottossy, A. P. (2005). Teacher selection: A Delphi study. Virginia Polytechnic Institute and State University.

Ujang, Z. (2017). Contribution of research for Sustainable Social Development. Paper presented at the International Symposium on Disaster Risk Managemet 2017, Kuala Lumpur.

UN. (2015). Transforming Our World: The 2030 Agenda for Sustainable Development. Resolution adopted by the General Assembly on 25 September 2015. (A/RES/70/1). New York: United Nation Retrieved from http://www.un.org/ga/search/view_doc.asp?symbol=A/RES/70/1&Lang=E.

Van der Molen, P. (2009a). Cadastres and climate change. PositionIT, 19 - Jan/Feb 2010, 17-21.

Van der Molen, P. (2009b). The Evolving Function of Land Administration in Society. Paper presented at the 7th FIG Regional Conference, Hanoi, Vietnam.

Warth, J., Heiko, A., & Darkow, I.-L. (2013). A dissent-based approach for multi-stakeholder scenario development—the future of electric drive vehicles. Technological forecasting and social change, 80(4), 566-583.

Waweru, P. K., & Omwenga, J. (2015). The Influence of Strategic Management Practices on Performance of Private Construction Firms in Kenya. International Journal of Scientific and Research Publications, 5(6).

Williamson, I., Enemark, S., Wallace, J., & Rajabifard, A. (2010). Land administration for sustainable development. Paper presented at the FIG Congress 2010-Facing the Challenges-Building the Capacity.

Yang, Y. N. (2003). Testing the Stability of Experts' Opinions between Successive Rounds of Delphi Studies.

Yousuf, M. I. (2007). Using experts' opinions through Delphi technique. Practical assessment, research & evaluation, 12(4), 1-8.

Yusoff, M. Y. M., & Halim, N. Z. A. (2012). Country Report: Unleashing the Full Potential of eKadaster on The Cadastral System of Malaysia (E/CONF.102/CRP.3). Retrieved from Bangkok, Thailand:

# 23

# CADASTRAL DATABASE POSITIONAL ACCURACY IMPROVEMENT

N. M. Hashim [a*], A. H. Omar [a*], S. N. M. Ramli [a*], K.M. Omar [a*], N. Din [a, *]

[a] Dept. of Geomatic Engineering, University Technology Malaysia, Skudai, Johor Bahru, MALAYSIA –
*norshahrizan@perlis.uitm.edu.my, *abdullahhisham@utm.my, *nadhirahramli09@gmail.com,
*kamaludinomar@utm.my, *Fizul21@yahoo.com

**KEY WORDS:** Positional Accuracy Improvement, Legacy Dataset; Cadastral Database Modernization

**ABSTRACT:**

Positional Accuracy Improvement (PAI) is the refining process of the geometry feature in a geospatial dataset to improve its actual position. This actual position relates to the absolute position in specific coordinate system and the relation to the neighborhood features. With the growth of spatial based technology especially Geographical Information System (GIS) and Global Navigation Satellite System (GNSS), the PAI campaign is inevitable especially to the legacy cadastral database. Integration of legacy dataset and higher accuracy dataset like GNSS observation is a potential solution for improving the legacy dataset. However, by merely integrating both datasets will lead to a distortion of the relative geometry. The improved dataset should be further treated to minimize inherent errors and fitting to the new accurate dataset. The main focus of this study is to describe a method of angular based Least Square Adjustment (LSA) for PAI process of legacy dataset. The existing high accuracy dataset known as National Digital Cadastral Database (NDCDB) is then used as bench mark to validate the results. It was found that the propose technique is highly possible for positional accuracy improvement of legacy spatial datasets.

## 1. INTRODUCTION

Many countries around the world have recognized and appreciated the value of accurate digital cadastral database. Theoretically, an accurate, efficient and updated cadastral database offers the better basis for planning and implementation of variety of real estate application (Durdin, 1993; Effenberg et al. 1999; Ting et al., 1999). The advancement in spatial based technology like Geographic Information System (GIS) also suggested an urgent need to maintain the spatial data in high accuracy to represent the real world.

Principally, numerous spatial datasets were previously digitized from hardcopy maps. The legacy datasets in use today are a combination of data from different sources, of technologies and measurement techniques. As a result, these legacy datasets have relatively low positional accuracy caused by errors resulting from the production and measurement method employed according to the technological and legal changes over time (Sisman, 2014). In addition, the common error in digitizing process such as distortion of source map, digitizing operational errors and ground coordinate system constituted a combination of systematic and random errors (Tong, Shi, & Liu, 2009). However, the urgency of updating legacy dataset is extremely crucial, the needs for combine spatial data from different sources and accuracy has dramatically increased. This process is crucial to allow different datasets to be jointly presented and analyzed. The process requires an understanding about the positional accuracies of the geometries in the datasets to avoid mismatches and misinterpretations of geospatial data.

Generally, resurvey, reprocessing the existing survey data and upgrading the existing cadastral datasets are the potential approaches for improving the legacy cadastral datasets. Resurvey all the cadastral parcel probably is the best technique to generate the new accurate dataset, however this process require lengthy time and estimated cost is very high (Arvanitis et al., 1999). Meanwhile, Buyong et al. (1992) and Durdin (1993) suggested that maintaining the old measurement and incorporating the new measurements will suffice for upgrading legacy datasets. However, Perelmuter et al. (1992) stated that the incorporating is unsuitable due to weakness of the original control network (datum) and field book record system. Additionally, from the economic perspective, 20000 existing field books for instance, will require hundreds of operators and take many years to accomplish (Fradkin et al., 2002).

A possible alternative for improving the legacy datasets with realistic cost is transforming the legacy dataset into new system. This high possibility supported by the advancement of satellite base technology especially Global Navigation Satellite System (GNSS). Possibly, the coordinate accuracy of the legacy dataset can be highly improved. Furthermore, the availability of high resolution aerial imagery offer high possibilities of using imagery as background when underlays to spatial dataset (Hope et al., 2008). The imagery which has high quality spatial resolution and absolute accuracy like Quick Bird can be used as a base map to check the relative position of the adjusted legacy datasets.

Essentially, the legacy dataset accuracy has to be improved line with the current high accuracy spatial technology. As a result, many organizations and researchers around the world used PAI approach for upgrading legacy datasets into high accuracy dataset (Donnelly et al.2006; Fradkin et al., 2002; Hesse et al., 1990; Morgenstren, 1989; Tamim, 1995). Based on the

---

* Corresponding author

importance of PAI, the focus of this paper is to propose the PAI method using angular based Least Square Adjustment (LSA). The detail discussion on PAI is explained in the following section.

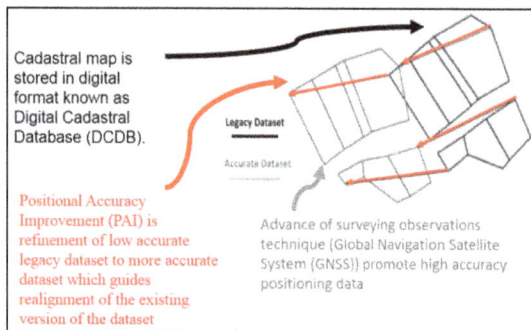

Figure 1. PAI concept (Hashim et al, 2016)

## 2. POSITIONAL ACCURACY IPROVEMENT

PAI is a process of improving the position of the geometric coordinates of a feature in a geospatial dataset to represent its actual position (Rönsdorf, 2008). The PAI can be classified as the improvement of low accurate legacy dataset to more accurate dataset. Figure 1 illustrates the concept of PAI (Norshahrizan et al, 2016).

PAI is commonly applied in two situations, PAI of Reference Data and PAI of User Data (Rönsdorf, 2008). The PAI of Reference Data links with improving the position of geometries in a reference dataset that describes physical or abstract features of the earth. The features position relate to the absolute position in a standard Coordinate Reference System such as Geocentric Datum of Malaysia 2000 (GDM 2000) in case of Malaysia or WGS-84 in global coordinate system. Meanwhile, the PAI of User Data describes the successive synchronization of legacy datasets with the already positionally improved reference dataset in order to retain the relationships between geometries.

To achieve an optimal solution in PAI, the method of LSA is often employed towards improving the positional accuracy of spatial datasets (Tamim 1995, Wolf and Ghilani 2006, Merritt and Masters 1999, Gielsdorf et al. 2004, Merritt 2005, Tong et al. 2005, Casado 2006, Hope et al. 2008, Tong et al. 2009). The LSA method is a well-established technique for solving an overdetermined system of equations by minimizing a weighted quadratic form of the residuals. Its application in estimating parameters in coordinate transformation can be found in literature; for example, Mikhail and Ackemann (1982), Wolf and Ghilani (2006), and Koch (1999). Tamim (1995) presented a methodology to create a digital cadastral overlay through upgrading digitized cadastral data. Merritt and Masters (1999) and Merritt (2005) developed the spatial adjustment engine based on the least squares method and applied it to improve the accuracy of cadastral data in Australia. Tong et al. (2005) presented a least squares adjustment model to resolve inconsistencies between the digitized and registered areas of cadastral parcels, and further improved the adjustment model by introducing scale parameters to reduce the influence of systematic error in the adjustment (Tong et al. 2009). Felus (2007) presented a workflow of three steps used to enhance the spatial accuracy of digital cadastral maps: a global

transformation from an old local system to a GPS-based WGS-84 system; a rubber-sheeting transformation for modifying boundary corners to fit existing ground features; and a LS adjustment with stochastic constraints to include additional cadastral information and geometric conditions. Hope et al. (2008) proposed a method of least squares with inequalities for data integration, in which topological relationships are modeled in the form of inequalities and optimal positioning solutions are obtained while preserving the spatial relationships among features.

Since the legacy dataset are less accurate in positioning, the integration of legacy datasets and highly accurate data such as those from GNSS is one of the most possible methods to improve the legacy datasets accuracy (Hope et al., 2008). However, it was found from Hope et al. (2008) that by simply replacing a sample of legacy dataset with more accurate version will lead to a distortion of the neighboring geometry (Figure 2). In addition, the relative geometry of the legacy datasets is often better than its absolute accuracy and supposed to be the spatial relationship or relative geometry between features must be preserved.

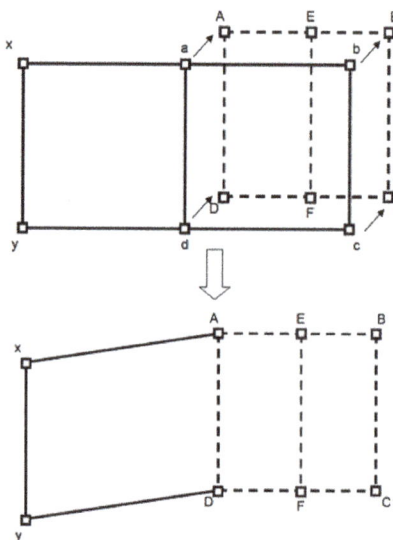

Figure 2. Cadastral boundaries (Solid Lines) and higher accuracy (Dashed Lines). Polygon xady is distorted if point abcd are simply replaced by ABCD (Hope et al., 2008)

The previous studies discussed above largely used the additional new observations which link with the legacy data for maintaining the geometry of adjusted data. In this study, the existing data alone from certified plan is used for maintaining the geometry of legacy data in the adjustment stage. The following section describes how the distance, bearings and angles are used in the LSA process.

### 2.1 Least square adjustment (LSA)

LSA is a model of solution widely used in the disciplines of surveying. The conventional Bowditch method is known as an arbitrary method since the corrections to the observations are applied irrespective of their uncertainties, whereas the least square adjustment method is more advanced technique. It adjusts observations based on the laws of probability, which models the occurrence of random errors (Halim et al., 2001). In other words, the LSA principle is such that the squares of the residuals must be minimized. Equation (1) expresses the fundamental principle of least squares:

$$\sum v^2 = v_1^2 + v_2^2 + \ldots + v_n^2 = minimum \qquad (1)$$

where    v = residual

There are two mathematical models in LSA, stochastic and functional model. The stochastic model involves the determination of variances and the weights of the observations. A functional model is an equation or set of equations that represents or defines an adjustment condition. There are two types of functional models, the conditional and parametric adjustments. In a conditional adjustment, geometric conditions are compulsorily imposed on the measurement and their residuals, for instance, the sum of the angles in a closed polygon is (n-2) x 180. In the parametric adjustment, observations are expressed in terms of unknown parameters that were never observed directly. For example, the well-known coordinate equations are used to model the angles, directions, and distances observed in a horizontal plane survey. The adjustment yields the most probable values for the coordinates (parameters), which in turn provide the most probable values for the adjusted observations.

In this study, the parametric adjustment of functional model is applied. The three mathematical models of observations, horizontal angle, azimuth and distance were used in the adjustment process. The mathematical model of bearing, horizontal angle and distance observation are like equations (2), (3) and (4) respectively (Amat, 2007; Setan, 1995; Wolf and Ghilani (2006).

$$\theta_1 = \tan^{-1}\left(\frac{X_B - X_A}{Y_B - Y_A}\right) \qquad (2)$$

$$\theta_1 = F(X_{AO}, Y_{AO}, X_{BO}, Y_{BO}) + \left(\frac{Y_{BO} - Y_{AO}}{L_{AB'O}^2}\right).\partial X_A$$
$$+ \left(\frac{Y_{AO} - Y_{BO}}{L_{AB'O}^2}\right).\partial X_B + \left(\frac{X_{BO} - X_{AO}}{L_{AB'O}^2}\right).\partial Y_A$$
$$+ \left(\frac{X_{AO} - X_{BO}}{L_{AB'O}^2}\right).\partial Y_B$$

where    $\theta_1$ = Bearing
$X_{AO}, Y_{AO}$ = Point A estimated coordinate
$X_{BO}, Y_{BO}$ = Point B estimated coordinate
$L_{AB'O}$ = Estimated horizontal distance A-B

$$\Delta\theta = \tan^{-1}\left(\frac{X_D - X_A}{Y_D - Y_A}\right) - \tan^{-1}\left(\frac{X_B - X_A}{Y_B - Y_A}\right) \qquad (3)$$

$$\Delta\theta = F(X_{AO}, Y_{AO}, X_{BO}, Y_{BO}, X_{DO}, Y_{DO})$$
$$+ \left(\frac{Y_{AO} - Y_{DO}}{L_{AD'O}^2} + \frac{Y_{BO} - Y_{AO}}{L_{AB'O}^2}\right).\partial X_A$$
$$+ \left(\frac{Y_{AO} - Y_{BO}}{L_{AB'O}^2}\right).\partial X_B + \left(\frac{Y_{DO} - Y_{AO}}{L_{AD'O}^2}\right).\partial X_D$$
$$+ \left(\frac{X_{AO} - X_{DO}}{L_{AD'O}^2} + \frac{X_{BO} - X_{AO}}{L_{AB'O}^2}\right).\partial Y_A$$
$$+ \left(\frac{X_{AO} - X_{BO}}{L_{AB'O}^2}\right).\partial Y_B + \left(\frac{X_{DO} - X_{AO}}{L_{AD'O}^2}\right).\partial Y_D$$

where    $\Delta\theta$ = Horizontal angle
$X_{AO}, Y_{AO}$ = Point A estimated coordinate
$X_{BO}, Y_{BO}$ = Point B estimated coordinate
$X_{DO}, Y_{DO}$ = Point D estimated coordinate
$L_{AB'O}$ = Estimated horizontal distance A-B
$L_{AD'O}$ = Estimated horizontal distance A-D

$$L_{AB'} = \sqrt{(X_B - X_A)^2 + (Y_B - Y_A)^2} \qquad (4)$$

$$L_{AB'} = F(X_{AO}, Y_{AO}, X_{BO}, Y_{BO}) + \left(\frac{X_{AO} - X_{BO}}{L_{AB'O}}\right).\partial X_A$$
$$+ \left(\frac{X_{BO} - X_{AO}}{L_{AB'O}}\right).\partial X_B + \left(\frac{Y_{AO} - Y_{BO}}{L_{AB'O}}\right).\partial Y_A$$
$$+ \left(\frac{Y_{BO} - Y_{AO}}{L_{AB'O}}\right).\partial Y_B$$

where    $L_{AB'}$ = Distance
$X_{AO}, Y_{AO}$ = Point A estimated coordinate
$X_{BO}, Y_{BO}$ = Point B estimated coordinate
$L_{AB'O}$ = Estimated horizontal distance A-B

## 3. METHODOLOGY

The research study area is in Kampung Pasir, Mukim Pulai, Johor Bahru District as shown in Figure 3. The area covers 77 parcels and the details of the data sources are illustrated in Table 1. The raw data for adjustment process was acquired from DSSM Johor in the form of Digital Cadastral Data Base (DCDB) or commonly known as Pangkalan Data Ukur Kadaster (PDUK). In this paper, the existing technique applied by DSMM is declared as DSMM Technique. Meanwhile, the proposed technique to produce the new accurate dataset is known as PAI Technique.

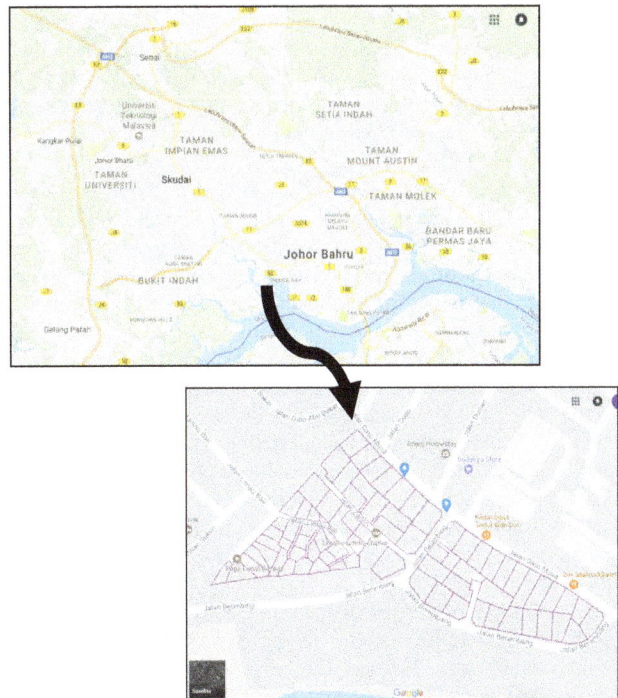

Figure 3. Study area

Table 1. Data sources

| Location | Kampung Pasir, Mukim Pulai, Daerah Johor Bahru, Johor, Malaysia |
|---|---|
| Parcel number | 4169-4189, 4199-4225, 119251-119277 & 119279 |
| Certified Plan (CP) | CP 40225 and CP 120293 |
| Data | PDUK (Block boundary line, CRM, Lot) |
| Data form | Bearing and distance |
| Data format | .tab |
| Adjustment Software | Star*net |

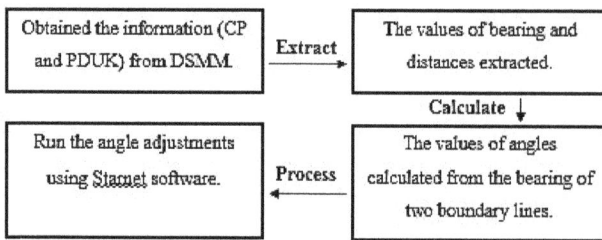

Figure 4. Interior angle extraction

The first step involves the calculation of interior angle of cadastral parcel. The existing raw data in LSA applied by DSMM are bearing and distance. The proposed method in this study is based on the angular and distance based LSA. The main justification of using angle method is to extract the independence or direct observation of raw data which contained minimum gross and systematic error. The bearing value is a very dependent data where it is largely based on the sources of the initial bearing. The common practice of the DSMM in the previous work is based on the solar observation and interior angle of three good boundary marks. As a result, it is extremely difficult to verify the quality of the reference bearing used in the previous work. In this study, the interior angle of these data is considered as an independent data which is calculated from the bearings of two boundary lines. Although the bearing control contains gross and systematic error, the interior angle between two boundary lines is free from the both factors.

The Star*net software has been used for the adjustment purposes. The Star*net is a rigorous least squares analysis software designed to adjust 2D and 3D survey networks. The output consists of a file of adjusted station coordinates and a statistical analysis of the adjustment. In addition, the graphical facilities are provided to allow the user to plot the network, including error ellipses of the adjusted points and relative error ellipses between stations. Star*net has the capacity to weight all input data both independently and by category which means that data can be defined as being FREE or FIXED and anything in between. The Star*net has the ability to control the weighting of input data where Giving accurately known measurements more weight than those measurements known to be less accurate.

The detail step of the processing in Star*net is shown in Figure 5. A new project is created which have two input file, bearing-distance and angle-distance. The bearing-distance input data is the current technique used by DSMM, meanwhile, the angle-distance input data is the new approach which is proposed in this study.

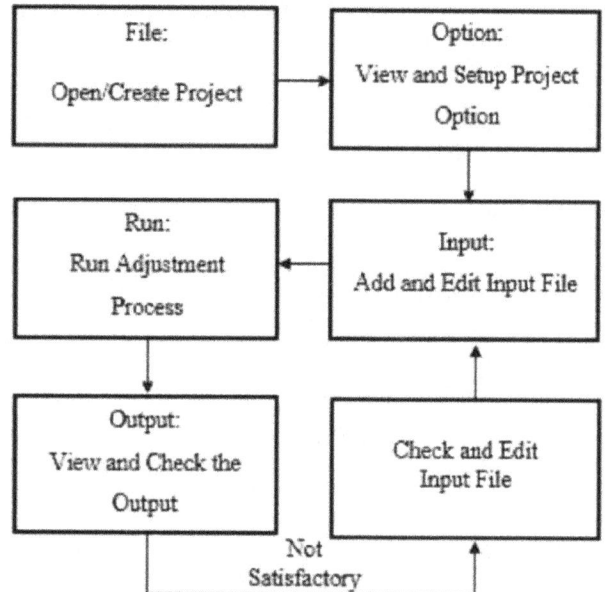

Figure 5. Processing stage in Star*net adjustment software

The final output of the adjustment in Star*net is the adjusted coordinate of entire boundary marks. The final coordinate will be exported to the dxf, coverage and shapefile format for further graphic and spatial based analysis. The AUTOCAD and ARCGIS software are the main tools for the processing. The final output of this process will be submitted in the accurate cadastral database. Figure 6 describes the entire process of suggested PAI as discussed above.

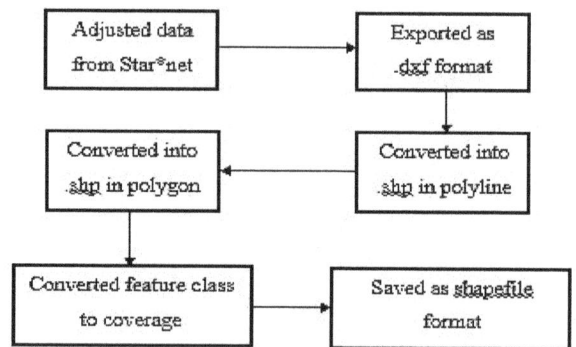

Figure 6. Accurate cadastral database development

## 4. RESULT AND ANALYSIS

In the verification stages, the output of the PAI Technique was compared to the result of DSMM Technique. The verification is based on four factors, area, bearing, distance and coordinate.

Table 2, 3, and 4 show the sample results of area comparisons by DSMM and PAI techniques. The actual area represents the area extracted from original certified plan (CP) and the adjusted area is obtained from the two different techniques. The results shown are selected from the minimum and maximum area differences of both methods.

Based on the DSMM Technique, the larger area differences given is 1.209 m$^2$ while the larger value given by PAI Technique is 0.725 m$^2$. In the meantime, the smallest differences of both techniques are 0.007 m$^2$ and 0.003 m$^2$ respectively.

Table 2. Minimum area difference

| TECHNIQUE | DSMM | PAI |
|---|---|---|
| LOT | 119258 | 4222 |
| ACTUAL AREA (m$^2$) | 212 | 595.9 |
| ADJUSTED AREA (m$^2$) | 211.993 | 595.897 |
| DIFFERENCE (m$^2$) | 0.007 | 0.003 |

Table 3. Maximum area difference

| TECHNIQUE | DSMM | PAI |
|---|---|---|
| LOT | 4189 | 4189 |
| ACTUAL AREA (m$^2$) | 875.3 | 875.3 |
| ADJUSTED AREA (m$^2$) | 874.271 | 874.575 |
| DIFFERENCE (m$^2$) | 1.029 | 0.725 |

In the entire data of area comparison analysis, the verification is based on the standard deviation value. The final result indicated that the standard deviation in PAI technique is 0.254 which is smaller than DSMM technique 0.292. The result proved that the PAI Technique is capable of retaining the area which indirectly evident in the geometry or position of relative boundary mark preserved.

Table 4 shows the differences in bearing values from CP, DSMM and PAI. The bearing values of DSMM are the adjusted bearing obtained from bearing adjustment method and bearing values of PAI are the adjusted bearing from angle adjustment method. The values selected represent the smallest, medium and largest bearing differences.

Table 4. Bearing comparison

| BM ID | | BEARING (dms) | | | DIFFERENCE (dms) | |
|---|---|---|---|---|---|---|
| FROM | TO | CP | DSMM | PAI | CP-DSMM | CP-PAI |
| 5135308322 | 5147708142 | 34-30-40 | 34-30-40 | 34-30-20 | 0-00-00 | 0-00-20 |
| 5267708003 | 5270207952 | 26-16-40 | 26-16-40 | 26-15-16 | 0-00-00 | 0-01-24 |
| 5268607586 | 5280707604 | 98-17-00 | 98-16-53 | 98-18-06 | 0-00-07 | -0-01-06 |
| 5314907946 | 5334808145 | 134-59-50 | 135-00-21 | 135-00-06 | -0-00-31 | -0-00-16 |
| 5360506938 | 5373807060 | 132-30-40 | 132-30-49 | 132-30-44 | -0-00-09 | -0-00-04 |
| 5514808060 | 5530008139 | 117-30-30 | 117-30-08 | 117-30-29 | 0-00-22 | 0-00-01 |

The bearing differences were calculated using the original bearings obtained from CP as reference. For DSMM Technique, the smallest differences is 0°0'0" and the larger value given is 0°0'31". The PAI result shows the maximum and minimum value of bearing differences to be 0°1'24" and 0°0'1" respectively. In terms of bearing comparison, 100% of bearing differences were under 30" in DSMM technique, while 88.8% were below 30". The standard deviation of bearing comparison between DSMM and PAI technique are 5.8" and 20.2'" respectively. In the PAI technique, the bearings are calculated directly from adjusted coordinate of boundary marks. Based on the result, although there are nonexistence of bearing data in the

adjustment process, the calculated bearings remain accurate showing values that are closed to the original values from CP. In addition, systematic and gross errors of input bearing can be removed when the PAI technique is applied.

Table 5 shows the distance values from CP, DSMM and PAI. The NDCDB distances are the adjusted distances obtained from bearing adjustment method and PAI distances are the adjusted distances from angle adjustment method.

Table 5. Distance comparison

| BM ID | | DISTANCE(m) | | | DIFFERENCE(m) | |
|---|---|---|---|---|---|---|
| From | To | CP | DSMM | PAI | CP-DSMM | CP-PAI |
| 5135308322 | 5147708142 | 21.793 | 21.793 | 21.79 | 0 | 0 |
| 5255108204 | 5236208200 | 18.907 | 18.918 | 18.92 | -0.011 | -0.009 |
| 5282507207 | 5299707376 | 24.156 | 24.166 | 24.18 | -0.01 | -0.023 |
| 5391408127 | 5403507966 | 20.115 | 20.136 | 20.12 | -0.021 | -0.009 |

The smallest difference of 0.000 m was found when there are no changes between the two methods. The largest differences for DSMM and PAI Technique are 0.021m and 0.023m respectively. The standard deviation of both method is 0.004m. From the overall data, 100% of both techniques are within the allowable tolerance of re-fixation which is not more than 0.050 m as stated in KPUP Circular Vol. 6/2009.

The PAI technique was applied to generate the high accuracy database. The generated cadastral database then has to be compared with the DSMM new dataset (NDCDB). Table 6 shows the result obtained from the adjustments.

Table 6. Coordinate and displacement

| BM ID | NDCDB COORDINATE | | PAI COORDINATE | | DIFFERENCE | | DISPLACEMENT |
|---|---|---|---|---|---|---|---|
| | EASTING | NORTHING | EASTING | NORTHING | ΔE | ΔN | |
| 5219607465 | 15222.438 | -60744.031 | 15222.440 | -60743.998 | -0.002 | -0.033 | 0.033 |
| 5222507761 | 15225.294 | -60773.583 | 15225.294 | -60773.587 | 0.000 | 0.004 | 0.004 |
| 5280607603 | 15283.431 | -60757.879 | 15283.442 | -60757.869 | -0.011 | -0.010 | 0.015 |
| 5385507589 | 15388.305 | -60756.402 | 15388.303 | -60756.403 | 0.002 | 0.001 | 0.002 |
| 5252607863 | 15255.320 | -60783.768 | 15255.320 | -60783.768 | 0.000 | 0.000 | 0.000 |

For the total 200 points of the boundary marks, it shows that 100% are under 0.050 m having the lowest and highest values of 0.000 m and 0.033 m respectively as shown in 6. The standard deviation of coordinate difference is 0.009m. Based on the result, it indicates that PAI method yielded datasets that are within tolerance level compared to the NDCDB coordinate and at the same time the area and geometry preservation of original parcel is improved.

## 5. CONCLUSION

This study proposed an angle based LSA in the process of enhancing legacy dataset towards an accurate dataset. The proposed method is also capable of to reducing the systematic and gross errors obtained during bearing observation obtained from low accuracy datum. The interior angle data offer an independent raw data which is directly calculated from two bearing without take into account the quality of the observed bearing. The proposed technique demonstrated geometric fitting transformed efficiently in the PAI process due to the less standard deviation of area comparison between PAI and original CP parcel area. In addition, the adjusted distance, bearing and coordinate also fall within the reasonable tolerance compared to the original data in CP and new database NDCDB. Finally, this study provides alternative tools for the enhancement of digital cadastral maps. These tools expectantly will assist the DSMM in managing the NDCDB toward implementing a more accurate digital cadastre in the next decade.

## ACKNOWLEDGEMENTS

The author would like to thanks and express sincere appreciation to DSMM State of Johor for providing valuable data and information regarding to the research done and Triple Axis for the funding.

## REFERENCES

Amat, M. A. C. (2007). *Implementasi pengoptimuman komputer dalam pembangunan perisian analisis pelarasan kuasa dua terkecil.* Universiti Teknologi Malaysia.

Arvanitis, A., & Koukopoulou, S. (1999). *Managing data during the update of the hellenic cadastre.* Paper presented at the FIG Com3 Annual Meeting, Budapest.

Buyong, T., & Kuhn, W. (1990). *Local adjustment for a measurement-based multipurpose cadastre systems.* Paper presented at the XIX Congress of FIG.

Casado, M. L. (2006). *Some basic mathematical constraints for the geometric conflation problem.* Paper presented at the Proceedings of the 7th international symposium on spatial accuracy assessment in natural resources and environmental sciences.

Donnelly, N., & Hannah, J. (2006). An Assessment of the Precision of the Observational Data Used in New Zealand's National Cadastral system. *Survey Review, 38*(300).

Durdin, P. M. (1993). Measurement-Based Databases: One Approach to the Integration of Survey and GIS Cadastral Data. *Surveying and Land Information Systems, 53*(1), 41-47.

Effenberg, W. W., Enemark, S., & Williamson, I. P. (1999). Framework for Discussion of Digital Spatial Data Flow within Cadastral Systems. *Australian surveyor, 44*(1), 35-43.

Felus, Y. A. (2007). On the Positional Enhancement of Digital Cadastral Maps. *Survey Review, 39*(306), 268-281. doi: 10.1179/175227007x197183.

Fradkin, K., & Doytsher, Y. (2002). Establishing an urban digital cadastre: analytical reconstruction of parcel boundaries. *Computers, Environment and Urban Systems, 26*(5), 447-463.

Gielsdorf, F., Gruendig, L., & Aschoff, B. (2004). *Positional accuracy improvement-A necessary tool for updating and integration of GIS data.* Paper presented at the Proceedings of the FIG working week.

Setan, H. (1995). *Functional and stochastic model for geometric detection of deformation in engineering: a practical approach.* City University.

Hashim, N., Omar, A., Omar, K., Abdullah, N., & Yatim, M. (2016). Cadastral Positioning Accuracy Improvement: A Case Study in Malaysia. International Archives of the

Photogrammetry, Remote Sensing & Spatial Information Sciences, 42.

Hesse, W. J., Benwell, G. L., & Williamson, I. P. (1990). Optimising, maintaining and updating the spatial accuracy of digital cadastral data bases. *Australian surveyor, 35*(2), 109-119.

Hope, S., Gordini, C., & Kealy, A. (2008). Positional accuracy improvement: lessons learned from regional Victoria, Australia. *Survey Review, 40*(307), 29-42. doi: 10.1179/003962608x253457

Koch, K.-R. (2013). *Parameter estimation and hypothesis testing in linear models*: Springer Science & Business Media.

Merrit, R., & Masters, E. (1999). *Digital cadastral upgrades—A progress report.* Paper presented at the Proceedings of the First International Symposium on Spatial Data Quality.

Merritt, R. (2005). *An assessment of using least squares adjustment to upgrade spatial data in GIS.* University of New South Wales.

Mikhail, E. M., & Ackermann, F. E. (1982). *Observations and least squares*: Univ Pr of Amer.

Morgenstern, D., Prell, K., & Riemer, H. (1989). Digitisation and Geometrical Improvement of Inhomogeneous Cadastral Maps. *Survey Review, 30*(234).

Perelmuter, A., & Steinberg, G. (1992). *LIS and Cadastre in Israel.* Paper presented at the Proceedings of the Israeli GIS/LIS'92.

Rönsdorf, C. (2008). Positional Accuracy Improvement (PAI) *Encyclopedia of GIS* (pp. 885-891): Springer.

Sisman, Y. (2014). Coordinate transformation of cadastral maps using different adjustment methods. *Journal of the Chinese Institute of Engineers, 37*(7), 869-882.

Tamim, N. (1995). A Methodology to Create a Digital Cadastral Overlay Through Upgrading Digitized Cadastral Data. *Surveying and Land Information Systems, 55*(1), 3-12.

Ting, L., & Williamson, I. P. (1999). Cadastral Trends: A Synthesis. *Australian surveyor, 44*(1), 46-54.

Tong, X., Shi, W., & Liu, D. (2005). A least squares-based method for adjusting the boundaries of area objects. *Photogrammetric engineering & remote sensing, 71*(2), 189-195.

Tong, X., Shi, W., & Liu, D. (2009). Introducing scale parameters for adjusting area objects in GIS based on least squares and variance component estimation. *International Journal of Geographical Information Science, 23*(11), 1413-1432.

Wolf, P., & Ghilani, C. (2006). Adjustment Computations Spatial Data Analysis: Hoboken.

# 24

# SINKHOLE SUSCEPTIBILITY HAZARD ZONES USING GIS AND ANALYTICAL HIERARCHICAL PROCESS (AHP)

Mohd Asri Hakim Mohd Rosdi, Ainon Nisa Othman, Muhamad Arief Muhd Zubir
Zulkiflee Abdul Latif & Zaharah Mohd Yusoff

Centre of Studies for Surveying Science and Geomatics, Faculty of Architecture, Planning and Surveying, Universiti Teknologi MARA, 40450 Shah Alam, Selangor Darul Ehsan, MALAYSIA.

**KEY WORDS:** Geographical Information System, Analytical Hierarchical Process, Sinkhole Susceptibility Hazard Zones

**ABSTRACT:**

Sinkhole is not classified as new phenomenon in this country, especially surround Klang Valley. Since 1968, the increasing numbers of sinkhole incident have been reported in Kuala Lumpur and the vicinity areas. As the results, it poses a serious threat for human lives, assets and structure especially in the capital city of Malaysia. Therefore, a Sinkhole Hazard Model (SHM) was generated with integration of GIS framework by applying Analytical Hierarchical Process (AHP) technique in order to produced sinkhole susceptibility hazard map for the particular area. Five consecutive parameters for main criteria each categorized by five sub classes were selected for this research which is Lithology (LT), Groundwater Level Decline (WLD), Soil Type (ST), Land Use (LU) and Proximity to Groundwater Wells (PG). A set of relative weights were assigned to each inducing factor and computed through pairwise comparison matrix derived from expert judgment. Lithology and Groundwater Level Decline has been identified gives the highest impact to the sinkhole development. A sinkhole susceptibility hazard zones was classified into five prone areas namely very low, low, moderate, high and very high hazard. The results obtained were validated with thirty three (33) previous sinkhole inventory data. This evaluation shows that the model indicates 64% and 21% of the sinkhole events fall within high and very high hazard zones respectively. Based on this outcome, it clearly represents that AHP approach is useful to predict natural disaster such as sinkhole hazard.

## 1. INTRODUCTION

Sinkhole or land subsidence is not a new phenomenon in Malaysia, especially surround Klang Valley. According to (Meng, 2005), sinkhole can be defined as on the ground surface depression due to the dissolving of the limestone near the surface or the collapse of an underground cave. Basically, Kuala Lumpur has two different geological formations, namely Kenny Hill Formation which consists of sedimentary rocks and Kuala Lumpur Limestone Formation with its famous highly erratic karstic constituents (Meng, 2005). Over 158 years of rapid development and rampant land use planning has led to specific changes in topography and geomorphology such as appearance of sinkholes. It can be disastrous and terrifying because of the condition is very unstable (Waltham, 2009). In urban areas such as Kuala Lumpur and Ampang city, the combination of industrial or development activities accelerated the process of sinkhole development. Overburden on the surface of earth for instance ex mining retention ponds, buildings, heavy traffic and changes in groundwater table induce to sinkhole process (Abidin, et al., 2002). Kuala Lumpur and Ampang city is located dominantly on the Kuala Lumpur Limestone Formation. Rapid development of these areas has had some impacts that are destructive to the environment. The cases can be originates from various places that having limestone bedrock formation, unpredictable and sudden. Often we heard recently in the newspaper or media about sinkhole tragedies and its effect to the human and infrastructure. Based on previous study, sinkhole only occurs in limestone bedrock areas (Abidin, et al., 2002). A sinkhole occurrence is seen as a result of high rainfall distribution and changing of groundwater levels in limestone areas. A stable and firm land surface is crucial for any urban development process in order to ensure public safety. This can increase country economic activity and stay competitive with others. In other word, any construction works must appropriately deal with the condition of karstic bedrock limestone. If not, many uncertainties and difficulties can be occurred in the future. Along with the globalization of technology, GIS become a wide information source especially in decision making for natural hazards. Since GIS had been implemented in many countries for natural hazard prediction, Malaysia is still not optimally used this technology to identify the conditions that may trigger sinkhole hazards. Thus, GIS were utilized to evaluate the potential sinkhole hazard areas using Analytical Hierarchy Process (AHP) technique. The AHP technique is developed by Saaty is based on three principles namely: decomposition, comparative judgment and synthesis of priorities (Malczewski, 1999). The use of GIS allows the combination of data from various sources gives the higher accuracy and time efficient (Kouri, et al., 2013). Inaccurate prediction will cause the human ignorance, then, more sinkhole to occur. An integration of GIS and AHP technique are needed to produce accurate models in order to produce potential sinkhole hazard maps.

## 2. METHODOLOGY

Generally, the research methodology framework is summarized in Figure 1. It describes the overall sequence of the analysis processes that consist of four phases including preliminary study, data collection, data processing as well as result and analysis in the final part. For the first phase, the problem statement and significant of research is determined within the

DBKL and MPAJ's area. Next, in the second phase all data were classified by primary and secondary data sources. In the third phase, data processing which involves weightage and software analysis determination is carrying out in order to achieve the research objectives. Next, the sinkhole hazard model is generated and used to map the susceptible location for sinkhole hazards as the last process in this research.

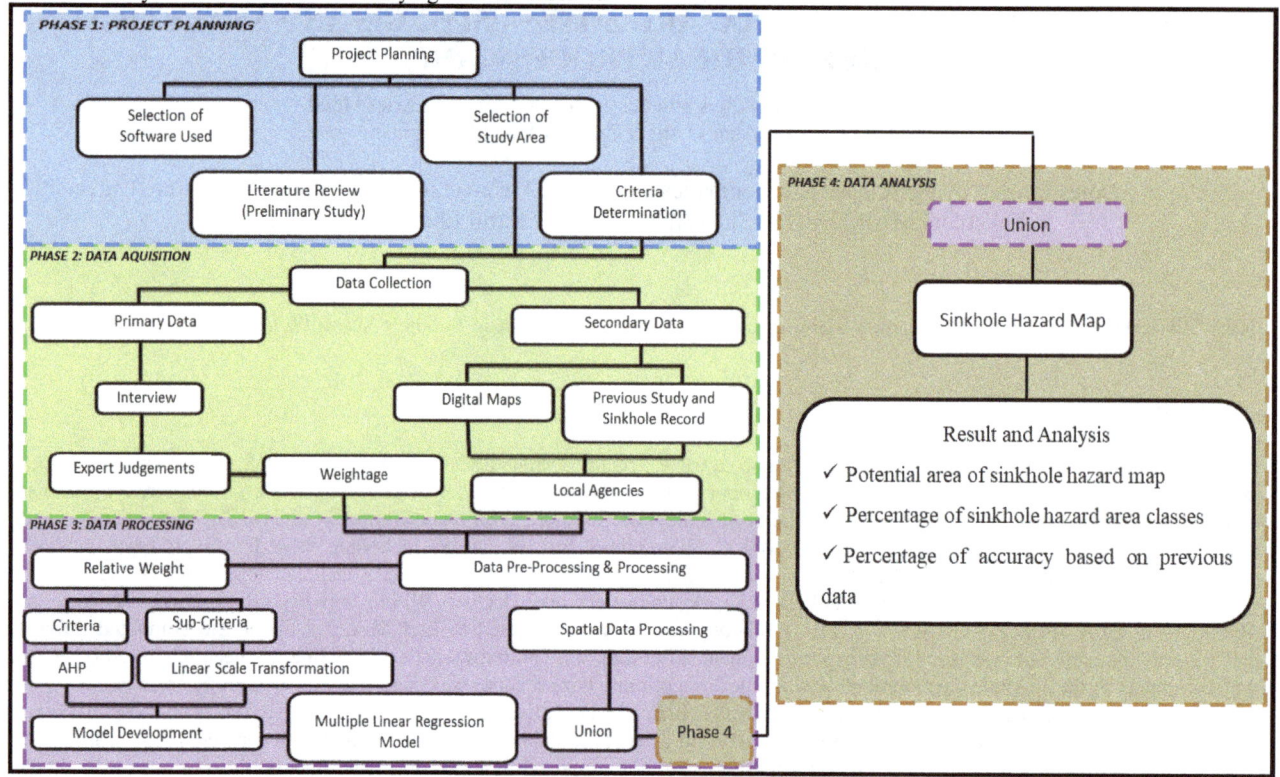

Figure 1. Research Methodology

## 2.1  Research Area

The area covers a whole part of DBKL and MPAJ administrative territories. The total area under DBKL is approximately 279,327 hectares, while the area for MPAJ is 14,350 hectares. This area was selected based on the geological setting, frequent issues occurred, the availability of previous sinkhole incident data and the availability of data. Basically, Kuala Lumpur lies on the extensive limestone bedrock which overlain by alluvial layer (Kong & Komoo, 1990). The formation of limestone covers the majority area of Kuala Lumpur and Ampang vicinity. The soil type mainly consists of urbanized land and forest. Figure 2 depicts the boundary and sinkhole location of the study area

## 2.2  Data Collection

The data used for this research are generally based on criteria determination process. A series of research papers and discussion with geological experts from various agencies supports the reliability of information. Five parameters were identified for sinkhole development in Malaysia namely bedrock lithology, soil type, water table decline, proximity to groundwater and land use (Table 1). The main source of primary data is by interviews with experts and some literature reviews from previous study. Geologist from local geotechnical engineering agencies was identified to acquire his judgment for relative weightage and some recommendations. Expert opinion is crucial because AHP technique is considered as heuristic method which is used expert judging concept that reduces the search activity in solving problem process.

Digital lithology and soil map acquired from Mineral and Geoscience Department considered as the main source. The topography of the area is analyzed using digital topographic map obtained from Department of Survey and Mapping Malaysia (JUPEM) and being extracted for the land use information. Groundwater level also was obtained from Mineral and Geoscience Department that includes groundwater well locations as well. For data validation, the sinkhole inventory data compiled by Mineral and Geoscience Department also being collected.

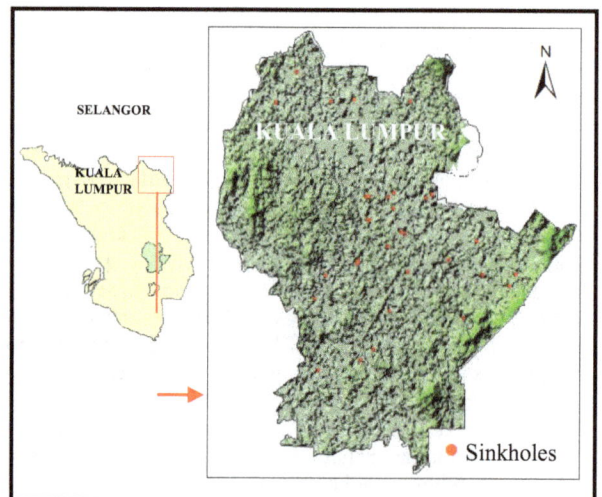

Figure 2.  Location of the research area

| Criteria | Unit |
|---|---|
| Lithology | Type |
| Soil properties | Type |
| Groundwater level decline | Meter Cubed (m³) |
| Land Use | Type |
| Proximity to groundwater well | Meter |

Table 1. Sinkhole criteria used and unit

## 2.3 Multi-criteria Decision Making Techniques (MCDM)

In the globalization era, solving problem becomes a difficult process when it required good decision to be made. It involves many procedures and parameters need to be concern in order to make a wise decision. Decision analysis is a set of systematic procedures for analyzing complex decision problems (Malczewski, 1999). A frequently applied approach is to decompose the problem into smaller, understandable parts that express relevant concerns (Alkema & Boerboom, 2012). The interpretation of an indicator as to whether its value is good with respect to its objective is a criterion (Ullman, 2006; Beinat, 1997). GIS Analyst needs to know the relative weight or importance of each factor in order to produce useful maps. In order to study the sinkhole formation, there are various techniques that can be used to the researchers to achieve their objectives goal. Ranking method, rating method, pairwise comparison method and AHP method are the suitable method that can be used to the researchers to study the disaster of sinkhole formation.

Integration of GIS and MCDM has been applied by different researchers in identifying sinkholes hazard area. The approach used is depending on the main goal of the study followed with the suitability of the region. There are several methods typically used locally and internationally are defined as heuristic, statistic and deterministic (Othman, et al., 2012).

Research by (Kouri, et al., 2013) used statistical method associated with GIS and remote sensing data to produce sinkholes hazard map in Kinta Valley, Perak. Eight causative parameters were used namely lithology, structure (lineament), soil cover, slope, land use mining, urban area features, ponds and rivers. Every parameter was calculated based on sinkhole location and a spatial database. Other paper by (Taheri, et al., 2015) using analytical hierarchy process (AHP) to determine sinkhole susceptibility map in Hamadan province, Iran. It combine with GIS environment considering eight causal factors namely distance to faults, water level decline, groundwater exploitation, penetration of deep wells into karst bedrock, distance to deep wells, groundwater alkalinity, bedrock lithology and alluvium thickness.

Therefore, the method illustrated in this research is the first contribution which explores the practicality of the AHP technique to determine potential sinkhole hazard area under DBKL and MPAJ administration. Based on literature reviews, the weightage of criteria is determined by AHP through normalized pairwise comparison matrix and linear scale transformation is used to calculate weight for sub-criteria. Thus, the result can be modelled by multiple linear regressions to map the sinkhole hazard zonation in the mention area.

## 2.4 Data Pre – Processing and Processing

All spatial and attribute data were processed throughout map digitizing, editing and conversion by using ArcGIS 10.1 software. The list of attributes weight of criteria and sub criteria are entered in the spatial data to classify the values. AHP and linear scale transformation techniques are used in this research to determine value of relative weight for criteria and sub criteria. The value of each criteria and sub criteria are derived from interviews and discussions with geologist expertise. The result of the weight is used to generate multiple linear regression models in order to produce sinkhole susceptibility hazard maps.

In this research, model development has been preliminary assessed considering relative weights assigned to five selected controlling factors (criteria) and to different classes of each one (sub criteria). A set of criteria have been weighted performing pairwise comparison matrices can be referred on Table 2. An important step of the AHP is to evaluate the consistency of the ratings. This can be carried out by calculating the consistency index (CI) and the consistency ratio (CR). The consistency index is defined by equation:

$$CI = \frac{\lambda - n}{n - 1} \qquad (1)$$

Where $\lambda$ is the average value of consistency vector of the preference matrix and n is the number of parameters. For the calculation of the consistency ratio (CR), the consistency index is compared with a random consistency index (RI):

$$CR = \frac{CI}{RI} \qquad (2)$$

The RI values have been tabulated by Saaty (1980). Consistency ratios higher than 0.1 suggest untrustworthy judgments, indicating that the comparisons and scores should be revised.

The development of model is mainly based on the final weight of criteria and subcriteria of the parameters. The expert judgment is represented in the series of mathematical models. In this research scope, multiple linear regression models are used to generate series of map of potential sinkholes hazard area. This model is named as Sinkhole Hazard Model (SHM) and consists of five (5) criteria which represented as follows:

$$SHM = (0.457 \times sc\_litho) + (0.109 \times sc\_soil) + (0.046 \times sc\_lu) + (0.299 \times sc\_wld) + (0.090 \times sc\_pg) \qquad (3)$$

Where sc_litho is standardized score for lithology sub criterion, sc_soil is standardized score for soil type sub criterion, sc_lu is standardized score for landuse sub criterion, sc_wld is standardized score for water level decline sub criterion and sc_pg is standardized score for proximity to groundwater sub criterion.

AHP technique is used to analyze complex decision problems taking into account a large number of factors or criteria. Each factor is evaluated on its importance with respect to another by applying Saaty's (1980) fundamental scale for pairwise comparison.

The potential sinkholes hazard zone has been initially evaluated considering the relative weights applied to five selected controlling parameters (criteria) and to different classification (sub-criteria). The pairwise comparison matrices in Table 2 are constructed to determine relative importance of each parameter for sinkhole development with respect to another one.

| Criterion | LT | ST | LU | WLD | PG |
|---|---|---|---|---|---|
| Lithology (LT) | 1 | 5 | 7 | 2 | 6 |
| Soil Type (ST) | 0.200 | 1 | 3 | 0.167 | 2 |
| Land Use (LU) | 0.143 | 0.333 | 1 | 0.200 | 0.333 |
| Water Level Decline (WLD) | 0.500 | 6 | 5 | 1 | 3 |
| Proximity to Groundwater (PG) | 0.167 | 0.500 | 3 | 0.333 | 1 |

Table 2. Pairwise comparison matrix of criteria

In Table 2, the relative scales factors have been entered by expert represent each variable involved. The variable comparisons are done in matrices form in order to enhance the weight computation process of potential sinkhole hazard. Prior to weightage calculation, every scale factor on each criterion must be converted into fraction in order to obtain the total column value for every cell. Then, the total scale factor is computed vertically by using this formula:

$$\sum = C^1 + C^2 + C^3 + \cdots + C^8 \qquad (4)$$

Where $\sum$ is total value of every columns variable and C is column variables. The normalized value is obtained as following:

$$N = C / \sum C \qquad (5)$$

Where N is Normalize Matrix, C is Criteria Comparison Matrix and $\sum C$ is total value of every columns variable. Relative importance or weight (W) is derived through eigenvector normalization process. The process is accomplished by averaging each normalized matrix by the sum of elements in the row. The same way goes to the other relative weight for criteria. Based on Table 3, the result of normalization weight, it can be determined that lithology has largest weight value 0.457, while landuse produces a smallest weight value 0.046.

| Criterion | Normalized Comparison Matrix | | | | | Weightage |
|---|---|---|---|---|---|---|
|  | LT | ST | LU | WLD | PG |  |
| LT | 0.498 | 0.390 | 0.368 | 0.541 | 0.486 | 0.457 |
| ST | 0.099 | 0.078 | 0.158 | 0.045 | 0.162 | 0.109 |
| LU | 0.071 | 0.026 | 0.053 | 0.054 | 0.027 | 0.046 |
| WLD | 0.249 | 0.468 | 0.263 | 0.270 | 0.243 | 0.299 |
| PG | 0.083 | 0.040 | 0.158 | 0.090 | 0.081 | 0.090 |

Table 3. Relative weightage value of each main criterion

The weightage of sub-criteria is derived by using linear scale transformation. Linear scale transformation is the most frequently used GIS based method from transforming input (subcriteria scores) data into subcriteria maps. The scale for the score is not fixed but depends on the $n^{th}$ value of the subcriteria in one parameters. Then, the weight is obtain from normalize the scores by dividing it with the total scores. The range score starts from 0 for the minimum value.

Generally, the total normalize weight must be 1. For example, soil type has five classes namely alluvium, steepland, sandy clay, clay loam and sand (mined land). The relative score for these five classes is 0, 1, 2, 3 and 4. Then, the standardized score must be computed by dividing the each relative score with the sum of all scores in the consecutive columns.

## 3. RESULTS AND ANALYSIS

### 3.1 Sinkhole Susceptibility Hazard Zonation Map

The sinkhole susceptibility hazard zonation maps generated from the SHM model is shown in Figure 4. The resulting maps data have been classified into five prone levels as: very low, low, moderate, high and very high risk. The result from this model have shown that the very low (Class 1), low (Class 2), moderate (Class 3), high (Class 4) and very high hazard (Class 5) zones constitute 14%, 24%, 21%, 31% and 10% of the study area respectively. It was found that the North West part can be categorized as high and very high hazard area. This area is mostly occurred in Kuala Lumpur Limestone Formation bedrock geology consisting limestone/marble and acid intrusive (undifferentiated) lithology (Taheri, et al., 2015). In this study area, most of sinkhole hazard occurred at the high value of water level decline which is -22 to -70 cubic meter of approximate yield. Furthermore, the alluvium type of soil can be considered as unsafe in some area. Besides, most of susceptible zones are covered by mining and urbanized land use. The sprawl of commercial and residential building in this location was erected on the ex – mined land which comprise of sands and clay properties. Meanwhile, for Ampang area, the relatively high and very high hazard falls at the center and western of the district. It covers the major part of Ampang city bounded by Kuala Lumpur territory. Most of the factors triggered are same with Kuala Lumpur as stated earlier. High dense urban areas plus surrounded by mined land vicinity can be classified identical with Kuala Lumpur.

### 3.2 Validation and Accuracy Assessment

In order to determine the accuracy of the sinkhole hazard map, this study was evaluated by overlying the previous sinkhole inventory data provided by The Malaysian Mineral and Geoscience Department. The totals of thirty three (33) location of previous sinkhole in Kuala Lumpur and Ampang have found located in the appropriate potential hazard classes. To validate the Sinkhole Hazard Model (SHM), the important classes that should be included are high and very high potential locations. Twenty (20) location of the tabulate data fall within high hazard areas and five locations are located in very high susceptible areas.

| Hazard Classes | Area (km$^2$) | Sinkhole Number | Sinkhole (%) |
|---|---|---|---|
| Very Low | 41.773 | None | 0 |
| Low | 70.212 | 2 | 6 |
| Moderate | 61.528 | 6 | 18 |
| High | 91.487 | 20 | 61 |
| Very High | 31.597 | 5 | 15 |
| Total | 296.597 | 33 | 100 |

Table 4. Accuracy percentage of the model

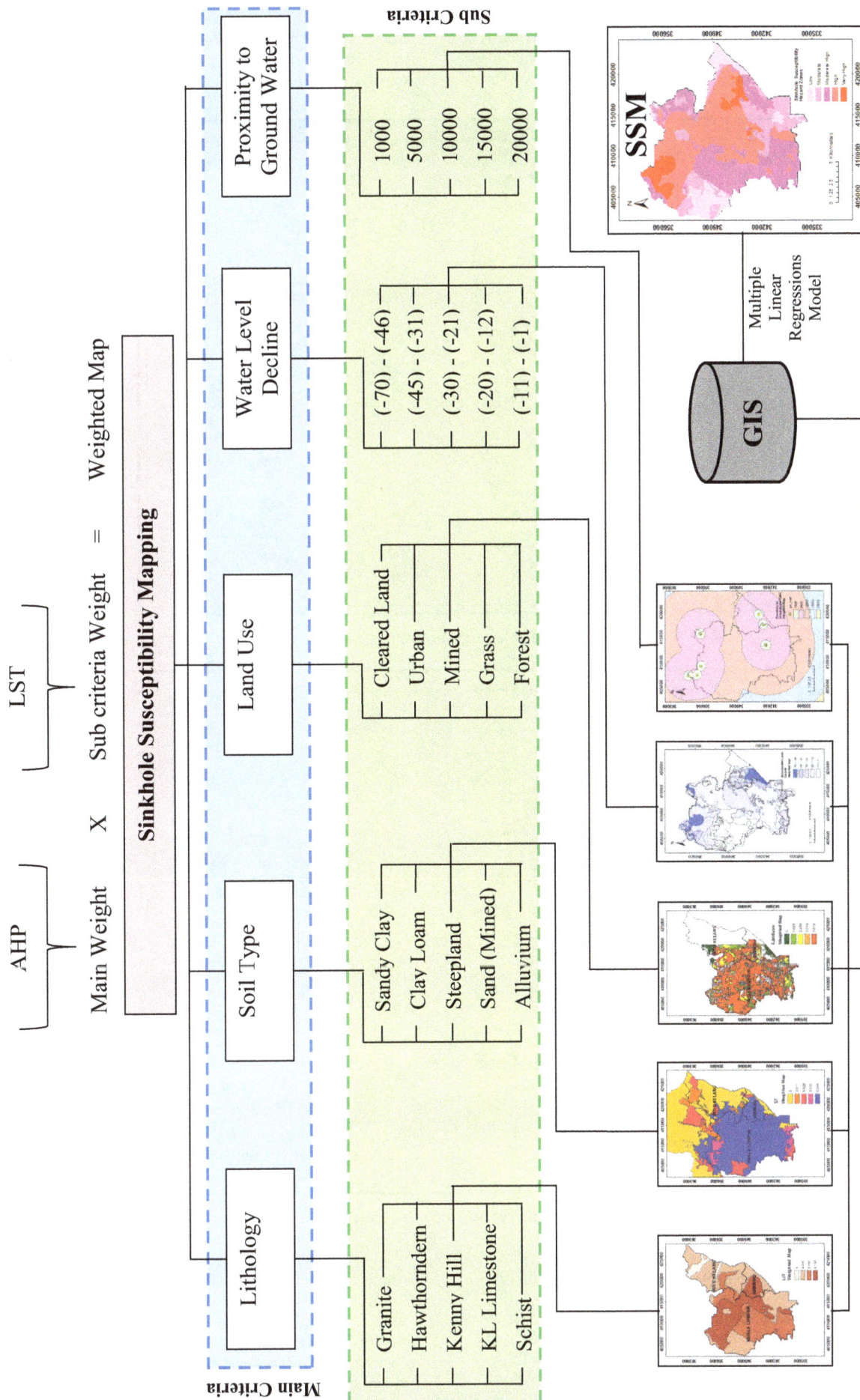

Figure 3. Flow diagram form a basis for constructing sinkhole susceptibility map

The hazard map shows in Figure 4 represents the potential sinkhole areas generated by model while Figure 5 represents the actual places of sinkhole cases and the potential sinkhole areas generated by model. From the result, it shows that accuracy of the model is 76% refer to the high and very high class while the remaining 24% fall within low and moderate class. Table 4 depicts the percentage of sinkhole based on actual previous data that had existed before.

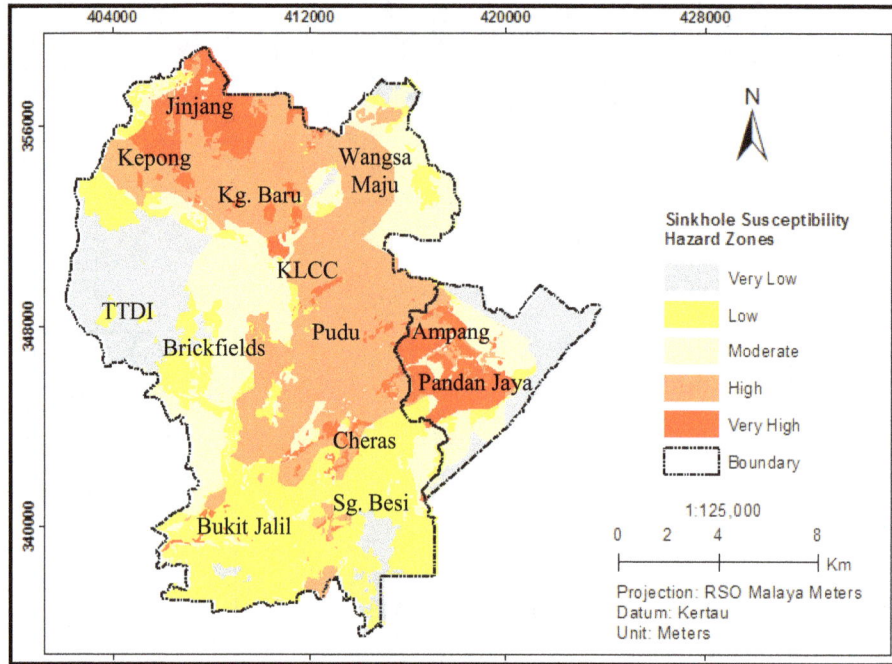

Figure 4. Sinkhole susceptibility hazard zones map

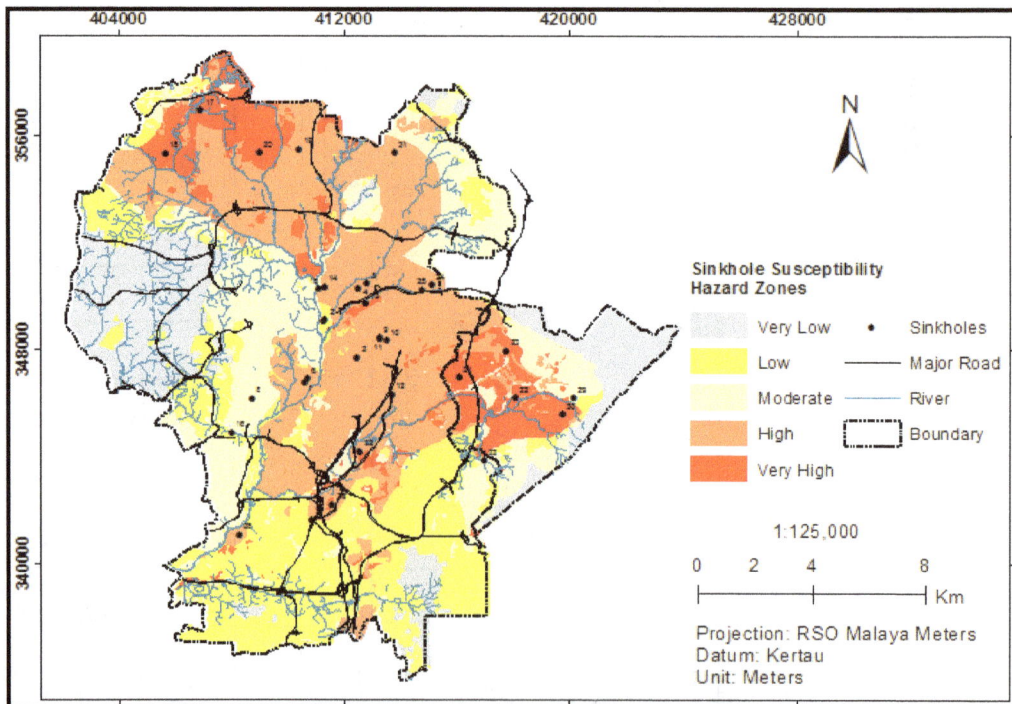

Figure 5. Sinkhole susceptibility map of the study area and location of the previous sinkhole

## 4. CONCLUSION AND DISCUSSION

Sinkhole hazard has increased dramatically since 1968 in Kuala Lumpur and Ampang areas. Rapid development of these areas has had some impacts that are destructive to the environment. Sinkhole can be classified as dangerous natural hazard that hard to predict when and where it will occur. However, sinkhole can be systematically managed even though cannot be completely prevented. The severity of impacts from sinkhole hazard can be minimized if the hazard zones can be predicted and mapped before any development activity takes place. Thus, accurate model need to be develop in order to produce reliable hazard maps. The model presented in this work has been constructed by integration of GIS and AHP approach. Results from this research can be used by the local authority to manage properly, systematically and plan development within their areas.

As the limitations of the study in order to improve this paper for the others researcher in the future, it suggests to use the others suitable techniques to detect the sinkhole phenomenon rather than AHP technique. Further study might be explore at the other different approach in multi-criteria decision making analysis such as Ranking, Rating, Fuzzy AHP or Weight of Evidence (WoE) method. Some suggestions could be also made for further study regarding of the triggering effect of the sinkhole incident. High resolution satellite images also might be used to obtain the latest land use and land cover classification of the areas for better time series. In addition, some limitations in this study of the model are partially correlated to the difficulty of acquiring data on some geological circumstances.

Another recommended technique that useful in predicted sinkhole hazard is the computation of the magnitude and frequency relationship. From the inventory data, the size and diameter of the sinkhole is recorded for estimation the possible sinkhole to occur in a year. This information is very crucial for local authority for planning and managing natural hazard in Malaysia.

## ACKNOWLEDGEMENT

Special thanks to Kementerian Pengajian Tinggi Malaysia (KPT) for the monetary fund under the research grant (FRGS/1/2016/WAB05/UITM/02/7), Dewan Bandaraya Kuala Lumpur (DBKL) and Department of Mineral and Geoscience for the data and expert opinion.

## REFERENCES

Abidin, R. Z., Jamalludin, D., & Arshad, M. F. (2002). *Sinkhole Physical Properties And Its Failure Risk Assessment.* Shah Alam, Selangor: Institute of Research, Development And Commercialisation UiTM.

Eisenberger, I. (1964). Genesis of Dimodal Distribution. *Technometrics*, 357-363.

Harries, K. (1999). *Mapping Crime: Principle and Practice.* Washington DC: U.S. Department of Justice Office of Justice Programs.

Kong, T. B., & Komoo, I. (1990). Urban Geology: Case Study of Kuala Lumpur, Malaysia. *Engineering Geology*, 71-94.

Kouri, O. A., Fugara, A. A., Rawashdeh, S. A., Sadoun , B., & Pradhan, B. (2013). Geospatial Modeling for Sinkhole Hazard Map Based on GIS & RS Data. *Journal of Geographic Information System*, 584-592.

Malczewski, J. (1999). *GIS and Multicriteria Decision Analysis.* New york: John Wiley & Sons.

Meng, T. S. (2005). Karstic Features of Kuala Lumpur Limestone. *Geological Society of Malaysia, Bulletin 46*, 447-453.

Saaty, T. L. (1980). *The analytic hierarchy process.* New York: McGraw-Hill.

Taheri, K., Gutierrez, F., Mohseni, H., Raeisi, E., & Taheri, M. (2015). Sinkhole susceptibility mapping using the analytical hierrarchy process(AHP) and magnitude - frequency realtiobships: A case study in Hamadan province Iran. *Geomorphology*, 64-79.

Waltham, T. (2009). Sinkhole Geohazards. *Geology Today*, 112-116.

# 25

# ASSESSMENT OF MULTIBEAM BACKSCATTER TEXTURE ANALYSIS FOR SEAFLOOR SEDIMENT CLASSIFICATION

Shahrin Amizul Samsudin and Rozaimi Che Hasan

UTM Razak School of Engineering and Advanced Technology,
Universiti Teknologi Malaysia, Kuala Lumpur
email: shahrinamizul@yahoo.com, rozaimi.kl@utm.my

**KEY WORDS:** Multibeam sonar, Acoustic backscatter, Acoustic classification, Grey Level Co-Occurrence Matrices, image analysis.

**ABSTRACT:**

Recently, there have been many debates to analyse backscatter data from multibeam echosounder system (MBES) for seafloor classifications. Among them, two common methods have been used lately for seafloor classification; (1) signal-based classification method which using Angular Range Analysis (ARA) and Image-based texture classification method which based on derived Grey Level Co-occurrence Matrices (GLCMs). Although ARA method could predict sediment types, its low spatial resolution limits its use with high spatial resolution dataset. Texture layers from GLCM on the other hand does not predict sediment types, but its high spatial resolution can be useful for image analysis. The objectives of this study are; (1) to investigate the correlation between MBES derived backscatter mosaic textures with seafloor sediment type derived from ARA method, and (2) to identify which GLCM texture layers have high similarities with sediment classification map derived from signal-based classification method. The study area was located at Tawau, covers an area of 4.7km2, situated off the channel in the Celebes Sea between Nunukan Island and Sebatik Island, East Malaysia. First, GLCM layers were derived from backscatter mosaic while sediment types (i.e. sediment map with classes) was also constructed using ARA method. Secondly, Principal Component Analysis (PCA) was used determine which GLCM layers contribute most to the variance (i.e. important layers). Finally, K-Means clustering algorithm was applied to the important GLCM layers and the results were compared with classes from ARA. From the results, PCA has identified that GLCM layers of Correlation, Entropy, Contrast and Mean contributed to the 98.77% of total variance. Among these layers, GLCM Mean showed a good agreement with sediment classes from ARA sediment map. This study has demonstrated different texture layers have different characterisation factors for sediment classification and proper analysis is needed before using these layers with any classification technique.

## 1. INTRODUCTION

Analysis and determination of physical properties of the seafloor is a crucial element for important marine activities, including coral reef management, fisheries habitat management and marine geology studies (Hedley *et al.*, 2016; Buhl-Mortensen *et al.*, 2015; Robidoux *et al.*, 2008; Hughes Clarke *et al.*, 1996). Over the last decades, the rapid developments in marine acoustic survey methods have revolutionised the formation of detailed maps of seafloor for the purpose of seabed habitat mapping (Brown *et al.*, 2011b). The use of high-resolution acoustic technique, in particular multibeam echosounder system (MBES) in providing full coverage topography (i.e. bathymetry) and acoustic backscatter (i.e. intensity returns) is vital for sediment and habitat types prediction (De Falco *et al.*, 2010; Medialdea *et al.*, 2008; Sutherland *et al.*, 2007). Backscatter returns from MBES is one of the potential dataset from acoustic technique that is seen to consist of important acoustic scattering information of the sediment types and offers huge possibility of remote identification of seafloor as well as proxy for habitat classes.

For sediment classification using backscatter from MBES, image analysis such as the use of image textural analysis is probably the most widely used technique in many studies (Herkül *et al.*, 2017; Lucieer *et al.*, 2016; Blondel *et al.*, 2015; Zhi *et al.*, 2014; Che Hasan, 2014; Hill *et al.*, 2014; Siwabessy *et al.*, 2013; Lucieer *et al.*, 2013; Fakiris *et al.*, 2012; Micallef *et al.*, 2012; Huang *et al.*, 2012; Lucieer *et al.*, 2011; Díaz, 2000). The technique, known as Grey Level Co-Occurrence (GLCM) method originated from textural analysis method of radar image using Haralick textures (Haralick *et al.*, 1973). As many texture layers can be derived from one image (in this case backscatter image), it is important to perform a detail assessment of which texture layers represent sediment classes. This is important because many habitat mapping process such as classification technique requires high spatial resolution data that can be incorporated with high spatial resolution bathymetry maps.

As backscatter data can be also represented by backscatter as a function of incidence angle, some studies have also used angular backscatter intensity (also known as signal-based backscatter) as one of the techniques to extract scattering information (Monteys *et al.*, 2016; Huang *et al.*, 2013; Lamarche *et al.*, 2011; Fonseca *et al.*, 2009; Parnum, 2007). Compared to backscatter image or mosaic, signal-based backscatter from MBES does not have high spatial resolution as the mosaic and thus might not be difficult to be integrated with other high spatial resolution maps such as bathymetry. However, one of the classification methods for signal-based backscatter, known as Angular Range Analysis (ARA) has been developed to automatically predict seafloor sediment types using acoustic inversion process (Fonseca and Mayer, 2007).

Consequently, the objectives of this paper are; (1) to investigate the correlation between MBES derived backscatter mosaic texture with seafloor sediment type, and (2) to identify which GLCM texture layers (i.e. from the image-based method) produce sediment classification map that have the highest similarities with signal-based classification method. Figure 2 shown the overall methodology flow chart conducted for this study.

## 2. METHODS

### 2.1 Study area

The study site is located in Tawau, Sabah, Malaysia which covers an area of 4.7km². It is situated off the channel in the Celebes Sea between Nunukan Island and Sebatik Island, East Malaysia (Figure 1). The site is adjacent to the international maritime border between Malaysia and Indonesia, located at about 1.5 km northwest of the Nunukan Island and 2.0 km southwest of the Sebatik Island.

mounted Kongsberg EM2040c multibeam bathymetric system. With a swath of seafloor ensonified four to five times the water depth on each survey line, offset for line spacing is set to three times water depth in order to provide ensonification overlap between adjacent survey lines. The positioning of the vessel during the survey was achieved by using C-Nav3050 DGPS system (horizontal accuracy: ± 0.45 m + 3 ppm and vertical accuracy: ± 0.90 m + 3 ppm) (Dubilier, 2016), integrated with an Kongsberg Seatex Motion Reference Unit MRU-5 (roll and pitch accuracy: 0.02° RMS at a ±5° amplitude)(Kongsberg, 2016), for roll, pitch and heave corrections. Multibeam data logging, Real-time navigation, display and quality control were using Seafloor Information System (SIS) software version 4.2.1 provided by Kongsberg. A sound velocity profile (SVP) through the water column in the survey area were daily collected in the beginning and at the end of survey process using Valeport

Figure 1: Location map of study area. The inset map shows the location of the study area relative to the location of Malaysia

### 2.2 Acoustic data acquisition

Acoustic data from MBES were acquired on the 26th of November 2017 until 1st of December 2016 using a hull-

### 2.3 Acoustic data processing

Backscatter data can be divided into two formats which are; (1) signal based data or backscatter intensity as a function of incidence angle, and (2) image-based data (i.e. backscatter mosaic). As a result, different classification methods have been established for each dataset (Brown et al., 2011a). The raw MBES backscatter data were processed in Fledermaus Geocoder Toolbox software version 7.4 (FMGT) to obtain (a) backscatter mosaic, and (b) prediction of sediment types using Angular Range Analysis (ARA) technique (Fonseca and Mayer 2007). An automated FMGT processing procedure was applied for both types of backscatter data (Quas et al., 2017). The corrections such as signal level adjustments and transmission loss, beam incidence angle, adjustments of beam footprint area and, Lambertian scattering adjustments were applied for each

Midas SVX2 equipment in order to obtain the speed of sound propagation in the water column at the survey area.

raw backscatter time series beam (QPS, 2016). Next, the backscatter intensity data were filtered based on beam angle , and then an anti-aliasing pass was run on the resulting backscatter swath data (QPS, 2016). For signal based seafloor classification, ARA technique was applied to the angular backscatter intensity to predict sediment types. This process produced estimated bottom sediment map by comparing the angular response/impedance estimates from calibrated backscatter values to empirical sediment models (Fonseca and Mayer 2007). The resulting seafloor characterisation map or ARA map was then exported to raster format for subsequent process. Note that in general, the spatial resolution from ARA map is low, with the size is half of the MBES swath width. Default ARA map yielded 30 sediment classes but then were regrouped to four (4) major sediment classes; sand, silt, clay, and gravel (Figure 3) as these classes are the dominant sediment types in ARA map. A set of random point was then generated from ARA map to be used as ground truth point. Along with ARA map, a backscatter mosaic image was produced at 1m pixel resolution for further analysis. In this study, sampling of ground truth was not available and therefore classes from ARA map were used as known classes to compare results with the classification map produced from texture layers of backscatter mosaic.

## 2.4  Derived GLCM and Image Statistics

Texture from image is an important characteristic for image classification such as used in many terrestrial remote sensing image processing and analysis. Eight (8) Haralick texture layers (Haralick et al., 1973) were derived from backscatter mosaic using ENVI 4.8 software; mean, variance, contrast, correlation, homogeneity, dissimilarity, entropy and angular second moment based on previous literature studies (Herkül et al., 2017; Diesing et al., 2016; Blondel et al., 2015; Lucieer et al., 2013; Huang et al., 2012; Lucieer et al., 2011). All texture layers were derived using Grey Level Co-occurrence Matrix (GLCM) method. GLCM calculates statistics by determining distinctive textural properties from an acoustic image showing the relationships between a given pixel and a specific neighbor (Díaz, 2000). For this study, Haralick texture layers were derived from GLCM calculated for a local rectangular window of 3x3 pixels.

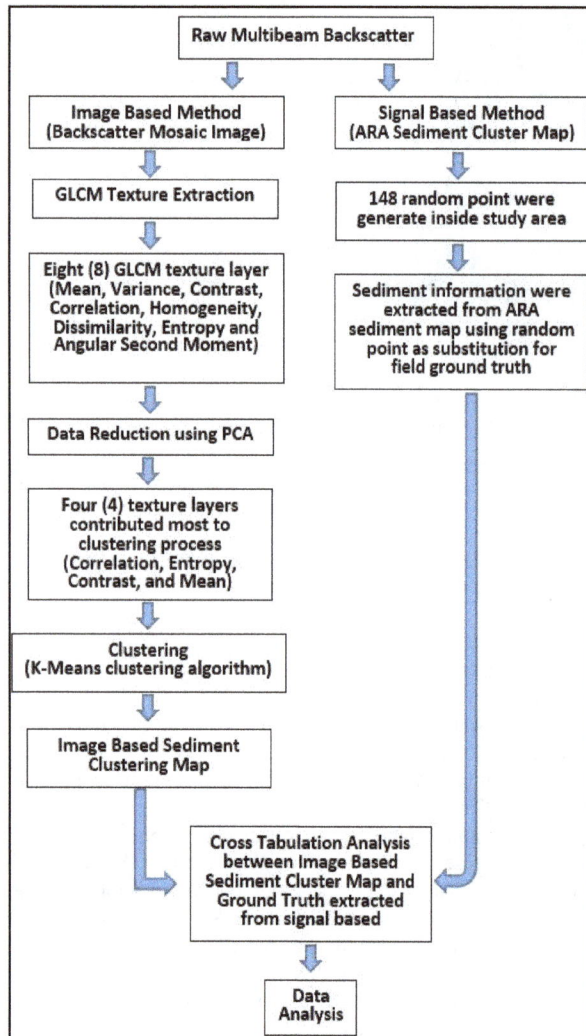

Figure 2: Study Flowchart

## 2.5  Principal Component Analysis (PCA)

Principal Component Analysis (PCA) method has been widely used in the previous study for data reduction and to avoid multicollinearity of the abiotic variables prior to clustering process (Ismail, 2016; Che Hasan, 2014; Verfaillie et al., 2009; Robidoux et al., 2008; Díaz, 2000). PCA has also been used to recognise which textural layers contributing most to the clustering map. PCA computes a set of new and linearly independent variables known as principal components (PCs) that account for most of the variance of the original variables. The PCs are produced from a linear combination of the original variables. PCA was used to determine; (a) which texture layers have the most contributions to the total variance of each rotated PC, and (b) correlations between different texture layers with each PC. Results from this will give a broad idea of which layers are important.

## 2.6  Clustering and comparison

After important texture layers have been identified, a K-Means clustering algorithm was applied to these texture layers. The K-Means clustering technique is widely being used for data segregating for terrestrial remote sensing and also in the marine environment. For this clustering process, the number of the cluster was set to be equal to the number of sediment classes in ARA map (i.e. four classes). A set of 148 points from ARA map were generated by creating random points inside the study area. Cross-tabulations of ground truth and clustered data for four PCs variables resulted from PCA were conducted to compare the occupancy of sediment within each cluster groups.

## 3.  Results

The results from PCA analysis produced four (4) PCs, explaining 98.77% of the total variance. The rotated component matrix (Table 1 and Table 2) shows the correlations between the rotated PCs and the original variables. The main GLCM variables that contributed to the highest variance of the PCA are Correlation (PCA1 -0.49%), Entropy (PCA1 -0.49%), Contrast (PCA2 0.57%), and Mean (PCA3 0.87%). This four GLCM layer obtained from PCA will be used for next cross tabulation analysis.

| PCA Layer | % Variance |
|-----------|------------|
| **1**     | **90.98**  |
| **2**     | **5.35**   |
| **3**     | **1.61**   |
| 4         | 0.83       |
| 5         | 0.52       |
| 6         | 0.43       |
| 7         | 0.22       |
| 8         | 0.06       |

Table 1. The contribution of all principal component analysis (PCA) bands to total variance.

| Texture Layer | PCA1 | PCA2 | PCA3 | PCA4 | PCA5 | PCA6 | PCA7 | PCA8 |
|---------------|------|------|------|------|------|------|------|------|
| Contrast      | -0.22 | **0.57** | 0.07 | 0.03 | 0.47 | -0.03 | -0.15 | -0.61 |
| Correlation   | **-0.49** | -0.26 | -0.21 | 0.50 | 0.06 | -0.62 | 0.06 | -0.01 |
| Dissimilarity | -0.30 | 0.41 | -0.02 | -0.09 | 0.38 | 0.03 | 0.08 | 0.76 |
| Entropy       | **-0.49** | -0.24 | -0.13 | 0.26 | 0.04 | 0.76 | 0.18 | -0.10 |
| Homogeneity   | 0.41 | 0.05 | 0.36 | 0.76 | 0.18 | 0.14 | -0.21 | 0.16 |
| Mean          | -0.35 | -0.26 | **0.87** | -0.20 | 0.02 | -0.08 | -0.03 | 0.00 |
| SecondMoment  | 0.21 | 0.08 | 0.15 | 0.08 | 0.14 | -0.08 | 0.94 | -0.11 |
| Variance      | -0.22 | 0.55 | 0.14 | 0.21 | -0.76 | 0.01 | 0.08 | 0.02 |

Table 2: Component matrix showing a correlation between rotated PCs and the original variables. Highest factor loads in each PC are highlighted in bold

The results from clustering map showed that, for GLCM Correlation and Entropy layers (Figures 4 and 5), the cluster map only showed two dominant classes. For the Contrast layer, the clustering was also showing poor cluster boundary although successfully produced four classes (Figure 5). Only clustering results from GLCM Mean layer showed cluster map with four classes and well delineated class boundary (Figures 6 and 7).

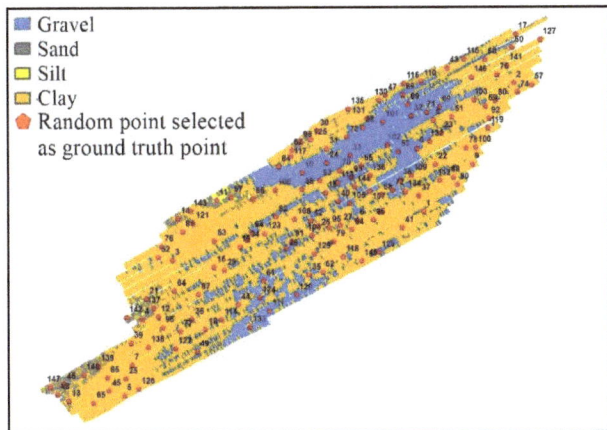

Figure 3: Sediment classes produced using Angular Range Analysis (ARA) and used for ground truthing

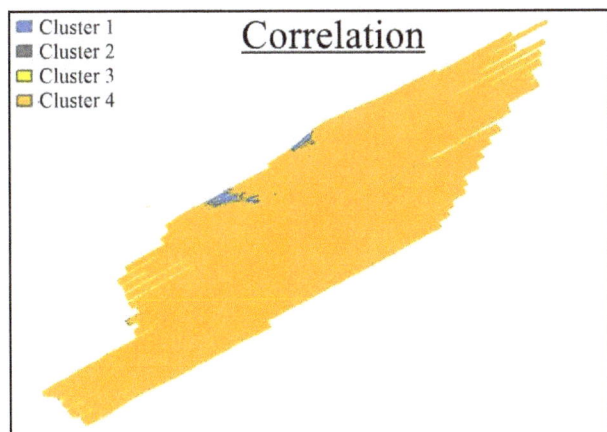

Figure 4: GLCM Correlation Cluster Map

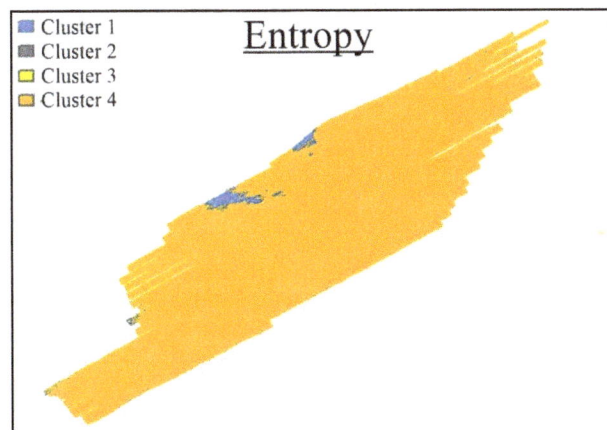

Figure 5: GLCM Entropy Cluster Map

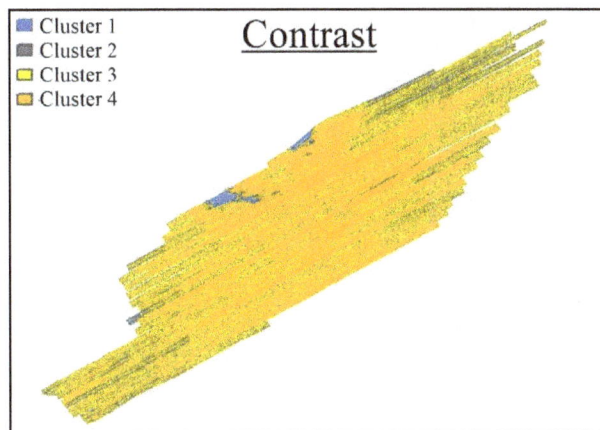

Figure 6: GLCM Contrast Cluster Map

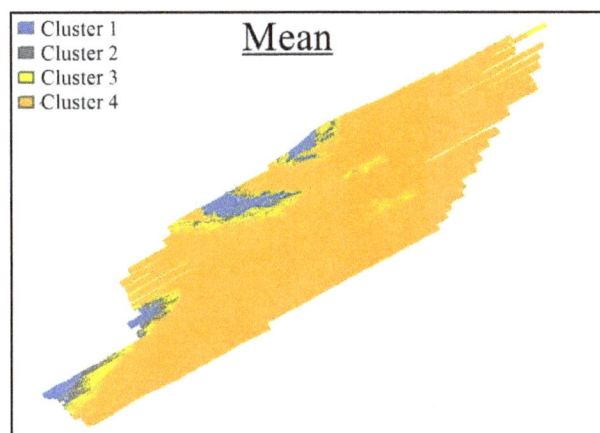

Figure 7: GLCM Mean Cluster Map

Cross tabulation analysis (Tables 3-6 and Figures 8-11) obtained from comparing sediment types and cluster maps in general produced inconsistent results except for GLCM Mean. For GLCM Entropy, only cluster 2 and cluster 4 have strong associations with a specified sediment class. For example, 100% of cluster 2 was related to Gravel and 67% of cluster 4 was identified as sand. For GLCM Contrast layer, three different clusters (clusters 2, 3 and 4) in the map showed high agreements with a single sediment type (i.e. sand), at 67%, 74% and 63% respectively. GLCM Correlation cluster map has identified two clusters with two different sediment types; cluster 1 with silt (67%) and cluster 4 with sand (65%).

However, for GLCM Mean, each cluster was showing relation to a unique sediment type, although there were some small percentage of other sediment types. For example, cluster 1 with gravel (42%), cluster 2 with silt (83%), cluster 3 with clay (43%) and cluster 4 with sand (74%).

| Sediment Type | Number of Ground Truth | Cluster 1(%) | Cluster 2(%) | Cluster 3(%) | Cluster 4(%) |
|---|---|---|---|---|---|
| Gravel | 36 | 17 | 100 | 50 | 24 |
| Sand | 94 | 17 | 0 | 0 | 67 |
| Silt | 13 | 33 | 0 | 50 | 7 |
| Clay | 5 | 33 | 0 | 0 | 2 |

Table 3: Cross tabulation between the GLCM Entropy cluster map and ground truth observations

| Sediment Type | Number of Ground Truth | Cluster 1(%) | Cluster 2(%) | Cluster 3(%) | Cluster 4(%) |
|---|---|---|---|---|---|
| Gravel | 36 | 0 | 11 | 23 | 30 |
| Sand | 94 | 0 | 67 | 74 | 63 |
| Silt | 13 | 50 | 19 | 3 | 5 |
| Clay | 5 | 50 | 4 | 0 | 2 |

Table 4: Cross tabulation between the GLCM Contrast cluster map and ground truth observations

| Sediment Type | Number of Ground Truth | Cluster 1(%) | Cluster 2(%) | Cluster 3(%) | Cluster 4(%) |
|---|---|---|---|---|---|
| Gravel | 36 | 0 | 0 | 0 | 25 |
| Sand | 94 | 0 | 0 | 0 | 65 |
| Silt | 13 | 67 | 0 | 0 | 8 |
| Clay | 5 | 33 | 0 | 0 | 3 |

Table 5: Cross tabulation between the GLCM Correlation cluster map and ground truth observations

| Sediment Type | Number of Ground Truth | Cluster 1(%) | Cluster 2(%) | Cluster 3(%) | Cluster 4(%) |
|---|---|---|---|---|---|
| Gravel | 36 | 42 | 0 | 14 | 24 |
| Sand | 94 | 8 | 17 | 14 | 74 |
| Silt | 13 | 33 | 83 | 29 | 2 |
| Clay | 5 | 17 | 0 | 43 | 0 |

Table 6: Cross tabulation between the GLCM Mean cluster map and ground truth observations

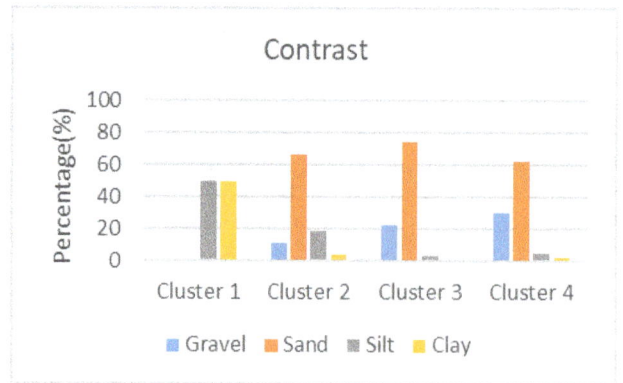

Figure 8: Per cluster sediment composition percentage for GLCM Correlation texture layer

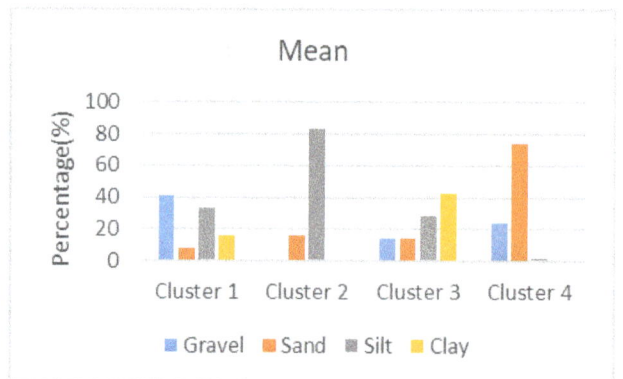

Figure 9: Per cluster sediment composition percentage for GLCM Entropy texture layer

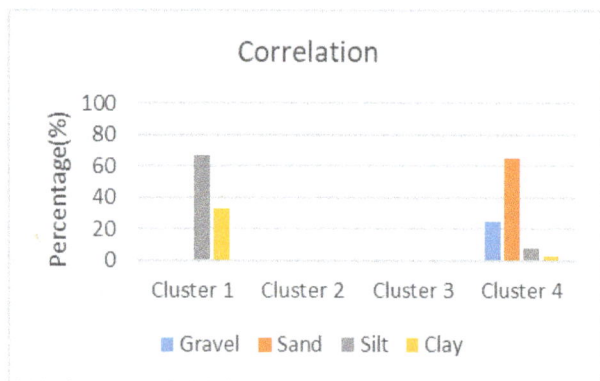

Figure 10: Per cluster sediment composition percentage for GLCM Contrast texture layer

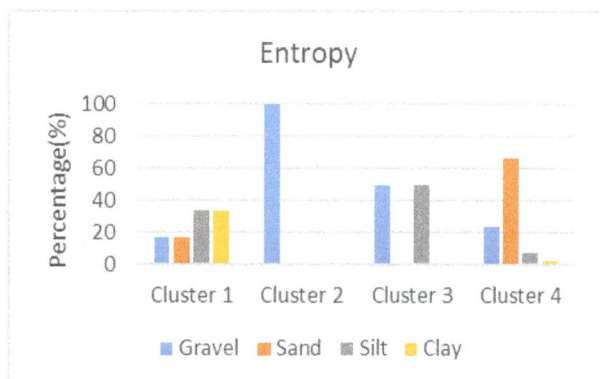

Figure 11: Per cluster sediment composition percentage for GLCM Mean texture layer

## 4. DISCUSSION

The approach of this study is to identify the correlation between MBES derived backscatter mosaic texture with seafloor sediment type and to identify the capability of texture based method to differentiate seafloor sediment classes. The research used sediment classes from ARA as substitute for ground truth and subsequently a set of random ground truth point was generated inside the study area. From the result obtained, it can be clearly seen that only clustering from GLCM Mean layer can provide significant discrimination compare with others three GLCM layers. Previous studies on texture-based sediment classification techniques have shown that the indices 'Mean' capture most of the textural variability within the data (Huvenne et al., 2007). Mean from backscatter has also been used in some of the sediment classification (Hill et al., 2014; Lucieer et al., 2013; Díaz, 2000).

The results from PCA analysis is able to identify some of the important layers in general. Principal component analysis has been broadly used to recognise which textural layers contributing most to the clustering map (Ismail et al., 2015; Che Hasan, 2014; Verfaillie et al., 2009; Robidoux et al., 2008; Díaz, 2000). GLCM Correlation, Entropy, Contrast and Mean are the main texture layer resulted from principal component analysis with percentage eigenvalue more than 1%. However, in this study, the PCA results of identifying the most significant layer disagree with the clustering map analysis. For example, PCA identified GLCM layers of Correlation and Entropy as the most influenced layers (PCA1), but the clustering map analysis has identified different layers (i.e. GLCM Mean). This is due to the small ratio of clay and silt within the study area. According to Che Hasan (2014); Müller and Eagles (2007), different sediment proportion within a sediment class also may cause backscatter intensity and texture analysis to diverge and unsupervised classification methods do not allow the control of such factors.

The study identified some relationships between the MBES backscatter mosaic and resulting clusters map with the backscatter derivatives GLCM Mean. Although GLCM Mean at the fourth place in the sequence of most contributing GLCM layer, previous researchers (Che Hasan, 2014) suggested that GLCM mean demonstrates the most significant layer for sediment clustering map.

## 5. CONCLUSION

A total of 4.7km$^2$ of multibeam sonar backscatter data from Tawau coastal area, Malaysia, was classified using GLCM and K-Means algorithms to find correlations between signal and image based backscatter. Notably, our approach is only using random ground truth point created in GIS software due to limitation during the survey. Hypothetically, if the ground-truth point of the survey had been carried out on a targeted K-Means clustering, the agreement observed may have been more convincing. However, on the basis of the comparisons with randomly created ground-truth data, the cross tabulation analysis conducted has shown encouraging results. In summary, only GLCM Mean texture layer show the significant similarities with signal based sediment classification map and demonstrate the ability to successfully delineating the major type of sediment. Overall, it can be concluded that image-based backscatter classification can assist the interpretation of multibeam backscatter data for the production of sediment maps.

## ACKNOWLEDGEMENT

The authors would like to thank the National Hydrographic Centre (NHC) and the Department of Survey and Mapping Malaysia (DSMM) for the acoustic data collection. Thanks to Universiti Teknologi Malaysia for providing research fund through Research University Grant (RUG) No. Q.K130000.2540.11H41 - 'Acoustic Classification Tool for Automatic Seabed Segmentation'. The authors are also grateful to Acburn Marine Automations Sdn. Bhd. for providing multibeam echosounder system and to all crew from *Bot Hidrografi 2* for their help during acoustic data acquisition.

## REFERENCES

Blondel, P., Prampolini, M. & Foglini, F. (2015). Acoustic textures and multibeam mapping of shallow marine habitats-Examples from Eastern Malta.

Brown, C. J., Smith, S. J., Lawton, P. & Anderson, J. T. (2011a). Benthic habitat mapping: A review of progress towards improved understanding of the spatial ecology of the seafloor using acoustic techniques. *Estuarine, Coastal and Shelf Science*, 92, 502-520.

Brown, C. J., Todd, B. J., Kostylev, V. E. & Pickrill, R. A. (2011b). Image-based classification of multibeam sonar backscatter data for objective surficial sediment mapping of Georges Bank, Canada. *Continental Shelf Research*, 31, S110-S119.

Buhl-Mortensen, L., Buhl-Mortensen, P., Dolan, M. & Gonzalez-Mirelis, G. (2015). Habitat mapping as a tool for conservation and sustainable use of marine resources: Some perspectives from the MAREANO programme, Norway. Journal of Sea Research, 100, 46-61.

Che Hasan, R. 2014. Multibeam backscatter for benthic biological habitat mapping. Deakin University.

De Falco, G., Tonielli, R., Di Martino, G., Innangi, S., Simeone, S. & Parnum, I. M. (2010). Relationships between multibeam backscatter, sediment grain size and Posidonia oceanica seagrass distribution. *Continental Shelf Research*, 30, 1941-1950.

Díaz, J. V. M. 2000. *Analysis of multibeam sonar data for the characterization of seafloor habitats*. University of New Brunswick, Department of Geodesy and Geomatics Engineering.

Diesing, M., Mitchell, P. & Stephens, D. (2016). Image-based seabed classification: what can we learn from terrestrial remote sensing? *ICES Journal of Marine Science*, 73, 2425-2441.

Dubilier, N. (2016). Trackline map and processing report for navigation sensors from Meteor M126.

Fakiris, E., Rzhanov, Y. & Zoura, D. (2012). On importance of acoustic backscatter corrections for texture-based seafloor characterization.

Fonseca, L., Brown, C., Calder, B., Mayer, L. & Rzhanov, Y. (2009). Angular range analysis of acoustic themes from Stanton Banks Ireland: A link between visual interpretation and multibeam echosounder angular signatures. *Applied Acoustics*, 70, 1298-1304.

Fonseca, L. & Mayer, L. (2007). Remote estimation of surficial seafloor properties through the application Angular Range Analysis to multibeam sonar data. *Marine Geophysical Researches*, 28, 119-26.

Haralick, R. M., Shanmugam, K. & Dinstein, I. H. (1973). Textural features for image classification. *IEEE Transactions on systems, man, and cybernetics*, 610-621.

Hedley, J., Roelfsema, C., Chollett, I., Harborne, A., Heron, S., Weeks, S., Skirving, W., Strong, A., Eakin, C., Christensen, T., Ticzon, V., Bejarano, S. & Mumby, P. (2016). Remote Sensing of Coral Reefs for Monitoring and Management: A Review. *Remote Sensing*, 8, 118.

Herkül, K., Peterson, A. & Paekivi, S. (2017). Applying multibeam sonar and mathematical modeling for mapping seabed substrate and biota of offshore shallows. *Estuarine, Coastal and Shelf Science*, 192, 57-71.

Hill, N. A., Lucieer, V., Barrett, N. S., Anderson, T. J. & Williams, S. B. (2014). Filling the gaps: Predicting the distribution of temperate reef biota using high resolution biological and acoustic data. *Estuarine, Coastal and Shelf Science*, 147, 137-147.

Huang, Z., Nichol, S. L., Siwabessy, J. P., Daniell, J. & Brooke, B. P. (2012). Predictive modelling of seabed sediment parameters using multibeam acoustic data: a case study on the Carnarvon Shelf, Western Australia. *International Journal of Geographical Information Science,* 26, 283-307.

Huang, Z., Siwabessy, J., Nichol, S., Anderson, T. & Brooke, B. (2013). Predictive mapping of seabed cover types using angular response curves of multibeam backscatter data: Testing different feature analysis approaches. *Continental Shelf Research,* 61, 12-22.

Hughes Clarke, J. E., Mayer, L. A. & Wells, D. E. (1996). Shallow-water imaging multibeam sonars: a new tool for investigating seafloor processes in the coastal zone and on the continental shelf. *Marine Geophysical Research,* 18, 607-629.

Huvenne, V. A., Hühnerbach, V., Blondel, P., Gómez Sichi, O. & LeBas, T. Detailed mapping of shallow-water environments using image texture analysis on sidescan sonar and multibeam backscatter imagery. Proceedings of the 2nd underwater acoustic measurements conference. Heraklion: FORTH, 2007.

Ismail, K. 2016. *Marine landscape mapping in submarine canyons.* University of Southampton.

Ismail, K., Huvenne, V. A. I. & Masson, D. G. (2015). Objective automated classification technique for marine landscape mapping in submarine canyons. *Marine Geology,* 362, 17-32.

Kongsberg, M. 2016. Data sheet MRU 5 (Ideal Marine Motion Sensor). Norway.

Lamarche, G., Lurton, X., Verdier, A. L. & Augustin, J. M. (2011). Quantitative characterisation of seafloor substrate and bedforms using advanced processing of multibeam backscatter-Application to Cook Strait, New Zealand. *Continental Shelf Research,* 31, S93-S109.

Lucieer, V., Hill, N., Barrett, N. & Nichol, S. Spatial analysis of multibeam acoustic data for the predictiion of marine substrates and benthic communities in temperate coastal waters. ISRSE 2011: International symposium of R 34th International Symposium on Remote Sensing of Environment, 2011. On USB.

Lucieer, V., Hill, N. A., Barrett, N. S. & Nichol, S. (2013). Do marine substrates 'look'and 'sound'the same? Supervised classification of multibeam acoustic data using autonomous underwater vehicle images. *Estuarine, Coastal and Shelf Science,* 117, 94-106.

Lucieer, V., Nau, A. W., Forrest, A. L. & Hawes, I. (2016). Fine-Scale Sea Ice Structure Characterized Using Underwater Acoustic Methods. *Remote Sensing,* 8, 821.

Medialdea, T., Somoza, L., León, R., Farrán, M., Ercilla, G., Maestro, A., Casas, D., Llave, E., Hernández-Molina, F. & Fernández-Puga, M. (2008). Multibeam backscatter as a tool for sea-floor characterization and identification of oil spills in the Galicia Bank. *Marine Geology,* 249, 93-107.

Micallef, A., Le Bas, T. P., Huvenne, V. A. I., Blondel, P., Hühnerbach, V. & Deidun, A. (2012). A multi-method approach for benthic habitat mapping of shallow coastal areas with high-resolution multibeam data. *Continental Shelf Research,* 39-40, 14-26.

Monteys, X., Hung, P., Scott, G., Garcia, X., Evans, R. L. & Kelleher, B. (2016). The use of multibeam backscatter angular response for marine sediment characterisation by comparison with shallow electromagnetic conductivity. *Applied Acoustics,* 112, 181-191.

Müller, R. D. & Eagles, S. (2007). Mapping seabed geology by ground-truthed textural image/neural network classification of acoustic backscatter mosaics. *Mathematical Geology,* 39, 575-592.

Parnum, I. M. (2007). Benthic habitat mapping using multibeam sonar systems.

QPS (2016). FMGeocoder Toolbox Online Manual. Fledermaus 7.7.x Documentation.

Quas, L., Church, I., O'Brien, S. J., Wiggert1, J. D. & Williamson, M. (2017). Application of high-resolution multibeam sonar backscatter to guide oceanographic investigations in the Mississippi Bight.

Robidoux, L., Fonseca, L. & Wyatt, G. A qualitative assessment of two multibeam echosounder (MBES) backscatter analysis approaches. Canadian Hydrographic Conference and National Surveyors Conference, Thursday, May, 2008.

Siwabessy, P. J. W., Daniell, J., Li, J., Huang, Z., Heap, A. D., Nichol, S., Anderson, T. J. & Tran, M. 2013. *Methodologies for seabed substrate characterisation using multibeam bathymetry, backscatter and video data: A case study from the carbonate banks of the Timor Sea, Northern Australia.*

Sutherland, T., Galloway, J., Loschiavo, R., Levings, C. & Hare, R. (2007). Calibration techniques and sampling resolution requirements for groundtruthing multibeam acoustic backscatter (EM3000) and QTC VIEW™ classification technology. *Estuarine, Coastal and Shelf Science,* 75, 447-458.

Verfaillie, E., Degraer, S., Schelfaut, K., Willems, W. & Van Lancker, V. (2009). A protocol for classifying ecologically relevant marine zones, a statistical approach. *Estuarine, Coastal and Shelf Science,* 83, 175-185.

Zhi, H., Siwabessy, J., Nichol, S. L. & Brooke, B. P. (2014). Predictive mapping of seabed substrata using high-resolution multibeam sonar data: A case study from a shelf with complex geomorphology. *Marine Geology,* 357, 37-52.

# 26

# RELIABILITY OF WIND SPEED DATA FROM SATELLITE ALTIMETER TO SUPPORT WIND TURBINE ENERGY

M. N. Uti[a], A. H. M. Din[a,b,c], and A. H. Omar[a]*

[a]Geomatic Innovation Research Group (GIG)
[b]Geoscience and Digital Earth Centre (INSTEG), Faculty of Geoinformation and Real Estate, Universiti Teknologi Malaysia, 81310 Johor Bahru, Johor, Malaysia
[c]Associate Fellow, Institute of Oceanography and Environment (INOS), Universiti Malaysia Terengganu, Kuala Terengganu, Terengganu, Malaysia, Email: amihassan@utm.my

**KEY WORDS:** Wind Speed Data, Wind Turbine Energy, Satellite Altimeter, Radar Altimeter Database System, Ground-truth buoy

**ABSTRACT:**

Satellite altimeter has proven itself to be one of the important tool to provide good quality information in oceanographic study. Nowadays, most countries in the world have begun in implementation the wind energy as one of their renewable energy for electric power generation. Many wind speed studies conducted in Malaysia using conventional method and scientific technique such as anemometer and volunteer observing ships (VOS) in order to obtain the wind speed data to support the development of renewable energy. However, there are some limitations regarding to this conventional method such as less coverage for both spatial and temporal and less continuity in data sharing by VOS members. Thus, the aim of this research is to determine the reliability of wind speed data by using multi-mission satellite altimeter to support wind energy potential in Malaysia seas. Therefore, the wind speed data are derived from nine types of satellite altimeter starting from year 1993 until 2016. Then, to validate the reliability of wind speed data from satellite altimeter, a comparison of wind speed data form ground-truth buoy that located at Sabah and Sarawak is conducted. The validation is carried out in terms of the correlation, the root mean square error (RMSE) calculation and satellite track analysis. As a result, both techniques showing a good correlation with value positive 0.7976 and 0.6148 for point located at Sabah and Sarawak Sea, respectively. It can be concluded that a step towards the reliability of wind speed data by using multi-mission satellite altimeter can be achieved to support renewable energy.

## 1. INTRODUCTION

### 1.1 Research Background

Malaysia has attempted to implement wind energy as one of the renewable energy (RE) source to accommodate the electrical power in Malaysia. Thus, many research and studies have been conducted by professionals and government agencies such as Malaysian Meteorological Department (MMD). Unfortunately, the study of wind speed for both onshore and offshores in Malaysia was deadlocked due to the location and method used (L. W. Ho, 1993). The geographical location of Malaysia at low wind-speed region has brought difficulties in wind speed study. Most of the research locations were placed at meteorological stations which are near/at airport and it is not suitable placed for wind speed study due to bias of wind speed observation caused by airplanes (L. W. Ho, 1993).

Besides that, MMD had conducting wind speed study at offshore areas by deployment of buoy and Volunteer Observing Ships (VOS). But, due to the shortage in data distribution by VOS members brought a great challenge to assess the wind speed data (L. W. Ho, 1993). Other than that, the existence of strong wind at a distinct time and location (low temporal and limited spatial coverage) becomes the biased factors of analysing the wind speed data by using buoy and VOS. Most of the wind speed observations conducted using the in-situ platforms such as ships and buoys have limited in spatial coverage and temporal. Since there are limitations for in-situ wind speed observation techniques in Malaysia to support the renewable energy thus, the aim of this research is to study the reliability of wind speed data using multi-mission satellite altimeter to support wind turbine generator over Malaysian seas. Thus, the scope of this research covers a 24 years (1993-2016) of wind speed data from satellite altimeter and moored buoy. Wind speed data from satellite altimeter are retrieved by using Radar Altimeter Database System (RADS). The first point of buoy is located at Sarawak offshore (5.15N, 111.82E) and the second point is located at Sabah Sea (5.83N, 114.39E). For altimeter wind speed data retrieval are covered from 0° to 14° of latitude and 95° to 125° of longitude including Malaysian Seas such as Malacca Straits, South China Sea, Celebes Sea and Sulu Sea, as shown in Figure 1. After that, data verification is conducted by comparing both of wind speed observation from satellite altimeter and buoy. Moored buoy will be acted as a ground-truth or bench-mark since it was point-based observation data. Verification of these data is focusing on the accuracy of the satellite altimeter measurements and its significance to provide a reliable data in terms of good spatial and temporal. These verification part also accounted with statistical analysis such as time-series graph, correlation and root mean square error (RMSE). Besides that, an assessment of the wind speed with respect to seasonal effect is approached by conducting an altimeter wind speed climatology data over 24 years from 1993 until 2016. The seasonal effect includes

Northeast Monsoon (November to February), Southwest Monsoon (May to August), First Inter-Monsoon (March and April) and Second Inter- Monsoon (September and October). The significant of this assessment is to identify and propose a potential location at/near offshores that received high wind speed for the wind turbine generator.

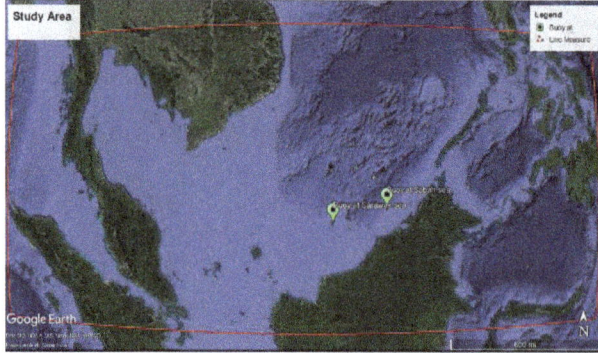

Figure 1. The map of study area (NOAA, 2010).

## 1.2 Satellite Altimetry Concepts

The basic principle of satellite altimeter is referred to the reflected pulse from the surface and backscattered of the wind and waves. Then, the pulse is received by altimeter sensor in a few milliseconds. There are several parameters for satellite altimeters which are returned signal from the sea; slant of the reflected signal; distance, $d$ from the sea level (measured by the time travel of the pulse); wind speed determination by the power from the impulse response. (Abdul Aziz et al., 2014)).

Since this technique acquired some parameters for calculation that mentioned above, several corrections for altimeter measurement have been accounted such as orbital errors, instrumental errors (offset antenna phase center, clock drift electronic time delay), atmospheric errors due to tropospheric and ionospheric effect, time lagging of measurements, clock drift and Doppler shift (Fu and Cazenave, 2001). Other components that can affect the signal such as inverted barometer effect, ocean tides, and electromagnetic bias as shown in Figure 2.

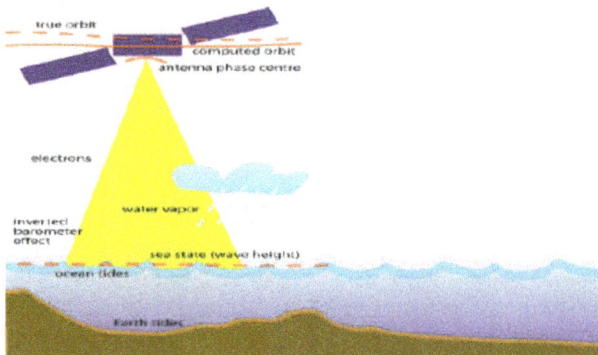

Figure 2. Corrections for altimeter range measurements

Corrected range ($R_{corrected}$) is related to the Observed range ($R_{obs}$).

$$R_{corrected} = R_{obs} - \Delta R_{dry} - \Delta R_{wet} - \Delta R_{iono} - \Delta R_{ssb} \qquad (1)$$

$R_{obs} = c\ t/2$ is the measured distance from the signal travel time, $t$ and $c$ is the speed of the echo pulse ignoring refraction.

where | $\Delta R_{dry}$ | : Dry tropospheric correction
| $\Delta R_{wet}$ | : Wet tropospheric correction
| $\Delta R_{iono}$ | : Ionospheric correction
| $\Delta R_{ssb}$ | : Sea-state bias correction

Altimetry measurement uses a nadir-pointing principle to calculate accurately the time taken of the emitted and returned signal of satellite altimeter. Figure 3 represents the diagram for the altimetry principles. $R$, is the measurement of slant from the angle, $\theta$, where $\theta$ is the antenna pointing angle, and $\theta'$ is the incidence angle, while $A_f$, is the the antenna footprint area (D.B. Chelton et al., 2001). Based on theory, with a specific energy the microwave pulse is permanently transmit from the on-board radar altimeter to the sea surface. The emitted signal from the satellite altimeter towards the sea surface will be reflected back to the altimeter sensor for the accurate measurement of the time taken of the signal between the satellite altimeter and ocean surface.

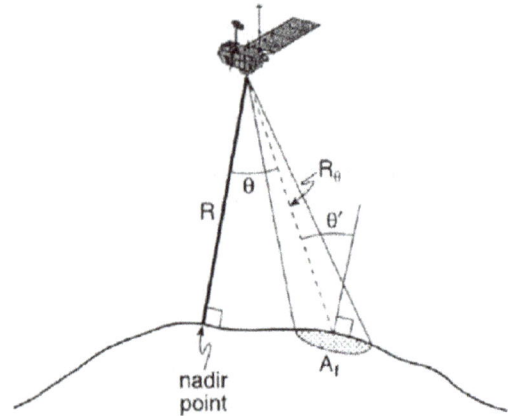

Figure 3. Altimetry principles

## 1.3 Buoy Concept

Buoy is one of the methods used for a meteorological and marine study. Nowadays, many countries in the world has moored and deployed the buoy on the surface of the ocean for meteorological study. Recently, United States of America already moored lots of buoys to form a buoy network that covering most of the west side of the USA coast, several Hawaiian Islands and the Gulf of Mexico (Mathisen, 2010). The used of marine buoy is to measure and collect the meteorological information such as ocean current, wave height, wind speed, humidity, temperature and the ocean $pH$. The measurement of these ocean parameters from the modern buoy because of the installation of the multipurpose sensors such as wind sensor, air pressure sensor, air temperature sensor, ocean temperature sensor, humidity sensor and others. The modern buoy also equipped with a GPS system in order to give coordinate of the buoy.

## 1.4 Previous Research

There were many research conducted with significant to study the reliability of wind speed measurement using the multi-mission satellite altimeter for the purpose of providing information to the renewable energy especially in wind turbine energy. Most of the previous study is carried out by assessing the observation of wind speed data from the satellite altimeter and buoy as shown in Table 1. Comparison of data from both techniques is conducted and analysed in order to verify the reliability of wind speed data from the satellite altimeter. Several methods of study are approached such as along-track analysis, cross-track analysis and statistical analysis to verify the reliability of wind speed data from multi-mission satellite altimeter.

| Previous Research | RMSE (m/s) | Correlation | Error biased | Sat |
|---|---|---|---|---|
| Ebuchi and Kawamura (1994) | 1.99 | NA | 0.56 | tx |
| Hwang et al., (1998) | 1.20 | 0.96 | NA | tx |
| Kshatriya et al., (2002) | 1.60 | 0.53 | NA | tx/pn |
| Abdalla et al., (2014) | 1.50 | 0.91 | -0.20 | j2 |
| Adballa (2014) | 1.48 | 0.86 | -0.32 | sa |
| Kumar et al., (2015) | 1.13 | 0.91 | -0.28 | sa |

Table 1. Previous research findings of the comparisons between altimeter and buoy measurements

Regarding to the previous research above, this paper will present the reliability of wind speed data by using multi-mission satellite altimeter for wind energy over Malaysian seas. Thus, this paper is focusing on the assessment of wind speed climatology and its reliability to support wind energy in Malaysia. The assessment is covered the Malaysian seas with two locations of moored buoy as shown in Figure 1. The altimetry data used were in the range of 24 years period of data starting from January 1993 until December 2016 by using nine different satellite altimeters which are ERS 1, ERS 2, SARAL, ENVISAT 1, CRYOSAT, TOPEX/POSEIDON, JASON 1, JASON 2 and JASON 3. Assessment of this study is done by verifying the altimeter with the ground-truth buoy by using cross-track analysis, statistical analysis and graphical presentation. A set of climatology data also analyzed in order to identify the potential location for wind energy implementation in Malaysia region

## 2 RESEARCH APPROACH

### 2.1 Research Approach

Recently, many researchers conducting their research by using the advance technology as a tool for data acquisition, processing or even data analysis. For marine and oceanographic study, satellite altimeter become one of the elevate technique used to provide a set of data for scientific research. Since 1970s, the launched of satellite altimeter has triggered the oceanographic community to use the altimetry data for marine studies and its advantage by providing good spatial and temporal observations (Fu and Cazenave, 2001). The evolution in space-based data observation such as altimeters has brought the wind speed measurement with a good accuracy comparable with the in-situ observations such as buoy (Dobson et al., 1987).

Wind speed data is retrieved from the satellite altimeter and selected moored buoy. For satellite altimeter, the data extraction divided into two phases; multi-mission satellite altimeter and single mission satellite altimeter. Wind speed data from selected

moored buoy that located at Sarawak and Sabah sea, is provided from oil and gas company. Buoy measurement only provide ocean information for particular area of interest due to limited spatial and temporal coverage (Wan et al., 2010). Other than that, these ground-truth buoy measurements are always been used as a benchmark to verify the ocean wind speed observations from the satellite altimeter.

Thus, the wind speed data from the satellite altimeter and buoy are compared to perform data validation and to provide a good continuity contemplate of wind speed data observation (Kumar et al., 2015). Satellite altimeter is an alternative way in providing a precise and stable data with a good coverage of both spatial and temporal. Fairly, many studies have been done in order to assess and validate the reliability of wind speed measurement from the satellite altimeter. Thus, this study is focusing on the reliability of the satellite altimeter for the appraisal of wind turbine energy in Malaysia.

### 2.2 Radar Altimeter Database System (RADS) Framework

Recently, altimetry data has been allocated through international agencies such as National Oceanic and Atmospheric Administration (NOAA) USA, AVISO, and EUMETSAT. RADS is a software produced by the collaborating between Delft Institute for Earth-Oriented Space Research (DEOS) and National Oceanic and Atmospheric Administration (NOAA). It is developed with integrated, approved and calibrated sea-level directory from the multi-mission altimeter. RADS is established to help in marine and oceanographic research by providing ocean information gathered from the satellite altimeter. RADS is operated using the scheme of an internet facility known as Netherlands Earth Observation NETwork (NEONET) to provide and support a remote sensing observation data. For this part, a special script is used in RADS software to extract a desired wind speed data from satellite altimeter.

The advantages of RADS is, users are able to access, process and verified their own altimetry data with accurate database and precise parameters (Andersen and Schrarroo, 2011). Currently, RADS capacitate all users to avulse data from several satellite altimeters such as ERS 1, ERS 2, JASON 1, JASON 2, ENVISAT 1, TOPEX, POSEIDON, GEOSAT and others. Table 2 shows the current status of satellite altimeter data from Radar Altimeter Database System (RADS).

| Altimeter | Phase | Time | Cycles |
|---|---|---|---|
| ERS 1 | C | 14 Apr 1992 - 20 Dec 1993 | 083 – 101 |
| | D | 24 Dec 1993 - 10 Apr 1994 | 103 - 138 |
| | E | 10 Apr 1994 - 28 Sep 1994 | 139 - 140 |
| | F | 28 Sep 1994 - 21 Mar 1995 | 141 - 143 |
| | G | 24 Mar 1995 - 02 Jun 1996 | 144 – 156 |
| TOPEX | A | 25 Sep 1992 - 11 Aug 2002 | 001 – 364 |
| | B | 10 Sep 2002 - 08 Oct 2005 | 369 - 481 |
| POSEIDON | A | O1 Oct 1992 - 12 Jul 2002 | 001 – 361 |
| ERS 2 | A | 29 Apr 1995 - 04 Jul 2011 | 000 – 169 |
| JASON 1 | A | 15 Jan 2002 - 26 Jan 2009 | 001 – 260 |
| | B | 10 Feb 2009 - 03 Mar 2012 | 262 - 374 |
| | C | 07 May 2012 - 21 Jun 2013 | 382 – 425 |
| ENVISAT1 | B | 14 May 2002 - 22 Oct 2010 | 006 – 094 |
| | C | 26 Oct 2010 - 08 Apr 2012 | 095 – 113 |
| JASON 2 | A | 04 Jul 2008 - 02 Oct 2016 | 000 – 303 |
| | B | 13 Oct 2016 - 04 Dec 2016 | 305 – 310 |
| CYROSAT 2 | A | 14 Jul 2010 - 04 Dec 2016 | 004 – 086 |
| SARAL | A | 14 Mar 2013 - 04 Jul 2016 | 001 – 035 |
| | B | 04 Jul 2016 - 04 Dec 2016 | 036 – 040 |
| JASON 3 | A | 01 Jan 2016 - 04 Dec 2016 | 000 – 035 |

Table 2. Status of RADS
(Source: http://rads.tudelft.nl/rads/status)

## 2.3 Methodology

The methodology for this research is involved convergence of satellite altimeter wind speed data with verification from two ground-truth buoys following with statistical analysis and graphical presentation to assess altimeter products with reference of ground-truth buoy. Numerous previous study had applied ground-truth buoy measurement to correlate and verify the satellite altimeter data considering buoy observations as reference data for wind speed study (Queffeulou, 2004). For spatial-temporal analysis of satellite altimeter and buoy observations, a track analysis of satellite altimeter corresponds to ground-truth buoy is approached. The altimeter measurement of spatial-temporal is using 9 days of moving-window, with resolution of 0.25°x 0.25°. Next, the collocated data sets are considered relative and further analysis is approached using statistical analysis by applying the correlation between both measurements and the root mean square error (rmse) calculation (Kumar et al., 2015). This procedure is applied the statistical regression to attain the probable relationship between the ground-truth buoy and the satellite altimeter measurements. This approach is useful to verify the reliability of the wind speed measurement from satellite altimeter correspond with buoys as a bench-mark. The graphical presentation is approached by production of time-series graph for both methods which is necessarily to evaluate the wind speed pattern for both measurements. Other than that, further climatology analysis of wind speed is approached to examine the seasonal effect towards the wind speed observations in Malaysian seas. The advantages of this analysis is to determine the pattern of the wind speed in Malaysian seas region and proposed the potential location for wind energy by selecting a specific place that received high average of wind speed corresponding with seasonal period.

## 2.4 Area of Interest and Wind Speed Data

This paper is conducted in order to provide and assess the wind speed data over Malaysian Sea as shown in Figure.1. The selected area of interest is covered the Malaysian Exclusive Economic Zone (EEZ). There are four zones of Malaysian seas that have been selected to assess the potential offshores/near offshores locations that possesses with a great wind speed. Every year, most of the zones faced with a great wind speed during the Northeast Monsoon (November to February) compare to the other monsoons.

### 2.4.1 Altimetry Data

For multi-mission satellite altimeter, the data extraction is from nine different types of satellite altimeter using Radar Altimeter Database System (RADS). For multi-mission satellite altimeter analysis, the data extraction is corresponded within the study area which is covered the Malaysian seas as shown in Figure. 1. A 24 years period of wind speed data is extracted starting from 1st January 1993 until 31st December 2016. The significant of these multi-mission satellite altimeter measurement, is to conduct a wind speed climatology assessment based on the daily and monthly solutions. Inside this processing, user will define the study area or geographical region in latitude and longitude also the related parameters. For this study, the satellite altimeter data is extracted between 0° N to 14° N for latitude and 95° E to 125° E longitude, which covers the four zones of Malaysian seas.

Next, is the single-mission satellite altimeter processing with significant to analyse the along-track of the satellite altimeter over the moored buoys and valuate the spatial coverage of the satellite altimeter. In this single–mission processing, the extraction of wind speed data from the satellite altimeter is based on the observation period from the ground-truth buoy. This study is used ground-truth buoys data from 1st October 1995 until 31st January 2012. Thus, a data selection of cycle and phase for each single-mission satellite altimeter will be referred to the ground-truth buoy period of time. After the selection, several satellite altimeter missions are chosen as shown in Table 3. These seven satellite altimeters are selected based on the time interval of the ground-truth buoys.

| Satellite | Phase | Cycle |
|---|---|---|
| ENVISAT 1 | B | 006-094 |
|  | C | 095-111 |
| ERS 1 | G | 150-156 |
| ERS 2 | A | 004-169 |
| JASON 1 | A | 001-260 |
|  | B | 262-371 |
| JASON 2 | A | 000-159 |
| POSEIDON | A | 113-361 |
| TOPEX | A | 112-364 |
|  | B | 369-481 |

Table 3. Single-mission satellite altimeter selected data from 1st October1995 until 31st January 2012

### 2.4.2 Buoy Data

In this research, wind speed data from the in-situ buoy (ground-truth), are used as a reference/benchmark for the data comparison and assessment with the satellite altimeter. Malaysia has deployed several numbers of buoys at the specific locations over Malaysian seas. For this study, the wind speed data from buoys located at Sabah and Sarawak sea are provided by the oil and gas company. Both of these buoys are installed with the sensors to provide oceanographic information. The buoys were install with the several sensors such as wave sensor, wind sensor, temperature sensor, air pressure sensor and ocean current sensor. For this part, we are used two moored buoy as a bench-mark as shown in Table 4.

| Buoy | Latitude | Longitude | Period |
|---|---|---|---|
| Sabah | 5.83 N | 114.39 E | Nov 04-Dec 07 |
| Sarawak | 5.15 N | 111.82 E | Oct 95-Jan 12 |

Table 4. Buoy information

## 3  RESULTS AND DISCUSSION

Wind speed measurements that derived from the multi-mission satellite altimeter is distinguished with the ground-truth buoy data to analyze the reliability of satellite altimeter wind speed data. In-situ measurements such as buoy have become a source of continuous measurements of oceanographic information at permanent points. Generally, in-situ measurements always become a bench-mark for the affirmation of the measurements from the satellite altimeter (Kumar et al., 2015). With the improvising of satellite altimeters, the respective blunder for the satellite altimeter has been downsized if compared with the ground-truth buoy, thus the linear-regression method is used for inter-comparison purposes (Durrant et al., 1987). Based on the previous study has concluded the relative error between the satellite altimeter and the ground-truth buoy are relatively equal

(Caires and Sterl, 2003). The Ground-truth buoy can be used as a bench-mark for quality control upon the instrument errors and for the affirmation part (Zhoa el al., 2012).

## 3.1 Validation of Wind Speed Data from Satellite Altimeter and Ground-truth Buoy

Previous research found that, the accuracy of the satellite altimeter measurements is commensurate with the measurements of the ground-truth buoy. Nowadays, satellite altimeter can provide a persistent and accurate wind speed measurements in a large coverage of both temporal and spatial. Previous study has concluded that the satellite altimeter measurements showing a good significant accuracy with the buoy measurements.

A collocation method for both techniques is used in order to assess and validate the reliability of the measurements from the satellite altimeter. A temporal and spatial assessment of the satellite altimeter is based on the specific parameters of each selected single-mission satellite altimeter as shown in Table 5.

| Satellite | Altitude (km) | Track spacing equator (km) | Repeat period (days) |
|-----------|---------------|----------------------------|----------------------|
| ENVISAT-1 | 796 | 80 | 35 |
| ERS-1 | 780 | 80 | 35 |
| ERS-2 | 785 | 80 | 35 |
| Jason-1 | 1336 | 315 | 10 |
| Jason-2 | 1336 | 315 | 10 |
| TOPEX | 1340 | 315 | 10 |
| CryoSat-2 | 717 | 75 | 28 |
| Saral | 800 | 75 | 35 |

Table 5. Satellite Altimeter Mission Parameters

## 3.2 Data Verification

Altimeter wind speed is measured from the energy of the reflected signal from the ocean surface roughness (Kumar et al., 2015). The comparison of monthly averages between collocated wind speed from the satellite altimeter and buoys are shown by using the graphical and statistical assessment. Based on Figure 4 and 5 showing that, the graphs of buoy-altimeter versus time to indicate the pattern of the wind speed observation for satellite altimeter and moored buoys at Sabah and Sarawak Sea, respectively.

Based on the Figure. 4, it illustrated a time-series graph by presenting a comparison between collocated satellite altimeter and buoy at point located at Sabah Sea. The graph shows, the wind speed pattern for collocated altimeter and buoy with respect to time is approximately the same. But there is a biased for each point of the pattern. In January 2005, the wind speed value for buoy is 6.17 m/s while the satellite altimeter with 7.07 m/s. From December 2005 until April 2006, the pattern of both lines is approximately equal and also happened for the lines pattern from June 2006 until May 2007. As we can see the value of the wind speed from satellite altimeter on July 2007 is higher compared to the buoy with 5.32 m/s for satellite altimeter and 3.75 m/s for buoy. Apart from that, most of the time these graph illustrated the satellite altimeter has indicated a higher value of wind speed observations compared to the moored buoy. As calculated before, the error biased for this Figure. 4 is positive 0.28 m/s accounted with the average difference of the wind speed value from the satellite altimeter and the moored buoy. It means that, the wind speed value from the satellite altimeter is higher compared to the wind speed measurement from the ground-truth buoy.

For Figure. 5, the graph illustrated the comparison of the wind speed pattern between the satellite altimeter and the buoy. As we can see, the wind speed value from both of the measurements showing the same pattern for a certain time during October 1995 until July 1996 and from October 1998 until October 1999. But, at a certain time there were large differences of the wind speed observation for both of the methods; such as for wind speed in December 1996 the satellite altimeter wind speed was 7.15 m/s while buoy was 3.81 m/s. It shows a huge difference value of the wind speed measurements between both of the techniques. This phenomenon also happened in September 2005, the monthly mean value of the wind speed from satellite altimeter was 6.87 m/s while buoy was 3.22 m/s. The graph also shows that, there was a data gap for the buoy during August 2001. These data gap is believed will affect the correlation measurements for both altimeter and buoy. The error biased for the wind speed observation from both techniques for point located at Sarawak Sea is 0.75 m/s, and these value is larger compared to the error biased of wind speed measurements at point located at Sabah Sea, which was 0.28 m/s.

Overall, these result has shown that the wind speed values from the satellite altimeter of both points were slightly larger compared to the wind speed values from the buoys. This may be due to the overestimation of wind speed from the satellite altimeter at low wind speed region (Lillibridge el al., 2014). This indicates the existence of swells in low wind speed region showing the ocean surface is rough causing to a strong reflected signal and over-estimation of wind speed measurements from the satellite altimeter (Bhownick et al., 2014).

For a further analysis, a statistical assessment on the monthly average for both of the satellite altimeter and buoy has been applied for each ground-truth buoys at Sabah Sea and Sarawak Sea. These statistical analysis is done to obtain a good comprehension of the wind speed between both of the measurements at a certain time for each point. The statistical parameter presented in this analysis is by looking up at the correlation and Root Mean Square Error (RMSE) for both measurements. As recalled from time-series analysis in the Figure 4 and 5, illustrated overall monthly mean of the wind speed from the satellite altimeter were slightly higher compared to wind speed from the buoys indicating the overestimation of the satellite altimeter wind speed. A further statistical analysis is approached to indicate the performance of the satellite altimeter measurements to show its reliability of providing the wind speed data.

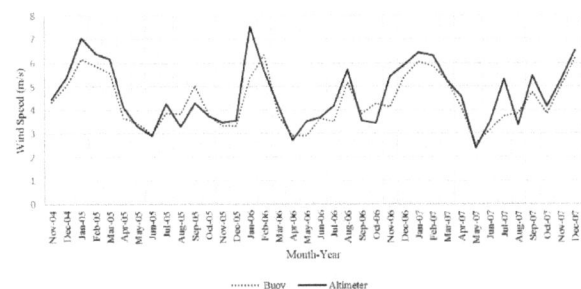

Figure 4. Time-series graph of inter-comparison altimeter and buoy for point located at Sabah Sea for monthly average.

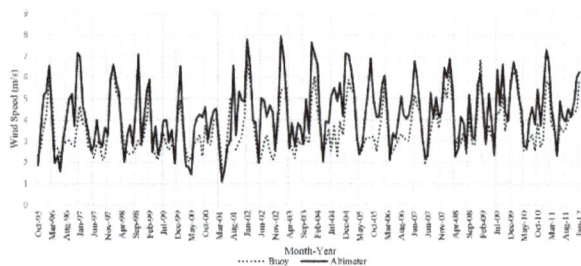

Figure 5. Time-series graph of inter-comparison altimeter and buoy for point located at Sarawak Sea for monthly average.

Based on the Figure 6(a) and (b), represented the correlation graph of the wind speed for point located at Sabah and Sarawak Sea, respectively. These correlation is produced by applying linear regression to both of the wind speed measurements. In Figure 6a), the graph shows the correlation of the wind speed measurements for both methods at Sabah Sea is positive 0.7976 with RMSE value ±0.65 m/s, while Figure 6b) illustrates the correlation graph for point located at Sarawak Sea is positive 0.6148 and RMSE ±1.25 m/s. Both of the graphs are shown with a good correlation results, means that the measurement of wind speed between satellite altimeter and buoy are good correlated. However, the RMSE value for point located at Sarawak Sea is much higher with ± 1.25 m/s compare to RMSE value for point located at Sabah Sea, ± 0.65. It can be related with the value of the correlation for both of the graphs, showing Figure 6a) has better positive correlation value compare to Figure 6b). This is happened because of several reasons, previous study explained if the grid point is located away from the coastal buoy, the land influence will affect the grid averaging. Indeed, the closer the satellite tracks to the buoy, the smaller the estimate of the bias between altimeter and the buoy (Shanas et al., 2014). Buoy that located at Sarawak sea is far more away from the land compare to the buoy located at Sabah sea as shown in Figure 7(a) and (b).

(b)

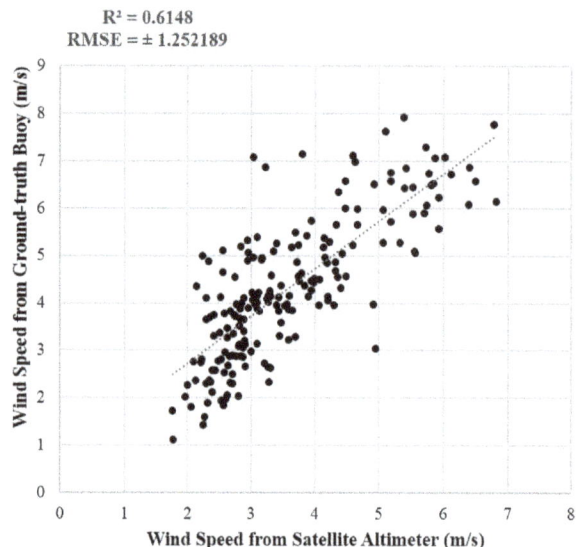

Figure 6. Correlation graph for point located at Sabah Sea (a) and Sarawak Sea (b)

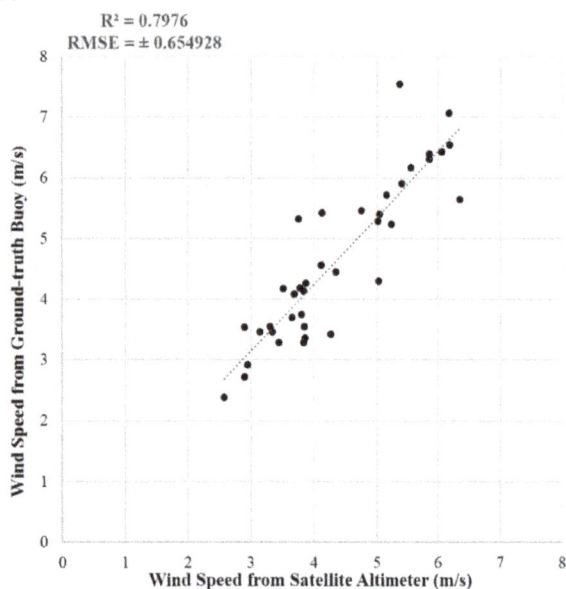

(a)

(a)

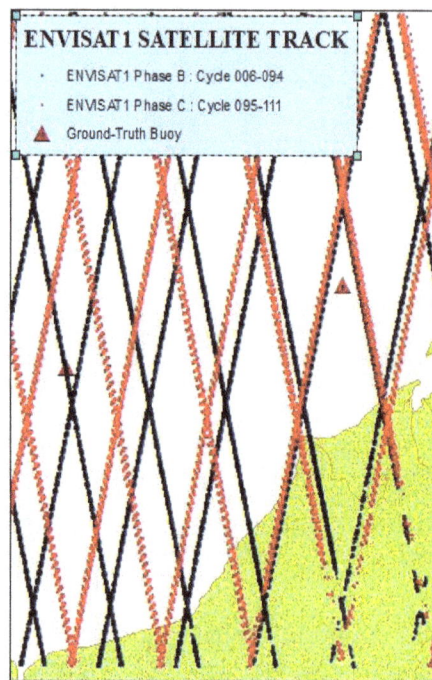

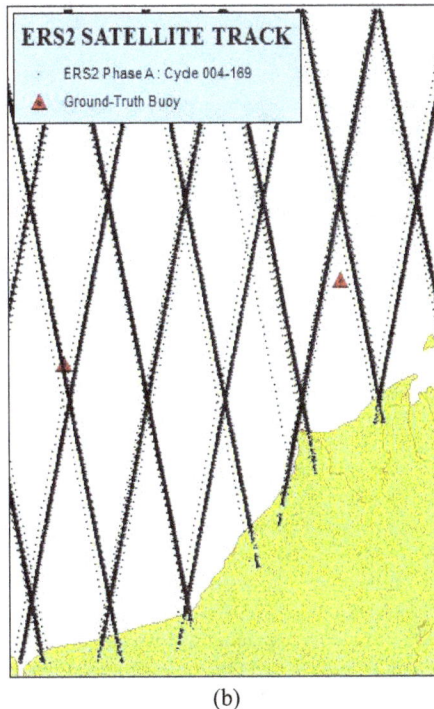

(b)

Figure 7. Satellite track over the ground truth buoy, (a) satellite track from ENVISAT-1 and (b) is satellite track from ERS-2

From the result of the satellite track presented in Figure. 7(a) and (b), has conclude the wind speed correlation and RMSE value for point located at Sarawak Sea should be better compared to the result from point at Sabah Sea. But, a reverse situation occurs explained the correlation and RMSE value for point located at Sabah Sea is much better compared to the point located at Sarawak Sea. This is happened because of the buoy at Sarawak Sea is located at shallow water compare to the buoy at Sabah Sea, where the depth of the water at Sarawak sea is 100m while the buoy at Sabah sea is 1032m. Sarawak point has a shallow water because of the location of the buoy is near with an island. Thus, the presence of land will influence the measurement of wind speed observations from the satellite altimeter. Since, the observation of wind speed from the satellite altimeter is estimated from the backscattered signal from the sea roughness, the shallow water of Sarawak Sea gives effect to the estimation of wind speed for satellite altimeter at Sarawak point. As a result, Sarawak point has a larger value of correlation and RMSE compared to the result with the point at Sabah Sea.

## 3.3  Climatology Analysis Due to Seasonal Effects

This research has utilized the altimetry data from nine satellite missions for 24 years period of observations from year 1993 until 2016. These altimetry data were combined to produce a wind speed climatology data using RADS software. The processing of these climatology data by using RADS is to evaluate the wind speed condition over Malaysian seas correspond with the seasonal variation. The seasonal analysis from climatology data set is based on the average of the monsoon season period which were Northeast Monsoon (November to February), Southwest Monsoon (May to August), First-inter Monsoon (March and April) and Second-inter Monsoon (September and October). The objective of this analysis is to assess the variation of the wind speed climatology data from the satellite altimeter and analyze the condition of the wind speed during each season. Henceforth, to propose a

suitable location in Malaysian offshores/near offshores for wind turbine energy is conducted by examined the particular location receiving a strong wind speed at a particular time.

### 3.3.1  Northeast Monsoon and Southwest Monsoon

The low pressure in the continent landmass can caused a seasonal warming phenomenon, which heats up quickly the air above the landmass compare to the air over the ocean (Masserah and Razali, 2016). This phenomenon will cause the monsoon season occurred. Previous research finding showing that, the wind speed during the Northeast Monsoon is stronger than the Southwest Monsoon. During the Northeast Monsoon (November to February), the Australian continent experienced with the summer season brought a low temperature in Asian region and produced a high pressure area compare to the Australian region. Thus, the wind circulates from the high pressure area of Asian continent to the low pressure area of Australian continent. The circulation of the wind across the equator has possess the wind blowing from the east-side of the Peninsular Malaysia are deflected to the Australian continents. During this Northeast Monsoon, wind from South China Sea is blown towards Malaysia region and forming a great wind speed compared to other monsoons (Masserah and Razali, 2016).

Based on the Figure. 8 presents the average of the wind speed during Northeast Monsson (November to February) for 24 years period of time. The map shows the average of the wind speed for the coastal areas is around 4m/s up to 5.7 m/s. While in Figure. 9, the average of the wind speed during the Southwest Monsoon is range from 2.0 m/s up to 4.5 m/s.

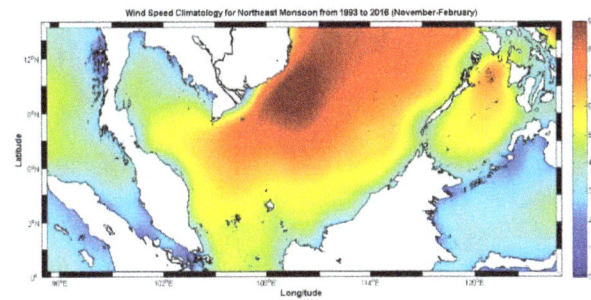

Figure 8. Average of wind speed during Northeast Monsoon from 1993 to 2016

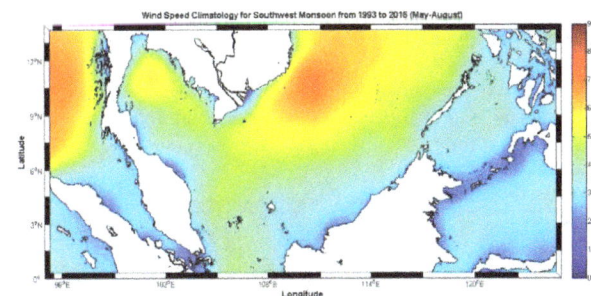

Figure 9. Average of wind speed during Southwest Monsoon 1993 to 2016

### 3.3.2  Monsoon Transition

Malaysia also experienced with two monsoons transition period which are First Inter-Monsoon (March to April) and Second Inter-Monsoon (September to October). These transition monsoons were occurred in short period of time in which the wind movement were facing the fluctuation and become slow with value not exceed 5.4 m/s (Masserah and Razali, 2016).

During the First Inter-Monsoon in Figure 10, shows the average of the wind speed near the coastal areas is range from 2.0 m/s up to 3.5 m/s. While, Figure 11 represents the wind speed condition during the Second Inter-Monsoon with the average of the wind speed range from 3.0 m/s up to 4.0 m/s. All the graphical result had showing the mean value of the wind speed for the coastal areas were maximum during the Northeast Monsoon up to 5.7 m/s, and minimum during the transition of the First Inter-Monsoon with the value of the wind speed were 3.5 m/s. Besides that, all the maps presented South China Sea area is received a stronger wind speed compare to Malacca Straits, Sulu Sea and Celebes Sea.

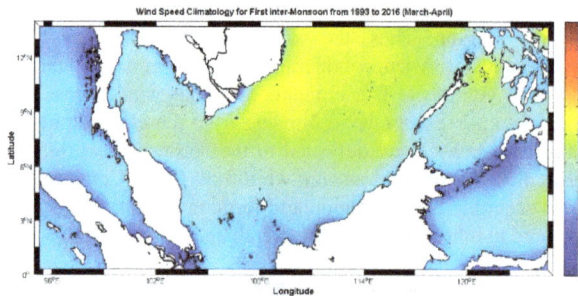

Figure 10. Average of wind speed during First Inter-Monsoon 1993 to 2016

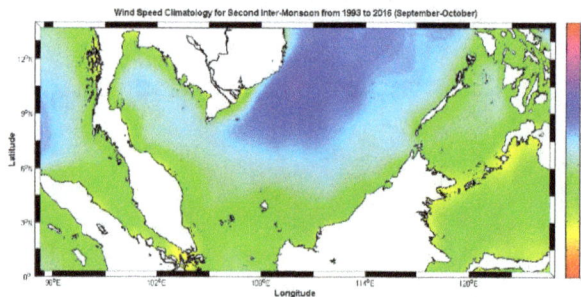

Figure 11. Average of wind speed during Second Inter-Monsoon 1993 to 2016

## 3.4    Wind Speed Climatology Using Time-Series Analysis

In this part, this paper explained the analysis of the wind speed climatology based on the monthly average of the satellite altimeter wind speed data from 1993 to 2016. The significant of this analysis is to analyze the wind speed condition for each sea at a particular time.

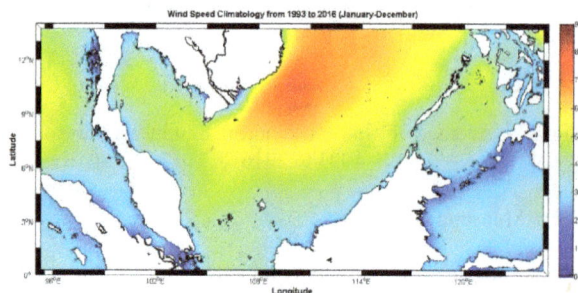

Figure 12. Average wind speed of twelve months for 24 years period 1993 to 2016

Based on Figure 12, the map illustrated the average wind speed for Malaysian seas over 24 years from 1993 until 2016. According to this map, it showed that South China Sea is received a strong wind speed compare to the other seas with maximum value of 7.5 m/s, followed by Sulu Sea with 5.0 m/s. While, both of Malacca Straits and Celebes Sea received a low average of wind speed with maximum value of 3.5 m/s. Based on Figure 13, the monthly assessment of the satellite altimeter wind speed for each Malaysian Seas plotted that South China Sea was dominantly received a strong wind speed, followed by Sulu Sea and two other seas which are Celebes Sea and Malacca Straits. Figure 13 also illustrates each of these Malaysian seas most likely have the same lines pattern of wind speed for each month. Commonly, each part of the sea received a strong wind during the Northeast Monsoon from November to February. South China Sea, Sulu Sea and Celebes Sea experienced a strong wind in January with value 7.5 m/s, 6.1 m/s and 3.5 m/s, respectively. While, Malacca Straits experienced a maximum wind speed in December with 3.3 m/s.

During the transition monsoons on March to April and September to October, each of the sea experienced with a fluctuated wind direction causing a slow wind speed at that time (Masserah and Razali, 2016). In the first transition monsoon, most of the seas facing with the slowest wind movement starting from March until the end of the April. Due to the location of Malacca Straits surrounded by a land mass especially the Sumatra Island make Malacca Straits received the worst wind speed in April with 2.4 m/s. Even though, there was a slow movement of wind during this first transition monsoon but South China Sea still receive a strong wind speed compared to the other sea with 3.5 m/s as shown Figure 13.

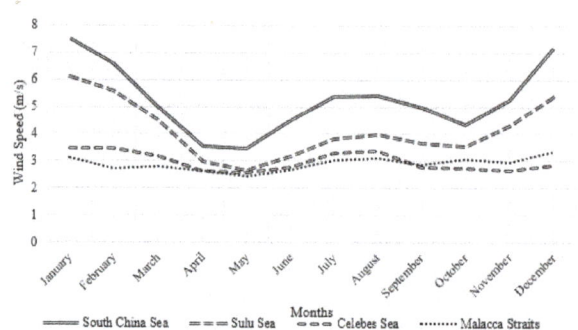

Figure 13. Monthly climatology of wind speed for each Malaysian seas over 24 years

The fluctuated wind movement also happened during the Second Inter-Monsoon (September to October) where all of the four seas facing with slow movement of wind speed. During that time, South China Sea only received maximum 4.9 m/s of wind speed in September, while Sulu Sea and Celebes Sea only experienced maximum wind speed with value 3.7 m/s and 2.8 m/s, respectively. In September, Malacca Straits received a minimum wind speed with 2.9 m/s and maximum wind speed with 3.1 m/s in October. According to the graph in Figure 13, shows the condition of wind speed during the second transition monsoon is slightly higher compare to the first transition monsoon. This is because, during October the wind-blown from the high pressure of Asia continent and the wind that blown towards South China Sea were start to move in the east direction towards the low pressure of Australian continent. In a result, forming a strong wind speed during Northeast Monsoon from November until February.

Figure 14 presents the graphical average of the wind speed for each months over 24 years of satellite altimeter measurements. According to the graph, the average wind speed is higher in January during the Northeast Monsoon with value 4.7 m/s and lower in April during the First Inter-Monsoon with value 2.6 m/s. Hence, it can be concluded that Malaysia will received a strong wind speed for four consecutive months from November until January with maximum value 4.7 m/s and minimum value 4.5 m/s.

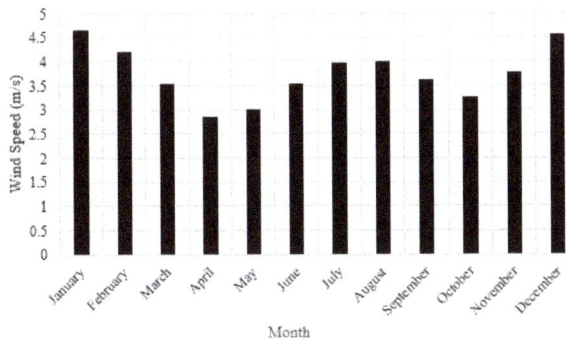

Figure 14. Average wind speed for each month over 24 years (1993-2016)

The analysis showed that, each part of the Malaysian Sea was experienced with a strong wind during the Northeast Monsoon and weak during the first transition monsoon. Apart from that, South China Sea becomes the region that received a strong wind speed compared to the other Malaysian Seas.

### 3.5 Proposed Location for Wind Turbine Energy

The location of Malaysia in low wind speed region has brought a great challenge to find the potential location in Malaysia for the implementation of wind energy and figuring the annual potential energy generated correspond to the strength of wind speed of the location (Mohamed et al., 2015). Thus, the climatology analysis gives an advantage in order to propose a suitable location that has good potential for wind turbine energy. Figure 15 shows three potential locations have experienced a strong wind speed during the Northeast Monsoon and received a strong wind blowing from South China Sea which are Tioman Island, Tenggol Island and Layang-layang Island. Tioman and Tenggol Island were located at the coast of East Peninsular Malaysia facing with strong wind from South China Sea with maximum value of wind speed 5.5 m/s. While Layang-layang Island is one of the island in Sabah Sea facing with maximum wind speed from South China Sea with the value of 4.5 m/s.

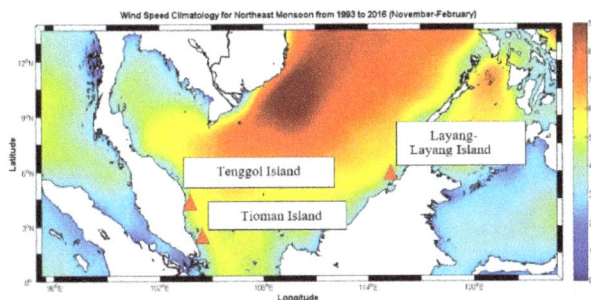

Figure 15. Shows the potential location ( ▲ ) for the Wind Turbine Generator (WTG)

These selected locations were based on the assessment of the ocean wind speed 10m above the sea levels. Since these selected islands facing with a strong wind speed during particular monsoon, the physical height for wind turbine generator need to be considered regarding the maximum value of the receiving wind from the sea towards each of the locations. The positioned of Malaysia in low-wind speed region make the coastal zone of Peninsular Malaysia, Sabah and Sarawak received maximum wind speed value below 7 m/s with maximum annual wind power less than 400 W/m² as shown in Table 7.

| Wind Class | 10m | | 80m | | 100m | |
|---|---|---|---|---|---|---|
| | Density (W/m²) | Wind Speed (m/s) | Density (W/m²) | Wind Speed (m/s) | Density (W/m²) | Wind Speed (m/s) |
| 1 | <100 | <4.4 | <240 | <5.9 | <260 | <6.1 |
| 2 | 100-150 | 4.4-5.1 | 240-380 | 5.9-6.9 | 260-420 | 6.1-7.1 |
| 3 | 150-200 | 5.1-5.6 | 380-490 | 6.9-7.5 | 420-560 | 7.1-7.8 |
| 4 | 200-250 | 5.6-6.0 | 490-620 | 7.5-8.1 | 560-670 | 7.8-8.3 |
| 5 | 250-300 | 6.0-6.4 | 620-740 | 8.1-8.6 | 670-820 | 8.3-8.9 |
| 6 | 300-400 | 6.4-7.0 | 740-970 | 8.6-9.4 | 820-1060 | 8.9-9.7 |
| 7 | >400 | >7.0 | >970 | >9.4 | >1060 | >9.7 |

Table 7. Wind classification (Oh et al., 2010)

### 4. CONCLUSION

In this study, the wind speed data from both satellite altimeter and buoy are compared for each point located at Sabah and Sarawak Sea. The study of 24 years wind speed data has proven its reliability to support the wind energy in Malaysia offshores. The result has demonstrated that the wind speed from the satellite altimeter are compared well with the ground-truth buoy and have a positive correlation for both points at Sabah and Sarawak sea with 0.7976 and 0.6148 respectively. Monthly mean statistics shows that the RMSE value for the satellite altimeter and buoy at Sabah Sea is ± 0.65 m/s, while for point located at Sarawak Sea with RMSE ± 1.25 m/s. The RMSE value is reasonable if compared with the previous study of collocated satellite altimeter and buoy as shown in Table 1.

The climatology wind speed data analysis shows that Northeast Monsoon received stronger wind speed from November until February with maximum value of 5.7 m/s. This is because of the circulation of ocean wind between two continents from low pressure to high pressure as mentioned in sub-section 3.3.1. During Southwest Monsoon, Malaysian Seas experienced a maximum wind speed with 4.5 m/s. Occasionally, due to the fluctuate movement of the wind in short period of time on First Inter-Monsoon and Second Inter-Monsoon brought a weak wind speed condition for each sea. Overall, Malacca Strait has possessed a week wind speed for each monsoon especially in the Second-Inter Monsoon with value of 2.3 m/s. From this result, most of the offshore and near coastal island in Malacca Straits were not suitable for wind energy implementation due to weak wind speed condition. Other than that, the location of Malacca Straits on the side of Sumatra Island has blocked the wind movement from the other continent. In reverse, for an offshores and near-coastal island facing the South China Sea have a great potential to provide a strong wind speed for wind energy in Malaysia. Most of the near-coastal Malaysian island on South China Sea receive a maximum wind speed of 5 m/s and stronger during the Northeast Monsoon with average 5.5 m/s. On behalf of that, three locations were proposed for wind energy site such as Tenggol Island (Terengganu), Tioman Island (Pahang) and Layang-layang Island (Sabah).

Based on the wind speed assessment, this paper has highlight the measurements of the wind speed data from the satellite altimeter is reliable to provide a useful and reliable wind speed information to support the wind energy in Malaysia for the future implementation of a green renewable energy.

## ACKNOWLEDGEMENT

The authors would like to thank to TU Delft, NOAA, Altimetrics LLC for providing altimetry data through Radar Altimeter Database System (RADS). We are grateful to the Ministry of Education (MOE) Malaysia, Fundamental Research Grant Scheme (Vot number: 4F706) and Universiti Teknologi Malaysia for funding this research under Research University Grant (Vot number: 12H99).

## REFERENCES

Abdul Aziz W.A.W, Omar K. M., Omar Y, Din A.H.M., 2014. Ocean wind speed characteristics over Malaysian Seas from multi-mission satellite altimeter during monsoon period, *Jurnal Teknologi (Science & Engineering)*, 4(71), pp. 79-82.

Andersen,O., B., Schrarroo, R., 2011. *Range and Geophysical Corrections in Coastal Regions: And Implications for Mean Sea Surface Determination*. In Coastal Altimetry.

Bhownick et al., 2015. Analysis of SARAL/Altika wind and wave over Indian Ocean and its real-time application in wave forecasting system at ISRO, *Mar. Geod*, 38(1), pp. 396-408.

Caires, S., Sterl, A., 2003. Validation of ocean wind and wave data using triple collocation, *J. Geophys. Res*, 108(C3), 3098.

D.B. Chelton, J.C Ries, B.J. Haines, L.L. Fu, P.S. Callahan., 2001. Satellite Altimetry, *Int. Geophys*.

Dobson et al., 1987. Validation of Geosat altimeter-derived wind speeds and significant wave heights using buoy data. *J. Geophys*, 92(10), 719-731.

Durrant et al., 1987. Validation of Geosat altimeter-derived wind speeds and significant wave heights using buoy data, *J. Geophys. Res*, 92(10), pp. 719-731.

Fu, L., Cazenave., 2001. *A Satellite Altimeter and Earth Sciences: A Handbook of Techniques and Application*. Academic Press.

J. P. Mathisen., 2010. *Metocean data*. Meteorological Technology International.

L. W. Ho., 1993. Wind Energy in Malaysia: Past, Present and Future, *Renewable and Sustainable Energy Review,* 53, pp. 279-295.

Lillibridge, J., Scharoo, R., Abdalla, S., Vandermark, D., 2014. One- and Two-dimensional wind speed models for Ka-Band altimetry, *J. Atmos. Ocean Tech*, 31, pp. 630-638.

Mohamed Nor, K., M. Shaaban., H. A. Rahman, Feasibility Assessment of Wind Energy Resources inMalaysia Based onNWP Model. *Renewable Energy*. Elsevier. pp. 147-154, Johor Bahru: Malaysia.

N.K Masserah, A.M. Razali., 2016. Modelling the wind direction behaviors during the monsoon seasons in Peninsular Malaysia, Renewable and Sustainable. *Energy Rev.,* 56, pp. 1419-1430.

P.R Shanas, V.S. Kumar, N.K. Hithin., 2014. Comparison of gridded multi-mission and along–track mono-mission satellite altimetry wave heights with in-situ near-shore buoy data, *Ocean Engineering*, 83, pp. 24-35.

Queffeulou, O,. 2004. Long Term validation of wave height measurements from altimeters. *Mar. Geod*, 27, pp. 495-510.

Satellite images (2016, November 20). Retrieved from Google Earth National Oceanic and Atmospheric Administration, U.S NAVY website: https://www.google.com/earth/.

T.H. Oh, S.Y. pang, S.C Chua., 2010. Energy Policy and Alternative Energy in Malaysia: Issues and Challenges for Sustainable Growth. *Renewable Sustain Energy Rev*, 14, pp. 1241-1252.

U.M Kumar, D. Swain, S.K. Sasamal, N.N Reddy, T. Ramanjappa., 2015. Validation of SARAL/Altika significant wave height and wind speed observations over North Indian Ocean, *Journal of Atmospheric and Solar-Terrestrial Physics*, 135, pp. 174-180.

W.B Wan Nik, M.Z. Ibrahim, K.B Samo., 2010. Wave Energy Potential of Peninsular Malaysia, *ARPN J. Appl* Sci, 5, pp. 11-23.

Zhoa, D., Li, S., Song, C., 2012. The comparison of altimeter retrieval algorithms of the wind speed and the wave period, *Acta Ocean. Sin*, 31(3), pp. 1-9.

# 27

# DETECTION OF STREET LIGHT POLES IN ROAD SCENES FROM MOBILE LIDAR MAPPING DATA FOR ITS APPLICATIONS

Siamak Talebi Nahr [a, *], Mohammad Saadatseresht [a], Jamshid Talebi [b]

[a] School of Surveying and Geospatial Engineering, College of Engineering, University of Tehran
[b] Pooya-Naghsh-Omid Consulting Engineers Company, 021-44095520, pooya.naghsh@gmail.com

**KEY WORDS:** MLS, light pole, Bhattacharya, SMRF, DBSCAN, ITS

**ABSTRACT:**

Identification of street light poles is very significant and crucial for intelligent transportation systems. Automatic detection and extraction of street light poles are a challenging task in road scenes. This is mainly because of complex road scenes. Nowadays mobile laser scanners have been used to acquire three-dimensional geospatial data of roadways over a large area at a normal driving speed. With respect to the high density of such data, new and beneficial algorithms are needed to extract objects from these data. In this article, our proposed algorithm for extraction of street light poles consists of five main steps: 1. Preprocessing, 2. Ground removal, 3. 3D connected components analysis, 4. Local geometric feature generation, 5. Extraction of street light poles using Bhattacharya distance metric. The proposed algorithm is tested on two rural roadways, called Area1 and Area2. Evaluation results for Area1 report 0.80, 0.72 and 0.62 for completeness, correctness and quality, respectively.

## 1. INTRODUCTION

Nowadays, high-density and high-accuracy Mobile Laser Scanning (MLS) data are becoming a primary source for highway mapping (Gong, Zhou et al. 2012), urban road distress assessment (Awrangjeb and Fraser 2014, Guan, Li et al. 2015), and road feature inventory (Pu, Rutzinger et al. 2011, Guan, Yu et al. 2015). Road feature inventory is a crucial necessity in intelligent transportation systems (ITS) applications. Pole-like road objects, including light poles, telecommunication poles and traffic signposts, located along roads/streets, are typical kinds of road infrastructure. For example, light poles provide illumination to pedestrians and vehicles at night for a clear visibility of the road environment. Traffic signposts, as a highly important transportation infrastructure, play a critical role in ITS, traffic safety, and route guidance. For ITS related applications, the information of pole-like objects can be used for road infrastructure maintenance, road safety analyses, advanced driver assistance, semantic mapping, and smart city applications. For instance, for a driver assistance system, the position of street light poles can be used to improve the stability of road tracking (Fleischer and Nagel 2001). Therefore, regular inventory and maintenance of street light poles is important. Due to the large number of street light poles on road, traditional manual survey methods are extremely time-consuming. A rapid and robust method is highly needed to obtain street light poles information.

Identification of street light poles is very significant and will make the detection of the attached objects easier. Regularly rounded and long, poles are made of different materials and they have different heights and radius. However, in the same scene, the type of poles is usually identical. Thus, recently, detecting street light poles has attracted increased attention in the literature.

MLS systems integrate laser scanner(s), a global navigation satellite system, an inertial measurement unit, a distance measurement indicator, and digital/video camera(s) (Murray, Haughey et al. 2011, Brogan, McLoughlin et al. 2013). MLS systems have been used to acquire three-dimensional (3-D) geospatial data of roadways over a large area at a normal driving speed (Figure 1).

The average density of the point clouds collected by a MLS system can reach up to 4000 points/m2 with a moving speed of approximately 50 km/h. Therefore, MLS systems provide a promising way to extract street lighting poles. In fact, automated extraction of street lighting poles from MLS point clouds has been an active research topic in recent years.

Figure 1. Street Mapper, the LiDAR mobile mapping system utilized in this research.

* talebi@ut.ac.ir

In this paper, we will propose a procedure to detect street light poles from rural road scenes. Our procedure consists of five main steps: 1. Pre-processing, 2. Ground removal, 3. 3D connected components analysis, 4. Local geometric feature generation, 5. Extraction of street light poles.

## 2.   RELATED WORKS

There are many articles for road object detection via MLS data. For example, based on eigenvalue analysis, principal component analysis (PCA) was a widely used method for detecting pole-like road objects from irregular point clouds(Yokoyama, Date et al. 2011, Yokoyama, Date et al. 2013). These methods detected linear pole-like structures by first constructing a covariance matrix for each point with its neighbours and then analysing eigenvalues decomposed from the covariance matrix (El-Halawany and Lichti 2011). Eigenvalue-based PCA methods show high computational efficiency. However, other objects (particularly tree trunks) in a road scene might cause a considerable number of false alarms.

With the prior knowledge of pole-like objects in shape and size, by using grammar rules, a voxel structure was applied to mobile LiDAR data (Cabo, Ordoñez et al. 2014). Through a 3-D neighbourhood analysis of voxel representations, pole-like objects were detected. In (Chen, Zhao et al. 2007, Kukko, Jaakkola et al. 2009, Lehtomäki, Jaakkola et al. 2010, Hu, Li et al. 2011), pole-like objects were extracted by analysing scan lines, rather than raw point clouds. To improve computational efficiency, some studies convert 3-D point clouds into 2-D representations (El-Halawany and Lichti 2013). Point density in 2-D representations was also exploited to detect light poles (Chen, Zhao et al. 2007).

## 3.   PROPOSED ALGORITHM

In this section, the proposed algorithm will be discussed. Therefore, in section 3.1 some pre-processing steps to prepare input data for the proposed algorithm are explained. Section 3.2 gives a brief explanation of ground removal algorithm. In section 3.3 a 3D connected components algorithm is used to find connected objects in Euclidian distance metric. In the following section, local geometric features are generated, and finally in section 3.5 Bhattacharya distance is used to detect street light poles.

### 3.1   Pre-processing

This section intends to reduce huge amount of MLS data. To do this, a buffer is set and the points out of it are eliminated. The centre of buffer is trajectory of sensor and its width is equal to two times of the road width. The reason we do this is because we know street light poles are placed in this range. After eliminating the points out of the buffer, the amount of data will reduce tangibly and the following parts of algorithm will run faster.

### 3.2   Ground Removal

After pre-processing section, the amount of data will be reduced, but because of the high density of MLS data, it is still high for computational purposes. To overcome this problem, a ground removal algorithm is used in this paper.

To remove ground points a Simple Morphological Removing Filter (SMRF) is used in this paper (Pingel, Clarke et al. 2013). The algorithm contains four theoretically distinct phases. The first is the creation of the minimum surface. The second is the

processing of the minimum surface, in which grid cells from the raster are identified as either containing bare earth or objects. This second stage signifies the heart of the algorithm. The third step is the creation of a DEM from these gridded points. The fourth step is the identification of the original MLS points as either bare earth or objects based on their relationship to the interpolated DEM.

### 3.3   3D Connected Components Analysis

After eliminating ground points, we need to group remaining points into detached objects based on Euclidian distance. On this basis, a 3D connected component extraction algorithm, which is called DBSCAN, is used (Tran, Drab et al. 2013). This algorithm is based on the number of points in a specified distance. The overall steps of this algorithm are shown in Figure 2.

```
for object i
    if the i-th object is not a member of a given cluster yet
        create a new cluster C
        while the neighboring objects satisfy the cluster
        condition
            expand cluster C
        end
    end
end
```

Figure 1. Overall steps of DBSCAN algorithm (Lehtomäki, Jaakkola et al. 2010).

At the end, some previously known properties of street light poles are used to filter unwanted objects. At the following steps, each of these connected objects will be processed separately.

### 3.4   Local Geometric Features Generation

In this paper statistical analysis of local geometric properties are used to generate street light poles features, as they are used in (Zai, Chen et al. 2015). On this basis, for each point of objects a 5D feature vector is calculated, which represents local geometric features for a point in a specified neighborhood. Equation (1) shows this feature vector:

$$F_p = \{N_x, N_y, N_z, dis, n_{var}\} \tag{1}$$

Where $N_x$, $N_y$ and $N_z$ are components of normal vector at a specified neighborhood of point $p$. $dis$ is distance between $p$ and center of gravity of all object points in XY plane. $n_{var}$ is local variance of normal vectors at a specified neighborhood of point $p$.

After generating $F_p$ for all points of an object, in the following step a 5D Gaussian pdf is estimated. If $\Phi_i = N_i(\mu_i, \Sigma_i)$ is 5D Gaussian function for object $i$, $\mu_i$ and $\Sigma_i$ are shown as:

$$\mu_i = \frac{1}{n} \sum_{j=1}^{n} f_{pj} \tag{2}$$

$$\Sigma_i = \frac{1}{n} \sum_{j=1}^{n} (f_{pj} - \mu_i)^T (f_{pj} - \mu_i) \tag{3}$$

### 3.5   Extraction of street light poles

3D connected components usually have similar geometric features. On this basis, probability density functions, which are estimated using these geometric features, should be also similar.

In this paper Bhattacharya distance is used to discriminate 5D Gaussian pdfs (Zai, Chen et al. 2015). This metric is computationally simple and can calculate distance between two n-dimensional normal Gaussian pdfs. If $N_p = \{\mu_p, \Sigma_p\}$ is the Gaussian pdf of a street light pole, its distance to another object's Gaussian pdf, $N_i = \{\mu_i, \Sigma_i\}$, could be computed as following equation:

$$d_{bhati} = \frac{1}{8}(\mu_i - \mu_p)^T \left[\frac{\Sigma_p + \Sigma_i}{2}\right]^{-1}(\mu_i - \mu_p) + \frac{1}{2}\ln\frac{\left|\frac{\Sigma_p + \Sigma_i}{2}\right|}{\sqrt{|\Sigma_p||\Sigma_i|}} \qquad (4)$$

Equation (4) shows the difference between two normal pdfs. The range for Bhattacharya distance is [0, +∞], in which, 0 means two normal pdfs are identical. Using this metric and a specified threshold similar objects can be extracted.

## 4. EXPERIMENTS

The algorithm was tested using a MLS dataset from the Riegl VMX-250. This system integrates two LIDAR sensors (VQ-250), and an IMU/GNSS unit, and it is deployed on the rear top of a van. The VQ-250 is a rotational sensor that acquires points with a 360° field of view on planes set, in this case, at 45° to the horizontal and 45° to the trajectory (i.e. driving direction).

The measurements were made along two rural roads in the north of Iran, called Area1 and Area2. Each of these roads are approximately 350m long. Area1 is placed at the out of the village and Area2 is a road in inside of the village. Therefore, Area2 is a little bit more complex than Area1. These roads have many objects such as bushes, lights, communication poles and trees. To implement proposed algorithm, a buffer with 25m width is used to filter unwanted points and decline the huge amount of MLS data. The Figure (3) shows two roads after applying buffer.

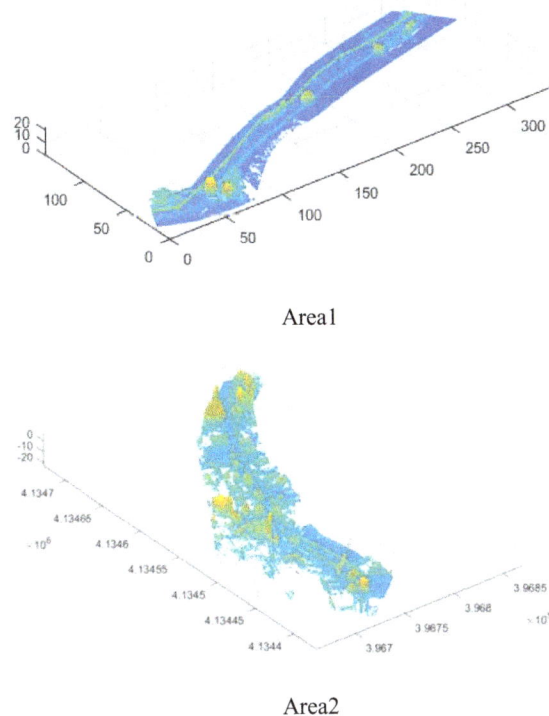

Area1

Area2

Figure 3. Test area before and after filtering using 25m buffer.

Although filtering unwanted points declines the amount of data, but it is still a high value and need to be reduced. This is because of high density of MLS points. To do this ground points are removed by SMRF algorithm. The SMRF parameters are shown in Table (1) and non-ground points are shown in Figure 4.

To generate 3D connected components DBSCAN algorithm is used. In this algorithm, search radius and minimum number of points in neighborhood are set to 0.30cm and 3, respectively. Generated connected components are shown in Figure (5).

| | |
|---|---|
| Cell size | 1m |
| Maximum expected slope | 0.2 |
| Maximum window radius | 16 |
| Elevation threshold | 0.45 |
| Elevation scaling factor | 1.2 |

Table 1. SMRF algorithm parameters.

Area1

Area2

Figure 4. Non-ground points generated using SMRF.

Number of connected components are 2975 and 3224 objects for Area1 and Area2, respectively. With respect to the primary properties of street light poles, we can filter these connected components to eliminate unwanted objects. On this basis, following criteria are used to filter out unwanted objects:

a. The number of points in each objects should be [100, 20000].

b. The height of the objects need to be more than 3 meters.

c. The maximum Z value of the objects need to be more than 6 meters.

These criteria conclude in 44 and 45 street light pole candidates for Area1 and Area2, respectively.

To generate feature vector (Eq. (1)) 8 neighbors of each point in an object are considered, then a 5D Gaussian pdf is estimated for each object. Next, one of the street light poles is chose manually. At last, Bhattacharya distance between desired light pole and other objects are calculated. Considering 0.4 as a threshold value, all of light poles are extracted. These objects are represented in red color in Figure (6).

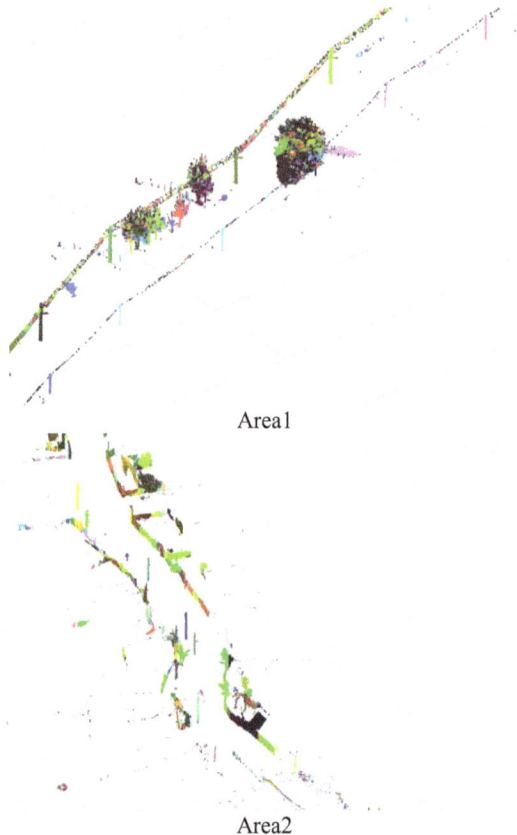

Area1

Area2

Figure 5. 3D connected components detected using DBSCAN.

## 5. EVALUATION

For evaluation purpose of results, we use completeness, correctness and quality indices presented in (Rutzinger, Rottensteiner et al. 2009). Equations 5 to 7 represent these indices. To analyze threshold impact on the results, evaluation indices are calculated for 0.3, 0.4, 0.5 and 0.6 values. Table (2) shows these results.

$$Comp = \frac{TP}{TP + FN} \tag{5}$$

$$Corr = \frac{TP}{TP + FP} \tag{6}$$

$$Corr = \frac{TP}{TP + FP + FN} \tag{7}$$

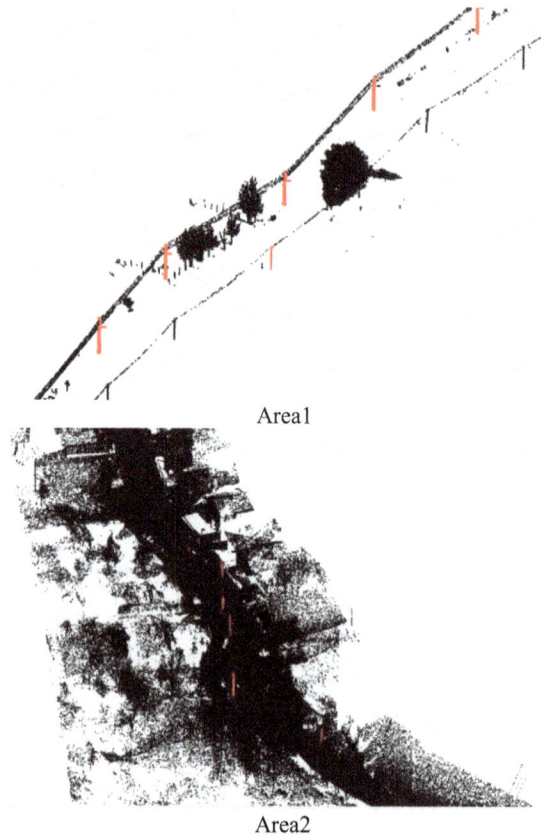

Area1

Area2

Figure 6. Final results of light poles extraction.

## 6. CONCLUSION

For evaluation purpose of proposed method, we used MLS data of two rural road with length of 350 m for each of them. There are many entities at the scene such as light poles, trees, telecommunication poles, bushes etc. To extract street light poles firstly SMRF algorithm is used to eliminate ground points. Next using DBSCAN algorithm 3D connected components are detected. Because of huge number of 3D connected components and with respect to the primary information of street light poles, these components are filtered out to get candidate street light poles. Local geometry features are generated at the next step. After estimating 5D Gaussian pdf on local geometry features, street light poles are extracted using Bhattacharya distance metric.

Completeness, correctness and quality indices are used to evaluate street light poles extraction results. As we can see in Table (2) proposed algorithm reached 0.80, 0.72 and 0.62 for Area1 and 0.79, 0.92 and 0.73 for Area2 as completeness, correctness and quality indices, respectively. These results represent a good performance of proposed algorithm for extraction of street light poles from MLS data in a rural road scene.

At last, considering some flaws and weaknesses of proposed algorithm, such as covered street light poles with trees, it seems using some segmentation algorithms after 3D connected components analysis could help improve results. Besides, manipulating local geometry features vector and using some other feature might be helpful. Finally determining Bhattacharya distance threshold is also a challenging task and could be considered in future researches.

### Area1

| | Ground Truth | | Extracted Results | | | Evaluation Results | | |
|---|---|---|---|---|---|---|---|---|
| Threshold | Light-poles | Other poles | TP | FN | FP | Completeness | Correctness | Quality |
| **0.3** | 10 | 12 | 6 | 4 | 0 | 0.60 | 1.00 | 0.60 |
| **0.4** | 10 | 12 | 7 | 3 | 1 | 0.70 | 0.88 | 0.63 |
| **0.5** | 10 | 12 | 8 | 2 | 3 | 0.80 | 0.72 | 0.62 |
| **0.6** | 10 | 12 | 8 | 2 | 6 | 0.80 | 0.57 | 0.50 |

### Area2

| | Ground Truth | | Extracted Results | | | Evaluation Results | | |
|---|---|---|---|---|---|---|---|---|
| Threshold | Light-poles | Other poles | TP | FN | FP | Completeness | Correctness | Quality |
| **0.3** | 14 | 15 | 6 | 8 | 0 | 0.43 | 1.00 | 0.43 |
| **0.4** | 14 | 15 | 8 | 6 | 0 | 0.57 | 1.00 | 0.57 |
| **0.5** | 14 | 15 | 9 | 5 | 1 | 0.64 | 0.90 | 0.60 |
| **0.6** | 14 | 15 | 11 | 3 | 1 | 0.79 | 0.92 | 0.73 |

Table 2. Evaluation results.

## REFERENCES

Awrangjeb, M. and C. S. Fraser (2014). "Automatic Segmentation of Raw LIDAR Data for Extraction of Building Roofs." Remote Sensing 6(5): 3716-3751.

Brogan, M., et al. (2013). "Assessment of stereo camera calibration techniques for a portable mobile mapping system." IET Computer Vision 7(3): 209-217.

Cabo, C., et al. (2014). "An algorithm for automatic detection of pole-like street furniture objects from Mobile Laser Scanner point clouds." ISPRS Journal of Photogrammetry and Remote Sensing 87: 47-56.

Chen, Y.-Z., et al. (2007). Amobile System Combining Laser Scanners and Cameras for Urban Spatial Objects Extraction. 2007 International Conference on Machine Learning and Cybernetics, IEEE.

El-Halawany, S. I. and D. D. Lichti (2011). Detection of road poles from mobile terrestrial laser scanner point cloud. Multi-Platform/Multi-Sensor Remote Sensing and Mapping (M2RSM), 2011 International Workshop on, IEEE.

El-Halawany, S. I. and D. D. Lichti (2013). "Detecting road poles from mobile terrestrial laser scanning data." GIScience & Remote Sensing 50(6): 704-722.

Fleischer, K. and H.-H. Nagel (2001). Machine-vision-based detection and tracking of stationary infrastructural objects beside inner-city roads. Intelligent Transportation Systems, 2001. Proceedings. 2001 IEEE, IEEE.

Gong, J., et al. (2012). Mobile terrestrial laser scanning for highway inventory data collection. Proceedings of International Conference on Computing in Civil Engineering, Clearwater Beach, FL, USA.

Guan, H., et al. (2015). "Iterative tensor voting for pavement crack extraction using mobile laser scanning data." IEEE Transactions on Geoscience and Remote Sensing 53(3): 1527-1537.

Guan, H., et al. (2015). "Deep learning-based tree classification using mobile LiDAR data." Remote Sensing Letters 6(11): 864-873.

Hu, Y., et al. (2011). A novel approach to extracting street lamps from vehicle-borne laser data. Geoinformatics, 2011 19th International Conference on, IEEE.

Kukko, A., et al. (2009). Mobile mapping system and computing methods for modelling of road environment. 2009 Joint Urban Remote Sensing Event, IEEE.

Lehtomäki, M., et al. (2010). "Detection of vertical pole-like objects in a road environment using vehicle-based laser scanning data." Remote Sensing 2(3): 641-664.

Murray, S., et al. (2011). "Mobile mapping system for the automated detection and analysis of road delineation." IET intelligent transport systems 5(4): 221-230.

Pingel, T. J., et al. (2013). "An improved simple morphological filter for the terrain classification of airborne LIDAR data." ISPRS Journal of Photogrammetry and Remote Sensing 77: 21-30.

Pu, S., et al. (2011). "Recognizing basic structures from mobile laser scanning data for road inventory studies." ISPRS Journal of Photogrammetry and Remote Sensing 66(6): S28-S39.

Rutzinger, M., et al. (2009). "A comparison of evaluation techniques for building extraction from airborne laser scanning." Selected Topics in Applied Earth Observations and Remote Sensing, IEEE Journal of 2(1): 11-20.

Tran, T. N., et al. (2013). "Revised DBSCAN algorithm to cluster data with dense adjacent clusters." Chemometrics and Intelligent Laboratory Systems 120: 92-96.

Yokoyama, H., et al. (2011). "Pole-like objects recognition from mobile laser scanning data using smoothing and principal component analysis." International Archives of Photogrammetry, Remote Sensing and Spatial Information Sciences 38(5/W12): 115-120.

Yokoyama, H., et al. (2013). "Detection and classification of pole-like objects from mobile laser scanning data of urban environments." International Journal of CAD/CAM 13(2).

Zai, D., et al. (2015). Inventory of 3D street lighting poles using mobile laser scanning point clouds. Geoscience and Remote Sensing Symposium (IGARSS), 2015 IEEE International, IEEE.

# 28

# AERIAL IMAGERY AND LIDAR DATA FUSION FOR UNAMBIGUOUS EXTRACTION OF ADJACENT LEVEL-BUILDINGS' FOOTPRINTS

S. Mola Ebrahimi [a], H. Arefi [a,*], H. Rasti Veis [a]

[a] School of Surveying and Geospatial Engineering, University of Tehran, Tehran, Iran

**KEY WORDS:** Footprint, Hough Transform, High Resolution Aerial Image, LiDAR Data, Edge Detector.

**ABSTRACT:**

Our paper aims to present a new approach to identify and extract building footprints using aerial images and LiDAR data. Employing an edge detector algorithm, our method first extracts the outer boundary of buildings, and then by taking advantage of Hough transform and extracting the boundary of connected buildings in a building block, it extracts building footprints located in each block. The proposed method first recognizes the predominant leading orientation of a building block using Hough transform, and then rotates the block according to the inverted complement of the dominant line's angle. Therefore the block poses horizontally. Afterwards, by use of another Hough transform, vertical lines, which might be the building boundaries of interest, are extracted and the final building footprints within a block are obtained. The proposed algorithm is implemented and tested on the urban area of Zeebruges, Belgium(IEEE Contest,2015). The areas of extracted footprints are compared to the corresponding areas in the reference data and mean error is equal to 7.43 m2. Besides, qualitative and quantitative evaluations suggest that the proposed algorithm leads to acceptable results in automated precise extraction of building footprints.

## 1. INTRODUCTION

As a consequence of rapid urbanization and population growth, having comprehensive and accurate information on urban structures and particularly buildings, is becoming essential for a proficient urban management. Such information is also a prerequisite for many other applications, such as urban risk management, disaster monitoring, real estate, and national security.

One of the most important features of buildings is their footprint that can be extracted directly using remote sensing data or field survey. Even though such information can be obtained using field survey which is costly, time consuming, and it cannot be updated automatically. In this regard, in order to expedite the extraction and update the process using remote sensing data would be evidently more effective. Building footprints can be extracted using remote sensing data manually or automatically. Since data updating is important in urban management, automatic methods are absolutely preferable.

This paper aims to present a valid and tractable method to extract building footprint using remote sensing optical image and LiDAR data. The proposed method is after automated building footprints' extraction, using Hough transformation in combination with optical imagery and LiDAR data. Footprint extraction for individual building within a building block in addition to isolated buildings is also considered in this paper.

There are two main remote sensing data sources available for building's attribute extraction including optical imagery (with three visible bands) and LiDAR data. Due to varying input data sources, there exist different techniques for building footprint extraction using remote sensing data. In the approach presented by d. San Koc (D.Koc San et al, 2010), buildings were extracted from high-resolution satellite images using SVM (Support Vector Machine) classifier and Canny edge detector. In another work, Yangeng Wei and his colleagues (Yangeng Wei et al 2004)

employed an unsupervised clustering method on panchromatic images to separate buildings from background and also take out the shadow class. Krishnamachari (Krishnamachari and Chellappa, 1994) used edge detector operator in order to extract edges in an image. Then the Markov chain model was utilized on extracted edges with a proper neighbourhood. Since there is no guarantee to always extract a complete polygon from the model, active curve model is subsequently implemented for this purpose. In another research area-based active contour model was further used to detect homogenous area (Nosrati and Saeedi, 2009). However, in their method a threshold were suggested for separating homogenous areas related to ground and vegetation from buildings. An approach proposed by Demir et al, incorporated four methods to extract buildings by combination of optical imagery and LiDAR data. In the first method, the buildings were extracted using DTM/DSM comparison as well as NDVI index analysis. The second method made use of a supervised classification on multispectral image that was refined by height information from LiDAR data. Last two methods were also founded upon LiDAR data. The former used voids in LiDAR DTM and NDVI classification, while the latter was based on the analysis of the density of raw LiDAR DSM data(Demir et al,2009). Siddiqui et al. also proposed a robust gradient-based method for building extraction from LiDAR and photogrammetric imagery. By transforming LiDAR information to intensity image without interpolation of point heights, and then analyzing the gradient information in the obtained image, their proposed algorithm could extract small single buildings with transparent roofs(Siddiqui et al,2016).

Aforementioned studies have proposed methods to extract footprint of small and large isolated buildings. In other words, these studies have not presented a way to extract buildings within a building block. Since building blocks contain several connected buildings with almost constant heights, edge analysis based on height or non-ground area classification in tree and building

[*] School of Surveying and Geospatial Engineering, University of Tehran, Tehran, Iran – Hossein.arefi@ut.ac.ir

class, cannot be considered as a proper way to extracts buildings within a block. Hence in this study, we propose a new method to extract buildings located in a building block as well as isolated buildings. The proposed method first recognizes the predominant leading orientation of a building block using Hough transform, and then rotates the block according to the inverted complement of the dominant line's angle. Therefore the block poses horizontally. Afterwards, by use of another Hough transform, vertical lines, which might be the building boundaries of interest, are extracted and the final building footprints within a block are obtained.

In the following sections, details on the proposed method are described and results obtained through its implementation are presented and discussed.

## 2.  PROPOSED ALGORITHM

As stated, this paper aims to extract building footprints automatically, using Hough transformation in combination with optical RGB image and LiDAR data. Figure (1) illustrates a workflow of our proposed method to this end. It should be noted that a first assumption to this method is that the RGB image and LiDAR data are quite well geo-referenced. In order to extract buildings, an initial step is to first distinguish buildings from other objects. Since the elevation information is one of the most important components in building detection, LiDAR image contains this source of useful information.

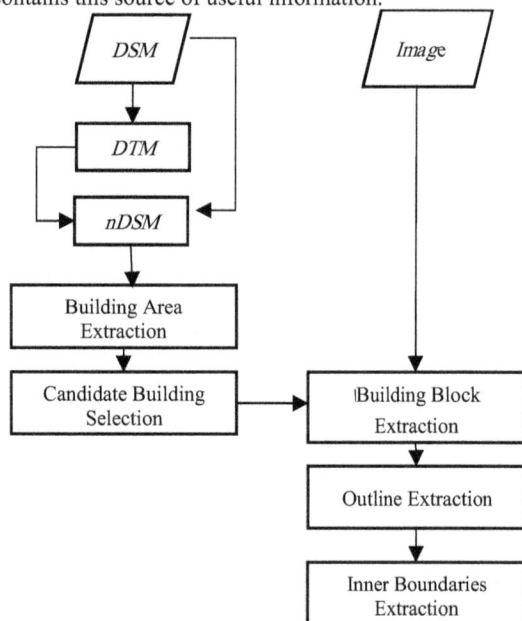

Figure 1. Workflow for building footprints using optical imagery and LiDAR data

In the first step of processing chain nDSM is created from DSM. The nDSM is selected so that includes objects that are high-rise but not vegetation. This (masked) nDSM can be considered as building map. Thus, buildings are obtained from LiDAR data, and each building whether isolated or in a building block, are projected on the optic image assuming that the optic image and LiDAR image are properly co-registered. Afterwards, if the candidate building is actually a building block, the outer and the inner boundaries are extracted. The following sections describe the proposed algorithm in detail.

### 2.1  nDSM Generation

Laser pulses emitted from a LiDAR system reflect from objects both on and above the ground surface. Objects above the ground surface can be extracted by subtracting bare ground or DTM (which is calculated using the geodesic morphological reconstruction algorithm (Arefi et al,2005)) form above ground or DSM as equation (1):

$$nDSM = DSM - DTM \qquad (1)$$

### 2.2  Building Extraction

In this step, non-ground points must be additionally classified into buildings and trees. First, buildings whose boundaries have high local variance and low vegetation index, are extracted from LiDAR data(Arefi et al,2003). Besides, there exists a big gap in the reflected pulses of LiDAR data (Arefi,2009), which can be considered as an index for distinguishing vegetation from buildings and hence is looked at as a novel index for vegetation discovery as equation (2), (3).

In our data set, vegetation was already eliminated, therefore only a threshold was adopted to remove garage and car segments. The output is a binary image showing building segments.

$$(VI)^{old\ technique} = FP - LP \qquad (2)$$

$$(VI)^{new\ technique} = FP - dilation(LP) \qquad (3)$$

Where      FP=First Reflected Pulse in LiDAR data
              LP = last reflected pulses in LiDAR data

Each region with the value of "1" in the output image is assumed as a candidate building. Candidate buildings are individually taken into account and projected to the geo-referenced optical image. Each candidate building in LiDAR image can be used as a mask in which building regions have a value of "1" and background and other buildings have a value equal to "0". This mask is projected to the optical image and as a result the image of candidate building is obtained. The resulting image is a colored image where buildings' pixels have the same value as pixel values of building image, and other pixels have the value of 0.

### 2.3  Buildings' Outline Detection

In order to determine building's footprint it is required to locate the outer boundary of the building. To this end, first the optic image of candidate building is transferred to gray value image, then an edge detector is used to extract patent changes throw boundary and the dark background. In other words, the goal of this step is to make the overall changes in the building surrounding against the black background more clear. Besides, it is essential to use a point reduction algorithm for simplification and elimination of bothersome points around the building of interest. Douglas-Pocker algorithm (Dilip K et al,2012) is used for this sake.

### 2.4  Inner Boundaries Extraction

As mentioned, for each candidate building, the dominant line must be extracted using Hough transformation. If the dominant line has a length more than a threshold (which is determined empirically by considering the length-to-width ratio of the smallest building in dataset), the so-called building might probably be a block and thus the inner boundaries must be detected. Figure (2) illustrates the flowchart of inner boundary extraction phase. The following subsections describe each step in details.

For each building that now has a dominant line, building block is rotated according to the inverted complement of the dominant line's angle, to have a horizontal orientation. Then, another Hough transformation is used to extract shorter length lines on the edge map extracted from RGB and LiDAR images.

By block rotation, inner boundaries can be detected as vertical lines and thus the following computations will be easier.

The lines with a slope equal to 10 degrees from vertical aspect are selected and considered as inner boundaries. Shorter lines, which belong to chimney edges, are eliminated by comparing the proportion of the vertical line's length to the distance from top and down boundaries along the so-called vertical line. If there exists more than one edge as inner boundary, a line should be fitted to them to get rid of the extra lines. Finally to ignore the ridge lines, local maxima are extracted.

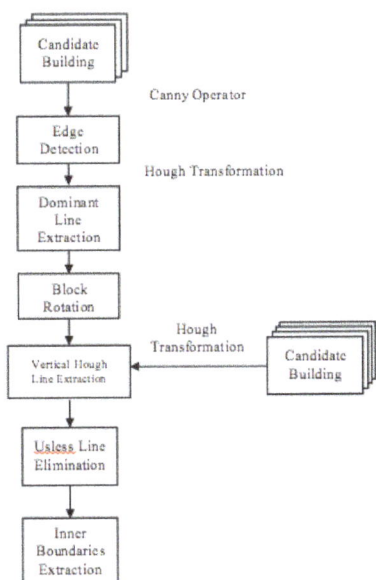

Figure 2. Workflow of Extraction of adjacent footprints in building block based on Hough transformation

## 2.5  Computing the Footprint of Each Building

The points of the outer boundary are categorized based on the inner boundary, and the intersections of the block outer boundary with the final segregating lines are proliferated at one pixel rate, and as the result all the points belonging to each footprint is designated to a specific group. As a final step, an inverse rotation is applied to each footprint in order to have the extracted buildings in the right orientation.

## 3.  IMPLEMENTATION AND RESULTS

The following subsections introduce case study and cover the evaluation results.

### 3.1  Case Study

The data we used in this study is consisted of an orthophoto RGB aerial image with 5cm spatial resolution captured from an urban area with 127.5*125 squared meters extent in Zeebruges, Belgium, as well as the LiDAR grid data of the same spot acquired in 10cm capture rate (IEEE data contest 2015).

This data is considered as a suitable evaluation case for building extraction algorithms due to the variety of building blocks in shape and size (fig 3).

Figure 3. Test data: (a) optic image of test data , (b) DSM of test data

### 3.2  Results

The aforementioned steps for extracting building footprints are graphically illustrated in following figures.

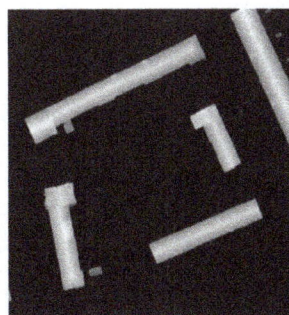

Figure 4 .Building area extracted from DSM

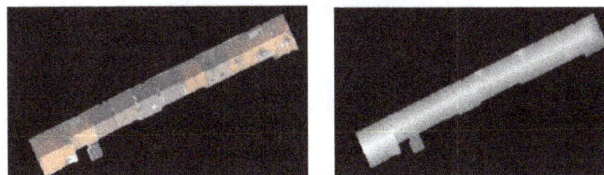

Figure 5. Building area extraction form a) optic imagery b)LiDAR data

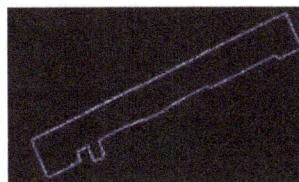

Figure 6. Building's Outline extraction by implementation of Canny operator (white polyline) and Douglas-Pocker algorithm (blue polyline)

Figure 7 . Rotation of building block to the inverted complement of angle of the dominant direction

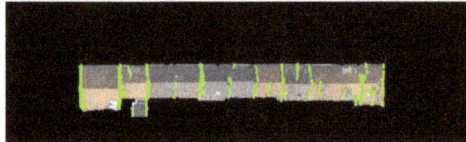

Figure 8 . Vertical or close to vertical Hough lines

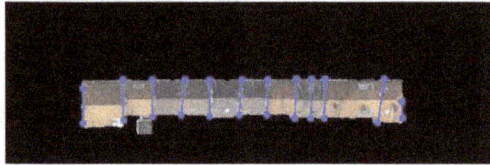

Figure 9 . Candidate inner boundaries

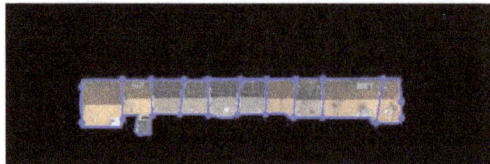

Figure 10 . Building boundaries by eliminating ridge lines

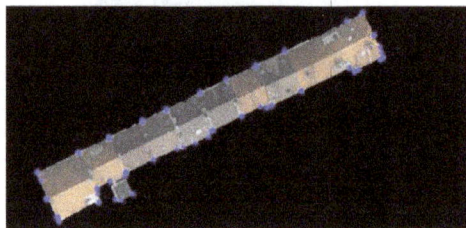

Figure 11 . Inverse rotation of building vertexes

The same procedure is applied for all other building blocks in the study area.
The RGB image of the case area is used for the results evaluation.

## 4. EVALUATION AND DISCUSSION

One straightforward strategy to assess the results of the proposed algorithm is to compare the building areas with their corresponding areas from the buildings extracted by human-operated digitization. To this end, we used ArcGIS software for building digitization (fig 12).

Figure 12 . Digitized footprints in ArcGIS (blue) and extracted footprints by algorithm (yellow)

Boundaries of the adjacent buildings in the eastern and western blocks are prone to less error due to closeness to optic nadir.

Besides, due to a small difference in color and height of adjacent buildings in the southern block, one of the inner boundaries is not detected.

Building footprints or what is called as *cadastral parcels* are considered as the foundation for all land administration systems (LIS). The area of a parcel is apparently one of the most important features, which can be calculated from those vertices. Besides, by having the length of the edges and the area of the polygon of interest, one can mathematically define the shape of the polygon(T. Xiaohua et al, 2001). Therefore, we can relate and estimate the errors of the proposed algorithm by RMSE index calculated for building areas(X. Tong et al, 2005). In the case dataset, we had 4 building blocks and a total number of 26 buildings for which we had their digitized footprints extracted. The areas of those footprints are compared to the areas calculated with the automated algorithm proposed in this paper.

| No. | Area calculated by ArcGIS($m^2$) | Area calculated by proposed algorithm($m^2$) |
|---|---|---|
| 1 | 111.42 | 112.53 |
| 2 | 61.83 | 70.84 |
| 3 | 58.85 | 62.94 |
| 4 | 54.73 | 55.35 |
| 5 | 66.22 | 69.47 |
| 6 | 61.32 | 61.33 |
| 7 | 56.92 | 56.03 |
| 8 | 66.05 | 65.57 |
| 9 | 130.12 | 144.59 |
| 10 | 72.17 | 54.40 |
| 11 | 74.37 | 68.28 |
| 12 | 58.66 | 60.17 |
| 13 | 54.31 | 58.06 |
| 14 | 53.55 | 54.74 |
| 15 | 61.97 | 64.07 |
| 16 | 57.71 | 59.06 |
| 17 | 59.59 | 86.94 |
| 18 | 56.94 | 0 |
| 19 | 115.67 | 142.78 |
| 20 | 60.47 | 59.32 |
| 21 | 55.59 | 57.94 |
| 22 | 58.17 | 60.58 |
| 23 | 58.79 | 60.65 |
| 24 | 62.81 | 66.02 |
| 25 | 58.76 | 70.82 |
| 26 | 102.44 | 101.15 |

As demonstrated in the table, the minimum difference between the manually-digitized area and the automatically-calculated area is 0.5 m$^2$ which is related to the building number 8. The maximum error of 54.96 m$^2$ is due to the building for which the proposed algorithm could not distinguish its boundary from the neighboring building. Regardless of specific cases, the average error and the RMSE of the proposed algorithm is equal to 7.43 and 14.45 m$^2$ respectively.

## 5. CONCLUDING REMARKS

The proposed method revealed a great potential for inner boundary distinction from the ridges. However, in the future studies it is better to free the algorithm from the restrictions of threshold settings. Since the core of this algorithm works by edge detection and is based on some assumptions about roof color and height, in cases where buildings have the exact same height and roof color, the algorithm cannot provide proper results. To solve this problem, we are working on neighborhood and adjacency information as extra sources of information to distinguish those buildings as well.

## REFERENCES

San D. Koc, and M. Turker, «Building extraction from high resolution satellite images using Hough transform,» *International Archives of the Photogrammetry, Remote Sensing and Spatial Information Science*, vol. 38, p. Part8, 2010.

Wei, Yanfeng, Zhongming Zhao, and Jianghong Song, «Urban building extraction from high-resolution satellite panchromatic image using clustering and edge detection,» *Geoscience and Remote Sensing Symposium,IGARSS'04,IEEE international,* vol. 3, 2004.

Krishnamachari, Santhana, and Rama Chellappa, «Delineating buildings by grouping lines with MRFs,» *IEEE Transactions on image processing*, vol. 5.1, pp. 164-168, 1996.

Nosrati, Masoud S., and Parvaneh Saeedi. "A Combined Approach for Building Detection in Satellite Imageries using Active Contours,» *IPCV,* 2009.

Demir, N., D. Poli, and E. Baltsavias, «Combination of image and lidar data for building extraction,» de *Proc. 9th Conference on'Optical 3D Measurement Techniques*, 2009.

Siddiqui, Fasahat Ullah, et a, «"A Robust Gradient Based Method for Building Extraction from LiDAR and Photogrammetric Imagery,» *Sensors,* vol. 16.7:1110, 2016.

Arefi, H., and M. Hahn, «"A morphological reconstruction algorithm for separating off-terrain points from terrain points in laser scanning data,» *International Archives of Photogrammetry, Remote Sensing and Spatial Information Sciences,* vol. 36.3/W19, pp. 120-125, 2005.

Arefi, H., M. Hahn, and J. Lindenberger, «LIDAR data classification with remote sensing,» de *Proceedings of the ISPRS Commission IV Joint Workshop :Challenges in Geospatial Analysis, Integration and Visualization II*, 2003.

H. Arefi, From LiDAR point clouds to 3D building models, Diss, DLR, 2009.

Hahn, M., H. Arefi, and J. Engels, «Automatic building outlines detection and approximation from airborne LIDAR data,» *Annual*, vol. 15, 2007.

Prasad, Dilip K., et al, «A novel framework for making dominant point detection methods nonparametric,» *Image and Vision Computing,* vol. 30.11, pp. 843-859, 2012.

T. Xiaohua, L. Dajie, and G. Jianya,, «A Methodology to Adjust Digital Cadastral Areas in GIS,» de *The 20th International Cartographic Conference. ICC*, 2001.

X. Tong, W. Shi, and D. Liu, «A Least Squares-Based Method for Adjusting the Boundaries of Area Objects,» ," *Photogrammetric Engineering & Remote Sensing,* vol. 71, pp. 189-195, 2005.

# 29

# DECISION LEVEL FUSION OF ORTHOPHOTO AND LIDAR DATA USING CONFUSION MATRIX INFORMATION FOR LNAD COVER CLASSIFICATION

S. Daneshtalab [a], H.Rastiveis [a]

[a] School of Surveying and Geospatial Engineering, College of Engineering, University of Tehran, Tehran, Iran -

(somaye.danesh, hrasti)@ut.ac.ir

KEY WORDS: Classification, LiDAR data, Orthophoto, SVM, Decision Level Fusion, Confusion matrix

ABSTRACT:

Automatic urban objects extraction from airborne remote sensing data is essential to process and efficiently interpret the vast amount of airborne imagery and Lidar data available today. The aim of this study is to propose a new approach for the integration of high-resolution aerial imagery and Lidar data to improve the accuracy of classification in the city complications. In the proposed method, first, the classification of each data is separately performed using Support Vector Machine algorithm. In this case, extracted Normalized Digital Surface Model (nDSM) and pulse intensity are used in classification of LiDAR data, and three spectral visible bands (Red, Green, Blue) are considered as feature vector for the orthoimage classification. Moreover, combining the extracted features of the image and Lidar data another classification is also performed using all the features. The outputs of these classifications are integrated in a decision level fusion system according to the their confusion matrices to find the final classification result. The proposed method was evaluated using an urban area of Zeebruges, Belgium. The obtained results represented several advantages of image fusion with respect to a single shot dataset. With the capabilities of the proposed decision level fusion method, most of the object extraction difficulties and uncertainty were decreased and, the overall accuracy and the kappa values were improved 7% and 10%, respectively.

## 1. INTRODUCTION

The performance of land cover classification using LiDAR data and Aerial imagery data separately has previously been analyzed and it was shown that superior results were achieved using LiDAR data (Jakubowski et al., 2013). However, simultaneous use of several remote sensing data from different sensors or methods of integration may be appropriate. In other words, the data obtained from different sources, each with aspects of the value that can be used together and are complementary (Esteban, Starr et al. 2005). Multiple sensors may provide complementary data, and fusion of information of different sensors can produce a better understanding of the observed site, which is not possible with single sensor (Simone, Farina et al. 2002).

Fusion of multiple data sets can be performed on signal, Pixel, feature and decision level (Pohl and Van Genderen 1998). In signal level fusion, signals from multiple sensors are combined together to create a new signal with a better signal-to-noise ratio than the input signals. In pixel level fusion, the information from different images, pixel by pixel, are merged to improve detection of objects in some tasks such as segmentation. Feature level fusion consists of merging the features extracted from different images. In this level of fusion, features are extracted from different sensors and combined to create a feature vector for classified using a classifier method (Abbasi, Arefi et al. 2015).

In decision level fusion, different datasets are combined at a higher level of integration. In this level of fusion, first the data from each single sensor is separately classified, then fusion consists of merging the output from the classification(Goebel and Yan 2004, Dong, Zhuang et al. 2009, Du, Liu et al. 2013).

Singh and Vogler investigated the impact of urban land use litter to detect classes of integration of Landsat imagery and lidar data began. By applying the supervised classification method showed that most like to merge the two data classification accuracy up to 32% compared to the separate use of lidar data has increased(Singh, Vogler et al. 2012).

Kim and colleagues used data integration aerial imagery and lidar ground cover gave way to improve classification accuracy. (Kim 2016). Gerke and Xiao proposed a combination of lidar and aerial image classification and automatic detection of effects used in urban areas. The method combines lidar data and aerial image of two dimensional geometry and spectral data liar image to extract four classroom buildings, trees, vegetation and land without vegetation land was used. The new classification method was introduced to the use of geometric and spectral information during the process of classification was defined (Gerke and Xiao 2013).

Another method for automatic extraction of buildings using lidar integration and optical image was presented by Li Wu. This procedure was completely data-driven, and also was suitable for any form of buildings(Li, Wu et al. 2013).

Bigdeli proposed a hyperspectral image classification system to integrate multiple fuzzy decision model based on lidar data

Decision template (DT. In this way, the characteristics of each data source separately extracted and the optimal properties were selected among them. Then, each of the data sources was classified separately using support vector machine. The output of this classification integrated using a Bayesian algorithm(Bigdeli, Samadzadegan et al. 2014).

Gilani, proposed a graph based algorithm for combination of multispectral images and airborne data lidar. The results showed enhancement in performance and improvement in the accuracy of the reconstruction and building recognition process(Gilani, Awrangjeb et al. 2015).Rastiveis Presented a Decision level fusion of lidar data and aerial color imagery based on Bayesian theory for urban area classification. This classification is performed in three different strategies: (1) using merely LiDAR data, (2) using merely image data, and (3) using all extracted features from LiDAR and image. The results of these classifiers were integrated in a decision level fusion based on Naïve Bayes algorithm (Rastiveis, 2015).

For the abovementioned researches, therefore, using color imagery with lidar data together in decision level of fusion may improve accuracy of classification in urban areas. In this study, we proposed a new decision level fusion of lidar data and orthophoto to improve the classification results. For this purpose, the paper will go on with describing the details of the proposed method. After that the study area would be introduced and, then, obtained results from the implementation of this method will presented.

## 2. PROPOSED METHOD

In this paper, an automatic object extraction system based on the integration of high resolution aerial orthophoto and LiDAR data is provided. Figure 1 shows the flowchart of the proposed algorithm.

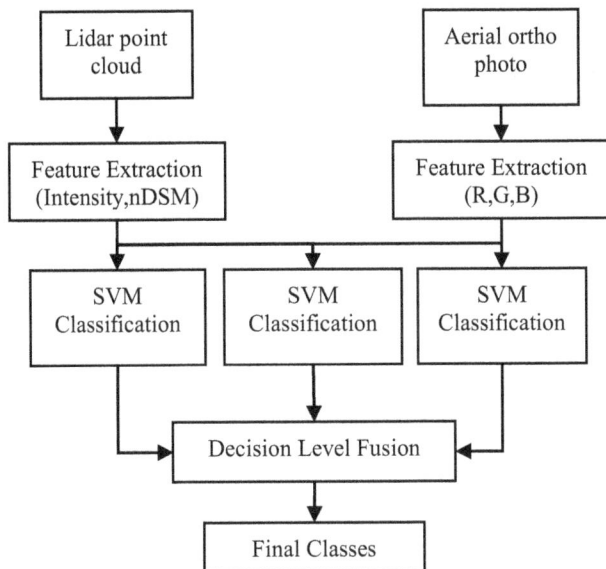

Figure 1.Flowchart of the proposed method for automatic object extraction from LiDAR data and orthophoto

As shown in Figure 1, after extracting proper features, both input data are classified through support vector machine classifier. Integrating all the extracted fetures from both sensors, another another classification are performed. Finally, according to the output from each of the classification and their error matrices, the results are integrated in a decision level fusion system to obtain the final classification result. The details of the proposed method are described in the following sections.

### 2.1 Data Preparation

In this step, as can be seen from Figure 1, different features are extracted from Lidar and ortho image data. These features must contain useful information to improve accuracy of classification process.

Intensity of Lidar as the primary attribute from lidar data recorded the reflectance on the surveyed area (Axelsson 1999). It is useful to discriminate asphalt areas from grassland or buildings in classification process (Buján, González-Ferreiro et al. 2012). Therefore, the intensity and is used as a feature in the classification of Lidar data. Besides, Normalized Digital Surface Model (nDSM) is extracted using LiDAR raw points cloud through Equation 1.

$$nDSM = DSM - DTM \qquad (1)$$

It should be noted that to simplify the fusion process the data set should have the same spatial resolutions. Therefore, a irregular format of the nDSM and the Intensity of LiDAR data are generated using interpolation techniques with the same resolution of the orthophoto.

From the orthophoto, three visible channels of Red, Green and Blue were considered as feature vector for the classification process.

Although different features may be used in classification process, however, increasing the number of features results more complexity in computations. Therefore, the most easy-access features for both data set are considered.

### 2.2 SVM Classification

There are a lot of classification algorithm that have been introduced during the last decades. In this paper, Support Vector Machine (SVM) classification algorithm is used to classify the data sets. Support vector machine is a machine learning method that is widely used for data analyzing and pattern recognition. The algorithm was invented by Vladimir Vapnik and the current standard incarnation was proposed by Corinna Cortes and Vladimir Vapnik. Classifying data has been one of the major parts in machine learning. The idea of support vector machine is to create a hyper plane between data sets to indicate which class it belongs to. The challenge is to train the machine to understand structure from data and mapping with the right class label, for the best result, the hyper plane has the largest distance to the nearest training data points of any class(Pal 2006). In this step, both data sets are classified using the aforementioned feature vectors.

### 4.2 Feature Level Fusion

In the orthophoto classification dataset, building was difficult to extract appropriately due to the similar spectral features with roads. But if only with orthophoto, it was not able to provide sufficient information. Meanwhile, in the lidar classification dataset, water and grass extraction also met problem caused by lidar data acquisition. Therefore, by integrating lidar and orthophoto, the problems could mitigate and the results of classification could be improved.

Here, all layers used in the previous classifications are integrated together as the input data of another SVM process. Orthophoto, including three spectral bands Red, Green, and Blue, and lidar derived Intensity image and nDSM image are employed in the classification process.

## 5.2 Decision Level Fusion

SVM classifiers are separately applied to each data. Then results of single classifiers are fused through a multiple classifier system classifier fusion is successfully applied on various types of data to improve single classification results. Classifier fusion can improve classification accuracy in comparison to a single classifier by combining results of classification algorithms on different data sets. The possible ways of integrating the outputs of classifiers in a decidedly fusion depend on what information can be obtained from the discrete members. In our proposed method, based on classification results in three lidar data, orthophoto and integration features, overall accuracy and omission, commision error for each class final image is created from the classification.

Finally, combining the previous classification results, the final classification are obtained. In this case, decision for each pixel are made according to the results of the overall classification from ortho photo, lidar data and the feature level fusion results. So that, first all pixels that the three classifications were agree about belonging them to the same class are remained without changing. The pixels that the classifications were disagree were labeled with regard to the overall accuracy and omission, commission error is a class act. In this case, the priority of a class in the final image is selected based on the lowest error rate or highest overall accuracy. Thus, all the pixels are checked and a decision is made on each.

## 3. EXPERIMENTAL RESULTS AND DISCUSSIONS

In this study, the standard data set from IEEE data fusion contest 2015 were used. The details of the data set are described in the following section. After that the obtained results from testing the algorithm using this data set are presented.

### 3.1 Data Set

The imaging data were acquired on March 13, 2011, using an airborne platform flying at the altitude of 300 m over the urban and the harbor areas of Zeebruges. The data were collected simultaneously and were georeferenced to WGS-84(IEEE, 2015).

The point density for the LiDAR sensor was approximately 65 points/m², which is related to point spacing of approximately 10 cm. Both the 3D point clouds and the resulting digital surface model (DSM) are provided. The color orthophotos were taken at nadir and have a spatial resolution of approximately 5 cm (IEEE, 2015). The ground truth of this data set five different land cover classes; in our proposed method these classes are Building, Grassland, Ground, Car and Tree. These features from LiDAR data, orthoimage and Ground trouth are displayed in Figure 2 (a-c).

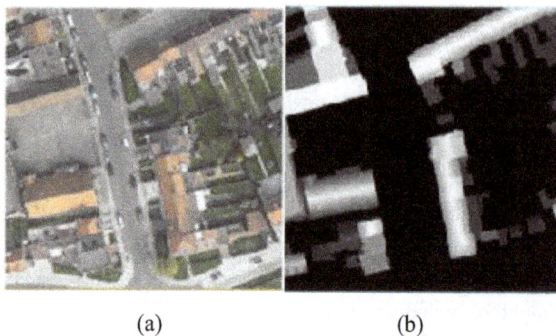

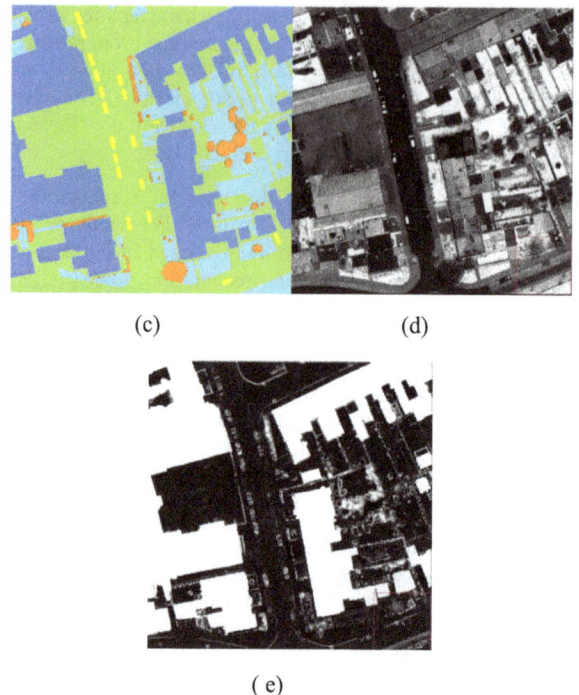

(a)                          (b)

(c)                          (d)

( e )

Figure 2. Applied data sets a) orthophoto, b) Lidar derived DSM, c) Ground trouth, d) Lidar pulse Intensity, e) nDSM

### 3.2 Results and Experiment

In the first step of the proposed method, feature spaces on orthophoto and Lidar data produced independently. Generate regular Lidar data with 5 cm spatial resolution, intensity image and nDSM. These features from LiDAR data are displayed in Figure ٢(d-e). In the next step, classification based on SVM is applied on orthophoto and Lidar data. This step by selecting the appropriate training and set dependent parameters were implemented. Five classes of "Buildings", "Cars, "Grass", "Ground" and "Tree" were considered. The main function of radial is the kernel used. Gamma parameter 1, parameter 100 penalty for lidar data and ortho photo by ٠,٣٣٣ and 100 has been set. In Figure 3 the outputs of the classifications are shown. After separately classifiying the orthophoto and the Lidar data, another classification based on fusing five extracted features were performed. Figure 4 represented the output of this classification. Also, figure 5 shows the output of the weighted majority voting method on these three categories. This method uses the correctness of each class as a weight in decision making. The Error matrix of each category in Table 1 to 5 is shown. The final classification was made from the decision level fusion method which is shown in Figure 6.

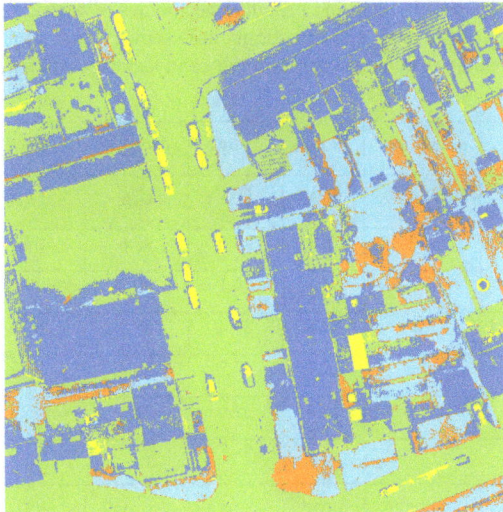

(a)

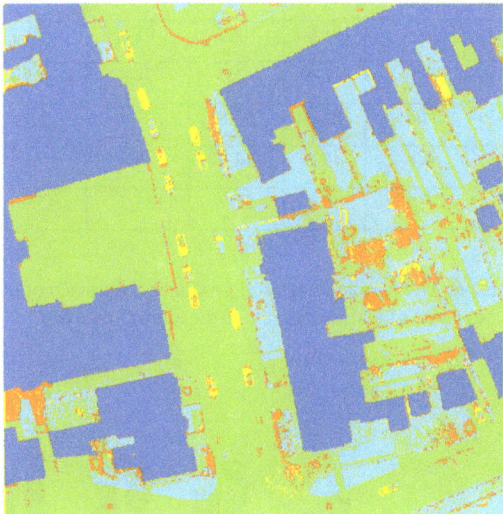

Figure 5. The output of the weighted majority voting classification.

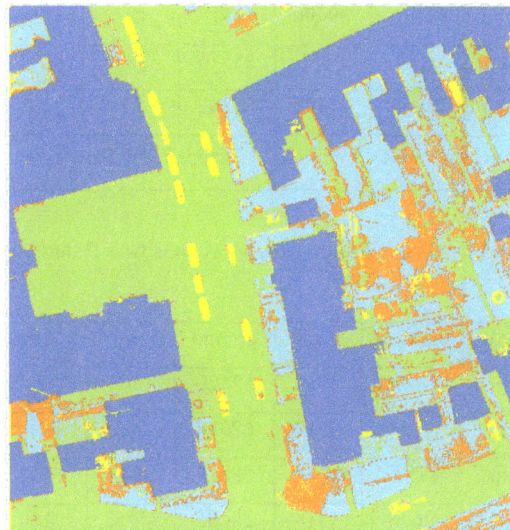

(b)

Figure 3. The output of the SVM classification,(a) orthophoto, (b) lidar data.

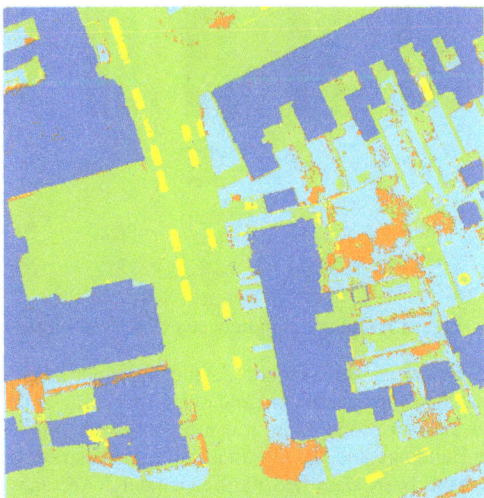

Figure 6. The output of the final classification using Decision Level Fusion.

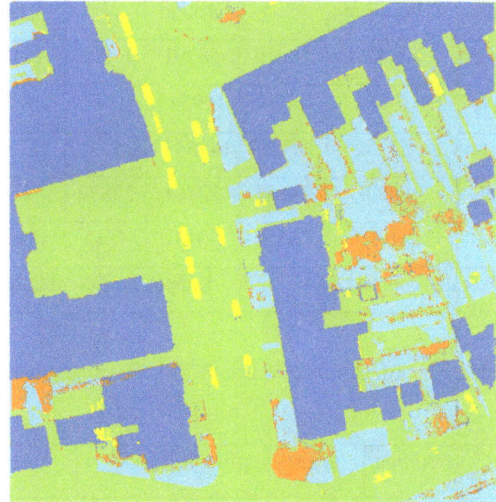

Figure 4. The output of the SVM classification from feature level fusion.

| Class | building | grass | ground | car | tree |
|---|---|---|---|---|---|
| Building | 896509 | 51260 | 331296 | 6696 | 11641 |
| Grass | 8333 | 475196 | 62784 | 105 | 47055 |
| Ground | 400187 | 10999 | 1344542 | 8911 | 2219 |
| Car | 25818 | 821 | 20305 | 31651 | 3 |
| Tree | 16559 | 97998 | 47727 | 4 | 63315 |
| OverallAccuracy 70.95 | | | KappaCoefficient 0.55 | | |

Table 1. Confusion Matrix from SVM classifiers orthophoto.

| Class | building | grass | ground | car | tree |
|---|---|---|---|---|---|
| Building | 1234290 | 10019 | 58201 | 0 | 7917 |
| Grass | 19887 | 400345 | 169546 | 4347 | 20771 |
| Ground | 63382 | 167759 | 1475474 | 15333 | 38177 |
| Car | 7860 | 14631 | 22701 | 22872 | 8414 |
| Tree | 21987 | 43520 | 80732 | 4815 | 48954 |
| OverallAccuracy 80.31 | | | KappaCoefficient 0.70 | | |

Table 2. Confusion Matrix from SVM classifiers lidar.

| Class | building | grass | ground | car | tree |
|---|---|---|---|---|---|
| Building | 1247403 | 7841 | 59441 | 137 | 4886 |
| Grass | 15225 | 477671 | 153113 | 97 | 35297 |
| Ground | 66546 | 89490 | 1515177 | 11872 | 10657 |
| Car | 6189 | 4151 | 28576 | 35228 | 411 |
| Tree | 12043 | 57121 | 50347 | 33 | 72982 |
| OverallAccuracy 84.51 | | | KappaCoefficient 0.76 | | |

Table 3 Confusion Matrix from SVM classifiers feature level.

| Class | building | grass | ground | car | tree |
|---|---|---|---|---|---|
| Building | 1237916 | 1762 | 52120 | 24 | 1787 |
| Grass | 8333 | 475196 | 62784 | 105 | 47055 |
| Ground | 87532 | 64805 | 1580796 | 9943 | 15028 |
| Car | 3676 | 1418 | 24605 | 37152 | 160 |
| Tree | 3874 | 20747 | 19256 | 3 | 47841 |
| OverallAccuracy 87.28 | | | KappaCoefficient 0.79 | | |

Table 4. Confusion Matrix from SVM classifiers decision.

| Class | building | grass | ground | car | tree |
|---|---|---|---|---|---|
| Building | 1249625 | 11741 | 64437 | 150 | 7395 |
| Grass | 6255 | 466702 | 82446 | 43 | 31717 |
| Ground | 78101 | 116290 | 1605405 | 12747 | 18209 |
| Car | 5683 | 2296 | 22612 | 34426 | 255 |
| Tree | 7742 | 39245 | 31754 | 1 | 66657 |
| OverallAccuracy 86.39 | | | KappaCoefficient 0.78 | | |

Table 5. Confusion Matrix from Weighted majority voting.

In Tables 6 (a-b) user and producer accuracies of each category for all performed classifications are shown, and Figure 7 depicted the resulted overall accuracy and kappa coefficient.

| Class | orthophoto | lidar | feature | decision | WMV |
|---|---|---|---|---|---|
| | U.A | U.A | U.A | U.A | U.A |
| Building | 69.1 | 94.1 | 94.5 | 95.6 | 93.7 |
| Grass | 80 | 65.1 | 70.1 | 80.0 | 79.4 |
| Ground | 76 | 83.8 | 89.4 | 89.9 | 87.6 |
| Car | 40.2 | 29.9 | 47.2 | 55.4 | 52.7 |
| Tree | 28.0 | 24.4 | 37.9 | 52.1 | 45.8 |

(a)

| Class | orthophoto | lidar | feature | decision | WMV |
|---|---|---|---|---|---|
| | P.A | P.A | P.A | P.A | P.A |
| Building | 66.5 | 91.6 | 92.5 | 91.8 | 92.7 |
| Grass | 74.6 | 62.9 | 75 | 74.6 | 73.3 |
| Ground | 74.4 | 81.6 | 83.8 | 87.4 | 88.8 |
| Car | 66.8 | 48.2 | 74.3 | 78.4 | 72.6 |
| Tree | 50.9 | 39.4 | 58.7 | 38.5 | 53.6 |

(b)

Table 6. The resulted user and producer accuracies of all classes. a) UserAccuracy , b)ProducerAccuracy

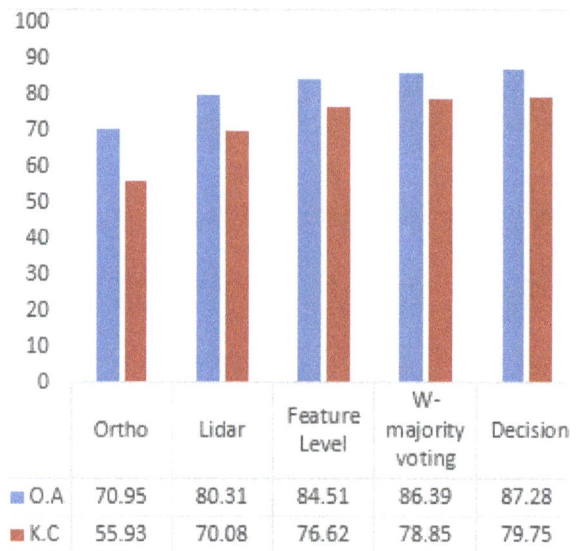

| | Ortho | Lidar | Feature Level | W-majority voting | Decision |
|---|---|---|---|---|---|
| O.A | 70.95 | 80.31 | 84.51 | 86.39 | 87.28 |
| K.C | 55.93 | 70.08 | 76.62 | 78.85 | 79.75 |

Figure 7. Comparison between classification results

As can be seen from Figure 7, the decision level fusion improved the kappa coefficient and overall accuracy of the three previous classifications. Also, it can be observed from Table 5 that the Buildings and the Ground classes, in relation to other classes, have been classified more accurately. Accuracy grass class in the other photo is higher of the other data. The building class in the lidar data and accuracy class of ground and tree in the feature fusion image data had the maximum accuracy in separately classification of the datasets. With the capabilities of the proposed decision level fusion method, most of the object extraction difficulties and uncertainty were decreased and, the overall accuracy and the kappa values were improved 7% and 10%, respectively.

## 4. CONCLUSION

In this paper, a new decision level fusion approach for the integration of high-resolution aerial imagery and Lidar data to improve the accuracy of classification in the city complications was proposed. The overall accuracy of the classification complications urban area in five different classes, for orthophoto 70.95%, 80.31% for lidar data, 84.51% at feature level and for the final image of integration in decision level 87.28% were obtained. The obtained results showed 7% and 10% improvement in the overall accuracy and the kappa values, respectively. the weighted majority voting is a fusion method at the decision level.  also  the results are compared with that method. Although the results were promising, however, more tests considereing different data ses, features, number of classes are suggested.

## REFERENCES

Abbasi, B., et al. (2015). "Fusion of hyperspectral and lidar data based on dimension reduction and maximum likelihood." The International Archives of Photogrammetry, Remote Sensing and Spatial Information Sciences 40(7): 569.

Axelsson, P. (1999). "Processing of laser scanner data—algorithms and applications." ISPRS Journal of Photogrammetry and Remote Sensing 54(2): 138-147.

Bigdeli, B., et al. (2014). "A decision fusion method based on multiple support vector machine system for fusion of hyperspectral and LIDAR data." International Journal of Image and Data Fusion 5(3): 196-209.

Buján, S., et al. (2012). "Land Use Classification from Lidar Data and Ortho-Images in a Rural Area." The Photogrammetric Record 27(140): 401-422.

Dong, J., et al. (2009" .(Advances in multi-sensor data fusion: Algorithms and applications." Sensors 9(10): 7771-7784.

Du, P., et al. (2013). "Information fusion techniques for change detection from multi-temporal remote sensing images." Information fusion 14(1): 19-27.

Esteban, J., et al. (2005). "A review of data fusion models and architectures: towards engineering guidelines." Neural Computing & Applications 14(4): 273-281.

Gerke, M. and J. Xiao (2013). "Supervised and unsupervised MRF based 3D scene classification in multiple view airborne oblique images." ISPRS Annals of the Photogrammetry, Remote Sensing and Spatial Information Sciences 2(3): 25-30.

Gilani, S., et al. (2015). "Fusion of LiDAR data and multispectral imagery for effective building detection based on graph and connected component analysis." The International Archives of Photogrammetry, Remote Sensing and Spatial Information Sciences 40(3): 65.

Goebel, K. and W. Yan (2004). Choosing classifiers for decision fusion. Proceedings of the Seventh International Conference on Information Fusion.

Kim, Y. (2016). "Generation of Land Cover Maps through the Fusion of Aerial Images and Airborne LiDAR Data in Urban Areas." Remote Sensing 8(6): 521.

H. Rastiveis, (2015). "Decision level fusion of LIDAR data and aerial color imagery based on Bayesian theory for urban area classification," The International Archives of Photogrammetry, Remote Sensing and Spatial Information Sciences, vol. 40, p. 589.

Li, Y., et al. (2013). "An improved building boundary extraction algorithm based on fusion of optical imagery and LiDAR data." Optik-International Journal for Light and Electron Optics 124(22): 5357-5362.

Pal, M. (2006). "Support vector machine-based feature selection for land cover classification: a case study with DAIS hyperspectral data." International Journal of Remote Sensing 27(14): 2877-2894.

Simone, G., et al. (2002). "Image fusion techniques for remote sensing applications." Information fusion 3(1): 3-15.

Singh, K. K., et al. (2012). "LiDAR-Landsat data fusion for large-area assessment of urban land cover: Balancing spatial resolution, data volume and mapping accuracy." ISPRS Journal of Photogrammetry and Remote Sensing 74: 110-121.

# 30

# OPTIMIZATION OF CLOSE RANGE PHOTOGRAMMETRY NETWORK DESIGN APPLYING FUZZY COMPUTATION

A. Sh. Amini[a]

[a] Dept. of Geomatic Engineering, Islamic Azad University – South Tehran Branch, Piroozi Street, Abuzar Boulevard, Dehhaghi Avenue, P. O. Box: 1777613651, Tehran, Iran - sh_amini@azad.ac.ir

**KEY WORDS:** Close range photogrammetry, Fuzzy computation, Network Design

**ABSTRACT:**

Measuring object 3D coordinates with optimum accuracy is one of the most important issues in close range photogrammetry. In this context, network design plays an important role in determination of optimum position of imaging stations. This is, however, not a trivial task due to various geometric and radiometric constraints affecting the quality of the measurement network. As a result, most camera stations in the network are defined on a try and error basis based on the user's experience and generic network concept. In this paper, we propose a post-processing task to investigate the quality of camera positions right after image capturing to achieve the best result. To do this, a new fuzzy reasoning approach is adopted, in which the constraints affecting the network design are all modeled. As a result, the position of all camera locations is defined based on fuzzy rules and inappropriate stations are determined. The experiments carried out show that after determination and elimination of the inappropriate images using the proposed fuzzy reasoning system, the accuracy of measurements is improved and enhanced about 17% for the latter network.

## 1. INTRODUCTION

Achieving high quality and low cost in production and dimensional quality control processes is an important aspect of industrial measurements (Fraser, 1997). As a non-contact, flexible, and accurate technique, close range photogrammetry can be used to facilitate the measurements in various applications (Luhmann et al., 2004). A very most important issue which affects the accuracy of measurements is the design of an appropriate image capturing network. Many researchers have been performed in this area to prove the results close range photogrammetry procedure.

Olague and Mohr (2002) investigated on the problem of where to place the cameras in order to obtain a minimal error in the 3D measurements. He posed the problem in terms of an optimization design and a global optimization process to minimize this criterion. In another investigation, he used of genetic algorithms for automating the photogrammetric network design process (Olague, 2001). Saadatseresht et al., (2004) proposed a novel method based on fuzzy logic reasoning strategy for the camera placement. He designed a system to make use of human type reasoning strategy by incorporating appropriate rules. Dunn et al., (2005) presented a novel camera network design methodology based on the Parisian approach to evolutionary computation. His experimental results illustrate significant improvements, in terms of solution quality and computational cost, when compared to canonical approaches. Fehr et al., (2009) investigated on several considerations for improving camera placement with the goal of developing a general algorithm that can be applied to a variety of surveillance and inspection systems. He presented an algorithm for placement problem in the context of computer vision and robotics.

In practice, due to existing environmental constraints and for more speed, network design is not fully applied and image capturing is performed experimentally. Based on the rule "more

image, higher accurate", many photogrammetrists prefer taking large number of images from objects, while many of which may not be necessary. Consequently, we will show that taking inappropriate images may lead to even decrease the accuracy of determined object points.

In this paper, a new fuzzy computation system is proposed that is able to determine unsuitable camera stations based on network design constraints that may have unfavorable effects on the result of the bundle adjustment. In this system, constraints related to distance are modeled based on fuzzy rules to decide whether or not a given image must be taken into account in bundle adjustment procedure. In the following, various fuzzy models developed in this paper are discussed and followed by experiments carried out to evaluate the accuracy of the results. The conclusions of the experiments are finally mentioned.

## 2. FUZZY MODELING OF NETWORK DESIGN CONSTRAINTS

Network design or camera placement involves with satisfaction of some vision constraints as well as optimization of measurement accuracy. On the other hand, one of the most important issues affected the quality of industrial photogrammetric network is image acquisition based on network design constraints (Atkinson, 1998). Image acquisition according these constraints leads to determination of the best accuracy on target positions of the object. Network design constraints have been shown in Figure 1. Network design constraints grouped in three classes of range, visibility, and accessibility-related constraints (Saadatseresht, 2004). Range or distance-related constraints include those applying to imaging scale, resolution, camera FOV, depth of field (DOF), number and distribution of points, and workspace. Visibility related constraints come from the visibility of a cluster of object points from a camera station which depends upon the constraints of target incidence angle, occluded areas, and camera FOV.

Accessibility-related constraints are typically dependent upon physical constraints of space, obstructions, and often the infeasibility of occupying certain geometrically favorable locations. Range constraints, as the most important constraints affected the results of close range procedure are divided in to two parts (Saadatseresht, 2004):

- Constraints related to minimum distance from camera to the object. These constraints determine minimum needed distance from each camera station. In other words, if the distance of the camera station is far from this value, appropriate accuracy of object coordinates in 3D modeling procedure is determined.

- Constraints related to maximum distance from camera to the object. These constraints determine maximum needed distance from each camera station. In other words, if the distance of the camera station is nearer from this value, appropriate accuracy of object coordinates in 3D modeling procedure is determined.

The most important factors affected the accuracy of exterior orientation of imaging stations and object coordinates are the distance between the camera and the object, and the angle between camera viewing direction and the object surface. Far distance between the image station and the object decreases the scale of the image and its resolution, consequently it makes recognizing the centre of the targets on the objects less accurate. Close distance of image capturing makes the targets blur in images, because the depth of field factor is decreases. Moreover, inappropriate distance and angle between camera station and the object deforms the targets seen in the image, consequently the centre of the targets is measured with error. These factors cause the exterior orientation of image stations and object coordinates having less accuracy in the bundle adjustment. According to the position of each camera station related to the object, for each constraint, a fuzzy value between 0 and 1 is specified. This value is specified according to the comparison between the distance of the station from object and the appropriate distance that is calculated from the constraint formula. Resultant of all specified fuzzy values defines the quality of the station and images captured from that. If the value is appropriate (usually more than 0.7), the images captured from that station are preserved, else (usually less than 0.7), the images are omitted from bundle adjustment procedure. In continue, fuzzy modeling procedure of mentioned constraints is introduced.

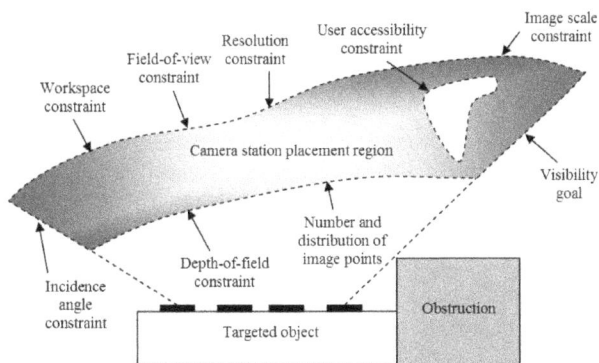

Figure 1. Constraints in photogrammetric network design

## 2.1 Applied Membership Functions

To fuzzy modeling of minimum distance constraints, "smf"

function is used (Menhaj, 1998). In this function, a value between 0 and 1 for distances around the minimum distance, value 1 for distances larger than this boundary and value 0 for distances less than this boundary is defined. To fuzzy modeling of maximum distance constraints, "zmf" function is used (Menhaj, 1998). In this function, a value between 0 and 1 for distances around the maximum distance, value 1 for distances less than this boundary and value 0 for distances larger than this boundary is defined. To combine these two constraint sets, "pimf" function is used (Menhaj, 1998). In this function, value 1 for distances in inner area, a value between 0 and 1 for distances in near boundaries and value 0 for distances in outer area is defined. An example of "smf", "zmf" and "pimf" functions have been shown in Figure 2.

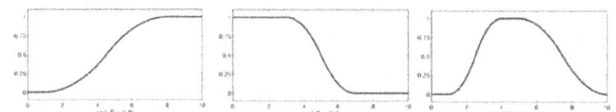

Figure 2. An instance of "smf" (left), "zmf" (middle) and "pimf" (right) functions from the toolbox of Matlab

## 2.2 Fuzzy Modeling of Range Constraints

In details, range constraints are included as (Saadatseresht, 2004):

-Minimum distance constraints: camera depth of field, and number and distribution of targets

-Maximum distance constraints: image resolution, image scale, and camera field of view

For each constraint, in addition to a value between 0 and 1, an attribute label is dedicated according to Table 1. Fuzzy modeling of each constraint is discussed in continue.

| Fuzzy value | Corresponding label |
|---|---|
| x = 0 | Unsuitable |
| 0 < x <= 0.7 | Weak |
| 0.7 < x < 1 | Appropriate |
| x = 1 | Robust |

Table 1. Relation between fuzzy values and corresponding labels for each constraint

*Camera Depth of Field Constraint*: Camera depth of field specified an area around the object that for a special distance between camera station and the object, a sharp image will be obtained (Saadatseresht, 2004). This constraint is appeared in equation (1):

$$D_{Depth}^{min} = \frac{d}{1 + \frac{d-f}{D_{HF}}} \qquad D_{HF} = \frac{f^2}{F_{stop}.\delta} \qquad (1)$$

where    $D_{Depth}^{min}$ = minimum depth of field distance,

$D_{HF}$ = ultra-focal distance
$d$ = initial distance between object and camera
$f$ = focal length
$F_{stop}$ = inner parameter of camera
$\delta$ = diameter of ambiguity circle

*Number and Distribution of Targets Constraint*: At least $k$ targets that have suitable distribution in image is an appropriate attribute for number of targets constraint and solving unknowns in adjustment (Saadatseresht, 2004). Equation (2) defines the appropriate distance to appear at least $k$ targets in each image:

$$D_{Point}^{min} = \frac{af\sqrt{k}}{d} \qquad (2)$$

where   $a$ = mean distance between targets in object space
$k$ = desired number of targets
$f$ = focal length
$d$ = frame size of the camera

*Image Resolution Constraint*: Image resolution constraint is mentioned to the ability of identifying the targets in an image (Saadatseresht, 2004). Equation (3) defines maximum distance between object and camera due to the image resolution constraint:

$$D_{Res}^{max} = \frac{f.D_t.Sin\varphi}{I_{Res}.T} \qquad (3)$$

where   $\varphi$ = angle between camera direction and object

$D_t$ = target dimension
$f$ = focal length
$T$ = minimum number of target pixels
$I_{Res}$ = dimension of each target

*Image Scale Constraint:* Image scale constraint determines the maximum distance that the accuracy decreases for more than that distance (Saadatseresht, 2004). Equation (4) defines this constraint:

$$D_{Scale}^{max} = \frac{D_0 f \sqrt{k}}{q S_p \sigma_i} \qquad (4)$$

where   $f$ = focal length
$k$ = number of images captured in each station
$S_p$ = relative accuracy value of measuring
$D_0$ = maximum diameter of the object
$\sigma_i$ = error of image coordinate measurement
$q$ = network stability factor

*Camera Field of View Constraint:* This constraint specifies maximum distance between object and camera that all or a part of the object appears in image area and the object is not appeared in a part of image space (Saadatseresht, 2004). Equation (5) defines this maximum distance:

$$D_{Fov}^{min} = \frac{D_o Sin(\alpha+\varphi)}{2 Sin(\alpha)} \qquad \alpha = tan^{-1}\left(\frac{0.9d_0}{2f}\right) \qquad (5)$$

where   $\alpha$ = half angle of vertex of camera pyramid
$\varphi$ = angle between camera direction and object

$D_0$ = length of maximum diameter of the object
$d_0$ = minimum of the frame size
$f$ = focal length in sequence

*Combining all Constraints:* In order to decide finally about the quality of the camera positions, it is necessary to combine all mentioned constraints. For this reason, the appropriate image capturing area is obtained according to equation (6) (Saadatseresht, 2004):

$$D_{min} = max(D_{Depth}, D_{Point})$$
$$D_{max} = min(D_{Res}, D_{Scale}, D_{Fov}) \qquad (6)$$
$$Range = D_{max} - D_{min}$$

where   $Range$ = appropriate image capturing area

According to the final fuzzy value, the system is decided whether each image is suitable for using in final bundle adjustment procedure or not. Providing that there is any unsuitable station, related image must be eliminated in adjustment procedure.

## 3. INPUT AND OUTPUT DATA

Input data in this procedure includes a data file of coordinates, camera information, object information, target characteristics and network design information. Output data includes the result of fuzzy modeling and decision about each image.

### 3.1 Input Data

In complete, data input includes:
*A position Data File:* Targets position coordinates, Cameras position coordinates.

*Camera Characteristics:* Focal length, F-stop parameter, Pixel size, Dimension of sensor (number of rows and columns).

*Object Characteristics*: Maximum length of the object, Expected accuracy on target positions.

*Target* Characteristics: Diameter of target, Number of pixels in each target, Number of expected targets in each image, Mean distance between each target.

*Network Design Information*: Network stability factor, Minimum angle between camera optical direction and the object surface.

### 3.2 Output Data

Output data includes a decision about the quality of each image position after fuzzy modeling of each constraint. Output information in details includes: Displaying targets and camera station positions, Membership functions of each constraint, Fuzzy value of each constraint , Final decision whether the image is appropriate for using in bundle adjustment procedure or not.

## 4. EXPERIMENTS

In three separate projects, a propeller of a plane (Amini, 2006), a car door (Opteka, 2004) and a nose of a plane (Amini, 2006) were modeled and measured by close range photogrammetry method that they have been shown in Figure 3.

Figure 3. The investigated propeller, the car door, and the plane nose

In investigation of the propeller, 3D modeling of its surface was implemented. The purpose of the investigation was determining the deformation between its two wings (Amini, 2006). In this investigation, 19 images were captured and after bundle adjustment procedure, the mean accuracy of x, y and z coordinates of the targets were determined.

The standard deviation of exterior orientation values of each station are shown in Table 2. Moreover, the fuzzy computation system was tested on each station and the results are also shown in Table 2.

| Station | Fuzzy Value | Exterior Orientation Error (mm) |
|---------|-------------|--------------------------------|
| 1 | 0.7464 | 0.2799 |
| 2 | 0.7474 | 0.2792 |
| 3 | 0.6484 | 0.3088 |
| 4 | 0.9263 | 0.2060 |
| 5 | 0.9189 | 0.2057 |
| 6 | 0.7905 | 0.2259 |
| 7 | 0.8435 | 0.2507 |
| 8 | 0.8938 | 0.2443 |
| 9 | 0.7663 | 0.2626 |
| 10 | 0.8856 | 0.2254 |
| 11 | 0.8530 | 0.2296 |
| 12 | 0.9155 | 0.1976 |
| 13 | 0.9244 | 0.1978 |
| 14 | 0.7804 | 0.2256 |
| 15 | 0.8319 | 0.2250 |
| 16 | 0.8824 | 0.2151 |
| 17 | 0.8619 | 0.2215 |
| 18 | 0.7551 | 0.3536 |
| 19 | 0.7471 | 0.3513 |

Table 2. Exterior orientation errors and the result of fuzzy value of each station in measuring the propeller

In Table 2, four maximum exterior orientation standard deviation values and four minimum fuzzy values are highlighted with the color of chromatic gray and also four minimum exterior orientation standard deviation values and four maximum fuzzy values are highlighted with the color of pallid gray. As Table 2 shows, the results for stations that have maximum/minimum exterior orientation standard deviation values are almost the same as the stations that have minimum/maximum fuzzy values. After omission of stations with low fuzzy values, bundle adjustment procedure was done with the rest of images. The result of bundle adjustment before and after omission of inappropriate images is shown in Table 3.

| Adjustment | δx (mm) | δy (mm) | δz (mm) | RMSE (mm) |
|------------|---------|---------|---------|-----------|
| Before omission | 0.0182 | 0.0265 | 0.0218 | 0.0388 |
| After omission | 0.0147 | 0.0232 | 0.0179 | 0.0328 |

Table 3. The results of bundle adjustment of the propeller before and after omission of images with low fuzzy values

The result of Table 3 shows that the accuracy of bundle adjustment proves after omission of stations with low fuzzy values. Moreover, camera station error ellipsoids for the propeller before and after omission of improper images are shown in Figure 4 in the same scale of 500 greater. Figure 4 shows that after omission of improper images, the rest of error ellipsoids growing smaller.

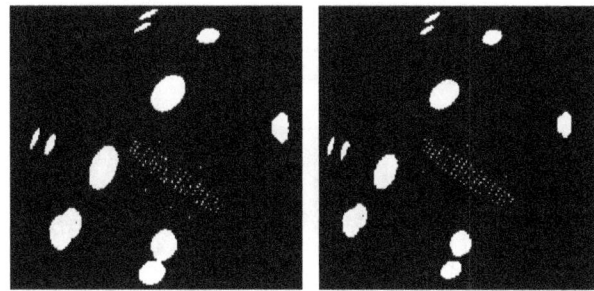

Figure 4. Camera station error ellipsoids for propeller in scale of 500 before (left) and after (right) omission of improper images

In investigation of the car door (Opteka, 2004), 9 images were captured and after bundle adjustment procedure, the mean accuracy of x, y and z coordinates of the targets were determined. The standard deviation of exterior orientation values of each station are shown in Table 4. Moreover, the fuzzy computation system was tested on each station and the results are also shown in Table 4. In Table 4, three maximum exterior orientation standard deviation values and three minimum fuzzy values are highlighted with the color of chromatic gray and also three minimum exterior orientation standard deviation values and three maximum fuzzy values are highlighted with the color of pallid gray.

| Station | Fuzzy Value | Exterior Orientation Error (mm) |
|---------|-------------|--------------------------------|
| 1 | 0.6414 | 0.3510 |
| 2 | 0.5541 | 0.3981 |
| 3 | 0.7941 | 0.3133 |
| 4 | 0.5543 | 0.3925 |
| 5 | 0.7486 | 0.4214 |
| 6 | 0.8000 | 0.3602 |
| 7 | 0.7993 | 0.3145 |
| 8 | 0.7856 | 0.3359 |
| 9 | 0.7994 | 0.3170 |

Table 4. Exterior orientation errors and the result of fuzzy value of each station in measuring the car door

As Table 4 shows again, the results for the stations that have maximum/minimum exterior orientation standard deviation values are almost the same as the stations that have minimum/maximum fuzzy values. After omission of stations with low fuzzy values, bundle adjustment procedure was done with the rest of images. The result of bundle adjustment before and after omission of inappropriate images is shown in Table 5.

| Adjustment | δx (mm) | δy (mm) | δz (mm) | RMSE (mm) |
|------------|---------|---------|---------|-----------|
| Before omission | 0.0099 | 0.0150 | 0.0148 | 0.0233 |
| After omission | 0.0125 | 0.0111 | 0.0092 | 0.0207 |

Table 5. The results of bundle adjustment of the car door before and after omission of images with low fuzzy values

The result of Table 5 shows that the accuracy of bundle adjustment after omission of stations with low fuzzy values proves better, but a little. The reason of small proven in results was decreasing the stations and consequently decreasing the corresponding rays' intersection. In Figure 5, camera station error ellipsoids for car door before and after omission of improper images are shown in the same scale of 700 greater. Figure 5 shows that after omission of improper images, the rest of error ellipsoids growing smaller.

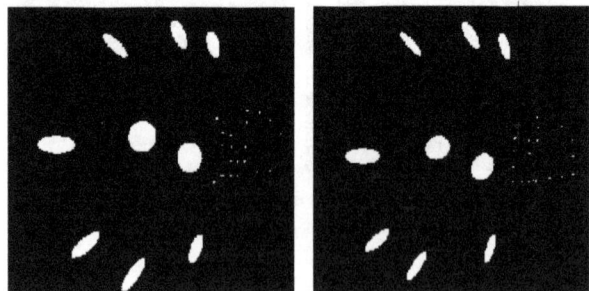

Figure 5. Camera station error ellipsoids for car door in scale of 700 before (left) and after (right) omission of improper images

In the third investigation, 3D modeling of a plane nose was implemented (Amini, 2006). In this investigation, 26 images were captured and after bundle adjustment procedure, the standard deviation of exterior orientation values of each station was determined that are shown in Table 6.

| Station | Fuzzy Value | Exterior Orientation Error (mm) |
|---------|-------------|----------------------------------|
| 1 | 0.8000 | 0.3948 |
| 2 | 0.6948 | 0.6332 |
| 3 | 0.8000 | 0.3878 |
| 4 | 0.6279 | 0.5844 |
| 5 | 0.9631 | 0.2671 |
| 6 | 0.8180 | 0.2982 |
| 7 | 0.9738 | 0.2645 |
| 8 | 0.8000 | 0.3704 |
| 9 | 0.8000 | 0.4016 |
| 10 | 0.5636 | 0.7945 |
| 11 | 0.8005 | 0.3951 |
| 12 | 0.6240 | 0.7100 |
| 13 | 0.8000 | 0.4304 |
| 14 | 0.8000 | 0.5492 |
| 15 | 0.8454 | 0.3162 |
| 16 | 0.9521 | 0.2563 |
| 17 | 0.9740 | 0.2403 |
| 18 | 0.9744 | 0.2648 |
| 19 | 0.8033 | 0.3971 |
| 20 | 0.8068 | 0.4254 |
| 21 | 0.8000 | 0.4832 |
| 22 | 0.8000 | 0.5819 |
| 23 | 1.0000 | 0.2116 |
| 24 | 1.0000 | 0.1866 |
| 25 | 0.8095 | 0.3026 |
| 26 | 0.8129 | 0.3154 |

Table 6. Exterior orientation errors and the result of fuzzy value of each station in measuring the plane nose

Moreover, the fuzzy computation system was tested on each station and the results are also shown in Table 6. In Table 6, four maximum exterior orientation standard deviation values and four minimum fuzzy values are highlighted with the color of chromatic gray and also four minimum exterior orientation standard deviation values and four maximum fuzzy values are highlighted with the color of pallid gray. As Table 6 shows again, the results for the stations that have maximum/minimum exterior orientation standard deviation values are almost the same as the stations that have minimum/maximum fuzzy values. After omission of stations with low fuzzy values, the bundle adjustment procedure was done with the rest of images.

The result of bundle adjustment before and after omission of inappropriate images is shown in Table 7.

| Adjustment | δx (mm) | δy (mm) | δz (mm) | RMSE (mm) |
|------------|---------|---------|---------|-----------|
| Before omission | 0.0550 | 0.0523 | 0.0424 | 0.0869 |
| After omission | 0.0415 | 0.0497 | 0.0363 | 0.0702 |

Table 7. The results of bundle adjustment of the plane nose before and after omission of images with low fuzzy values

The result of Table 7 shows that the accuracy of bundle adjustment proves after omission of stations with low fuzzy values. Camera station error ellipsoids for plane nose before and after omission of improper images are shown in Figure 6 in the same scale of 1000 greater. As Figure 6 shows, after omission of improper images, rest of error ellipsoids growing smaller.

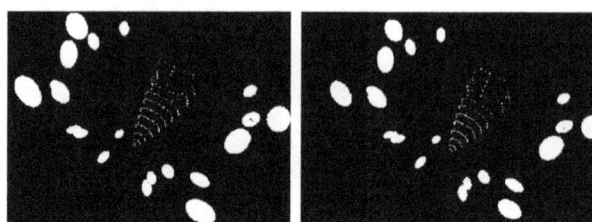

Figure 6. Camera station error ellipsoids for plane nose in scale of 1000 before (left) and after (right) omission of improper images

## 5. CONCLUSIONS

Close range photogrammetry is a suitable and efficient method in accurate modeling and measurement. The most important issue affected the accuracy in a close range photogrammetry procedure is an appropriate camera station network design. In practice, network design is not fully observed and image capturing is performed experimentally and consequently, some images may not suitable for using in bundle adjustment procedure. In this paper, a decision system is established based on fuzzy computation that can be able to specify unsuitable images based on network design constraints that may have unfavorable effect on the result of bundle adjustment. The program is experimented on three data sets determined from including a propeller of a plane, a car door, and a nose of a plane.

The results of three investigations showed that the quality of each image capturing station is almost related to fuzzy value of that station. Consequently, the fuzzy system truly is able to specify inappropriate images that have improper effects on the result of bundle adjustment. Moreover, after omission of stations with low fuzzy values in all three experiments, the accuracy of bundle adjustment proves. They are 15.5% average accuracy improvement for propeller of a plane (0.0388mm to 0.0328mm), 6.9% for car door (0.0233mm to 0.0217mm), and 7.7% for nose of a plane (0.0869mm to 0.0802mm). Moreover, dimensions of camera station error ellipsoids after omission of improper images becomes smaller that confirms applied fuzzy method. As a result, employing the suggested fuzzy system to detect and eliminate inappropriate images, it improves the accuracy of coordinates of targets in average about 17%.

## 6. REFERENCES

Amini, A. Sh., 2006. Investigation of Dimensional Quality Control of Industrial Equipments with Close Range Photogrammetry Method. *M. S. thesis, K.N. Toosi University of Technology, Tehran, Iran.*

Atkinson, K. B., 1998. Close Range Photogrammetry and Machine Vision. *Whittles Scotland Publishers.*

Dunn, E., Olague, G. and Lutton, E., 2005. Automated Photogrammetric Network Design using the Parisian Approach. *7th European Workshop on Evolutionary Computation in Image Analysis and Signal Processing, Springer-Verlag.*

Fehr, D., Fiore, L. and Papanikolopoulos, N., 2009. Issues and Solutions in Surveillance Camera Placement. *In the Proceedings of the IEEE/RSJ International Conference on Intelligent Robots and Systems.*

Fraser, C., 1997. Innovations in Automation for Vision Metrology Systems. *Intl. J. Photogrammetric Record, 15: 901-911.*

Luhmann, T., Robson, S., Kyle, S. and Harley, I., 2008. Close Range Photogrammetry: Principles, Methods and Applications, *Whittles England Publishers.*

Menhaj, M. B., 1998. Fuzzy Computation. *Elmo San'at Iran Publishing, Iran.*

Olague, G., 2001. Autonomous Photogrammetric Network Design using Genetic Algorithms. *In the Proceedings of the Evo Workshops on Applications of Evolutionary Computing, Springer-Verlag.*

Olague, G. and Mohr, R., 2002. Optimal Camera Placement for Accurate Reconstruction", *Pattern Recognition, 35: 927-944.*

Opteka, J., 2004. Precision Target Mensuration in Vision Metrology. *PhD thesis, University of Wien.*

Saadatseresht, M., 2004. Automatic Camera Placement in Vision Metrology Based on a Fuzzy Inference System, *PhD thesis, University of Tehran, Tehran, Iran.*

Saadatseresht, M., Samdzadegan, F., Azizi, A. and Hahn, M., 2004. Camera Placement for Network Design in Vision Metrology Based On Fuzzy Inference System. *International Conference of ISPRS, Istanbul.*

# 31

# WOODLAND MAPPING AT SINGLE-TREE LEVELS USING OBJECT-ORIENTED CLASSIFICATION OF UNMANNED AERIAL VEHICLE (UAV) IMAGES

A. Chenari [a], Y. Erfanifard [a, *], M. Dehghani [b], H.R. Pourghasemi [a]

[a] Dept. of Natural Resources and Environment, College of Agriculture, Shiraz University, Shiraz, Iran - afrooz.chenari71@gmail.com, erfanifard@shirazu.ac.ir, hr.pourghasemi@shirazu.ac.ir
[b] Dept. of Civil and Environmental Engineering, School of Engineering, Shiraz University, Shiraz, maryamdehghani@shirazu.ac.ir

**KEY WORDS:** Crown area, Object-oriented classification, Tree map, Unmanned aerial vehicle, Woodland.

**ABSTRACT:**

Remotely sensed datasets offer a reliable means to precisely estimate biophysical characteristics of individual species sparsely distributed in open woodlands. Moreover, object-oriented classification has exhibited significant advantages over different classification methods for delineation of tree crowns and recognition of species in various types of ecosystems. However, it still is unclear if this widely-used classification method can have its advantages on unmanned aerial vehicle (UAV) digital images for mapping vegetation cover at single-tree levels. In this study, UAV orthoimagery was classified using object-oriented classification method for mapping a part of wild pistachio nature reserve in Zagros open woodlands, Fars Province, Iran. This research focused on recognizing two main species of the study area (i.e., wild pistachio and wild almond) and estimating their mean crown area. The orthoimage of study area was consisted of 1,076 images with spatial resolution of 3.47 cm which was georeferenced using 12 ground control points (RMSE=8 cm) gathered by real-time kinematic (RTK) method. The results showed that the UAV orthoimagery classified by object-oriented method efficiently estimated mean crown area of wild pistachios ($52.09\pm24.67$ m$^2$) and wild almonds ($3.97\pm1.69$ m$^2$) with no significant difference with their observed values ($\alpha=0.05$). In addition, the results showed that wild pistachios (accuracy of 0.90 and precision of 0.92) and wild almonds (accuracy of 0.90 and precision of 0.89) were well recognized by image segmentation. In general, we concluded that UAV orthoimagery can efficiently produce precise biophysical data of vegetation stands at single-tree levels, which therefore is suitable for assessment and monitoring open woodlands.

## 1. INTRODUCTION

Biophysical characteristics of vegetation cover play an important role in exploring ecological and socio-economic effects of vegetation in open woodlands. In addition, these characteristics may influence on ecosystem patterns and processes since essential nutrients and moisture are concentrated beneath canopy of plants in these open woodlands with sparse vegetation cover (Korhonen et al., 2006; Wallace et al., 2008; Chianucci et al., 2016).

Similar to other woodlands, Zagros open woodlands in western Iran are very important on ecological (such as wildlife habitat, soil and water conservation) and socio-economic (for example, dependence of rural communities and nomads to by-products) attributes that makes it essential to prevent their existence to be at risk. An ongoing challenge to sustainable management of these valuable woodlands; however, is cost-effective and accurate monitoring of spatial and temporal changes in vegetation cover. Assessment and monitoring of species at single-tree level in such ecosystems typically concentrate on a number of parameters such as composition of species individuals, abundance of species offsprings, and biophysical characteristics such as crown area and branching architecture. As mentioned above, biophysical characteristics and in particular crown area and density (i.e., number of individuals per unit area), is strongly associated with establishment and growth of recruits in these open arid and semi-arid woodlands.

Regarding to the importance of the mentioned biophysical characteristics of vegetation, mapping these characteristics has become one of the major goals of open woodland mensuration (Chopping et al., 2008). There are various types of methods that have been developed to assess the mentioned characteristics by field measurements (Rautiainen et al., 2005; Korhonen et al., 2006; Hanberry et al., 2012). However, remotely sensed imagery from spaceborne and/or airborne platforms are preferred to quantify canopy cover and density in different forest ecosystems by previous researchers due to their less costs and time-consumption and more practical for large ecosystems, e.g. Zagros semi-arid woodlands in western Iran (Carreiras et al., 2006; Gleason and Im, 2011; Feret and Asner, 2012; Yang et al., 2012; Rezayan and Erfanifard, 2016). The outcome of classification of remotely sensed imagery is a thematic map exhibiting the spatial distribution of plants and their corresponding crowns within the studied environment (Schowengerdt, 2007; Panta et al., 2008).

Remote sensing datasets provide an outstanding way to obtain thematic maps of vegetation cover from local to global scales, while most of these datasets obtained by spaceborne platforms are not appropriate to map biophysical characteristics of single plants in open woodlands because of their low or medium spatial resolution as explored by Colomina and Molina (2014). On the one hand, it is so expensive to apply satellite data with very high spatial resolution (e.g., WorldView-3 with 30 cm spatial resolution), and on the other hand, datasets acquired by digital cameras (such as UltraCam-D) and airborne platforms are not available in all places (Chianucci et al., 2016). Therefore, it seems necessary to use remotely sensed datasets with appropriate spatial resolution for vegetation mapping of individual plants that are available with reasonable cost and complexity. Unmanned aerial vehicles (UAVs), as a result of recent advances in remote sensing, can combine temporal and spatial resolutions to provide appropriate datasets for mapping forested areas at single-tree levels. Diaz-Varela et al. (2014) mentioned that UAVs are attractive and flexible tools for mapping and monitoring of various aspects of environment. Progress made in internal sensors, reduction of costs, more flexibility compared to spaceborne and airborne platforms, and precise GPS embedded are some of unique advantages that have enhanced the application of UAVs in a wide range of subjects. Recent studies have exhibited the feasibility of using UAVs in monitoring (Torres-Sanchez et al., 2015; Vega et al., 2015; Mlambo et al., 2017) and measurement of characteristics plant species (Chianucci et al., 2016; McNeil et al., 2016; Ivosevic et al., 2017). In contrary, relatively few investigations have studied the applicability of orthoimages taken by UAVs for mapping biophysical characteristics of trees and shrubs at single-tree levels. Kuzmin et al. (2016) used UAV imagery to recognize tree species in a mixed boreal forest with overall accuracy of 82%. In addition, UAVs are very interesting instruments that hold great potential for mapping individual plants in open woodlands.

UAVs take images at very high spatial resolution which makes precise recognition of features possible. As the spatial resolution of imagery increases, the heterogeneity of pixels may lead to salt-and-pepper effect in pixel-based classification techniques. Previous studies revealed that it is so difficult to accurately distinguish features on images with very high spatial resolution by pixel-based techniques (Blaschke et al., 2000; Blaschke and Strobl, 2001). Taking into account the disadvantages of pixel-based techniques, many researchers have applied object-oriented classification techniques to recognize features on images with very high spatial resolution (Ouyang et al., 2011). To explore efficiency of object-oriented classification method to map single species in mixed stands on very high resolution datasets, it needs to be tested on UAV orthoimagery.

In this study, we hypothesize that UAV orthoimages can be implemented to map vegetation spatial arrangement of plant species in Zagros open woodlands. It was also hypothesized that object-oriented classification method on UAV orthoimagery is efficient to recognize type of species of individual plants in mixed stands. The main objective of the current study is to delineate crown area and spatial position of individual wild pistachio trees and wild almond shrubs in a mixed stand using object-oriented classification method on UAV orthoimagery.

## 2. METHODOLOGY

### 2.1 Study Area

Zagros open woodlands cover a vast area of Zagros Mountain ranges stretching from Piranshahr (Western Azerbaijan Province), northwestern Iran, to the vicinity of Firoozabad (Fars Province) in southwestern Iran, having an average length and width of 1,300 and 200 km, respectively (Fig. 1a). The study area classified as semi-arid based on Ambreje method, Zagros Woodlands cover 5 million hectares and consist 40% of Iran's forests.

The most important and widespread species is oak with three species (i.e. *Quercus brantii* var. *persica, Q. libani,* and *Q. infectoria*) that consist 85% of Zagros open woodlands. These oak species often form pure stands and sometimes, are mixed with wild pistachio (*Pistacia* spp.) and wild almond (*Amygdalus* spp.) (Owji and Hamzepour, 2012; Sagheb Talebi et al. 2014).

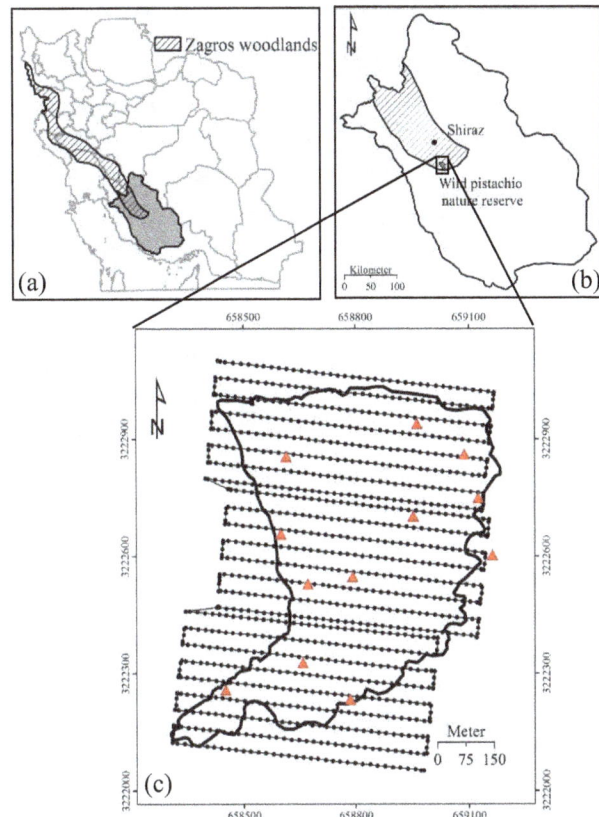

Figure 1. Map identifying the 45-ha study area in wild pistachio nature reserve (a) in Zagros woodlands of Fars province (a) and extent of Zagros woodlands in West Iran (b). The study area: filled circles and black lines show flight plan and red triangles indicate the location of ground control points (c).

The study site with 45 ha area is located in wild pistachio nature reserve in Fars Province, Iran (Fig. 1b), between 52° 30′ to 52° 40′ E and 29° 00′ to 29° 15′ N (Fig. 1c). The minimum and maximum elevations are 1,859 and 1,935 m a.s.l., respectively. The mean annual precipitation and temperature are 383 mm and 27.6 °C, respectively (Owji and Hamzepour, 2012). The study site was fully covered by wild pistachio (*Pistacia atlantica* var. *mutica*) (Fig. 2) as the second most frequent tree species in Zagros Woodlands, accompanied by wild almond shrubs (Fig. 3).

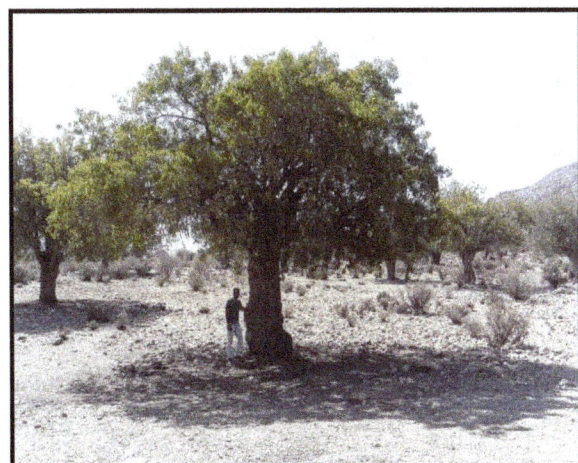

Figure 2. A wild pistachio tree within the study site (Photo taken by A. Chenari).

Figure 3. The study site was covered by wild pistachio trees and wild almond shrubs (Photo taken by Phantom-4 UAV).

## 2.2 UAV Imagery

The UAV applied in this study was DJI Phantom-4 quadcopter equipped with a digital camera with a 1/2.3 inch CMOS sensor and a size of 3,000 × 4,000 pixels (Fig. 4a). It has a weight of about 1,380 g including a digital camera, battery, and GPS. The maximum flight time length was about 28 min. If the necessary parameters, i.e., area of interest, desired ground resolution, longitudinal overlap are determined; the UAV automatically calculates the number of strips and flight height (Table 1). UAV take-off and landing are managed and monitored by operator through remote controller. In addition, operator can control the UAV during flight by means of remote controller up to maximum distance of 5 km from the vehicle. All RGB (Red-Green-Blue) images were acquired in jpeg format with the digital camera set in automatic mode in clear and calm climatic conditions at noon to reduce shadow effect and wind effects on the flight mission and collected images.

Figure 4. The DJI Phantom-4 unmanned aerial vehicle was used for image acquisition in this study (a), Leica GS10 global position system with dual frequency was applied to collect ground control points (b) (Photo taken by A. Chenrai).

| Parameters of flight plan | Values |
|---|---|
| Area of interest | 45 ha |
| Desired ground resolution | 3.5 cm |
| Longitudinal overlap | 70% |
| The number of strips | 25 strips |
| Flight height | 70 m |

Table 1. The parameters of flight plan of Phantom-4 UAV that was used in this study.

The collected images were automatically matched using an algorithm that identifies tie points which were also use to retrieve the parameters of orientation necessary in aerial triangulation. Moreover, 12 ground control points (GCPs) (Fig. 1a) collected by Leica GS10 GPS with dual frequency (Fig. 4b) were used in the triangulation process (www.dji.com).

## 2.3 Image Classification

The main goal of classification of a remotely sensed image is to automatically assign consistent groups of pixels into appropriate themes or classes. In general, the spectral content of each pixel is applied as the numerical basis for classification. Pixel-based classification techniques use spectral information of each pixel or digital numbers which are based on spectral reflectance and emittance of features to categorize classes (Schowengerdt, 2007). Although, pixel-based classification techniques have some advantages in analysis of remotely sensed datasets; but, they have limitations with the use of very high spatial resolution imagery. These techniques produce inconsistent results on very high resolution imagery which are beyond the expectations of discriminating objects of interest (Ouyang et al., 2011; Rougier et al., 2016; Xiao et al., 2016).

The development of object-oriented classification method as an alternative to pixel-based approach is suitable for remotely sensed datasets with high and very high spatial resolutions. Object-oriented classification method contributes in segmentation of imagery into groups of pixels, i.e. objects. It takes not only the spectral information, but also form and texture into account. The classification procedure begins with critical step of defining groups of neighboring pixels into significant areas which are called segments (Blaschke et al., 2000; Kuzmin et al., 2016). In this method, single pixels are not classified separately but homogeneous pixels representing objects of interest are extracted. Such segmentation step can be performed in multiple spatial resolutions to make it possible to discriminate several levels of objects. Segmentation is the first and most important step in object-oriented image classification of eCognition software (Version 8) (www.ecognition.com). The shape of each object (wild pistachio and wild almond in this study) is combined with its color and texture properties to delineate image objects properly.

Image classification starts with creating a hierarchical network of objects of interest through multi-resolution segmentations. Small objects are represented by upper-level segments, while large objects are shown by lower-level segments. In the next step, the neighborhood relationships were used to detect similar segments based on sample areas. The delineated objects are then grouped based on their physical properties, i.e. shape, color and texture of crowns of studies species. Finally, the classified objects are aggregated into homogeneous groups. The combined groups, i.e. segments or crowns of wild pistachio trees and wild almond shrubs, are then considered for further processes (Yan et al., 2006; Ouyang et al., 2011; Lu and He, 2017).

In segmentation step, it is necessary to determine three parameters of scale, color/shape, and smoothness/compactness as accurate as possible. Scale parameter influences the size of objects indirectly and determines the acceptable heterogeneity of objects. As scale parameter increases, objects become larger. Color/shape parameter assigns the effect of color vs. shape homogeneity on object delineation. When shape criterion increases, spectral information has less effect on object extraction. Smoothness/compactness parameter determines the compactness of object when the amount of shape criterion is considered larger than zero. The eCognition offers two separate procedures to delineate objects of interest. In this study, we used nearest neighbor and determined some samples to detect the objects, i.e. wild pistachio and wild almond.

The performance of object-oriented classification method on UAV orthoimagery was evaluated by a confusion matrix constructed by 200 randomly selected wild pistachios (100 trees) and wild almonds (100 shrubs). For validation of results, indices of accuracy (Eq. 1), precision (Eq. 2), sensitivity (Eq. 3), and specificity were applied (Eq. 4):

$$\text{Accuracy} = \frac{TP+TN}{P} \tag{1}$$

$$\text{Precision} = \frac{TP}{TP+FP} \tag{2}$$

$$\text{Sensitivity} = \frac{TP}{TP+FN} \tag{3}$$

$$\text{Specificity} = \frac{TN}{TN+FP} \tag{4}$$

where, P is the total number of samples (200 in here), TP is the number of samples correctly assigned to the class of interest, TN is the number of samples correctly assigned to other classes, FP is the number of samples that are incorrectly assigned to the class of interest, FN is the number of samples that are incorrectly assigned to other classes. Also, in this study Cohen's Kappa index ($\kappa$) was used to assess the reliability of the applied classification method.

$$\kappa = \frac{p_o - p_e}{1 - p_e} \tag{5}$$

$$p_o = TP + TN \tag{6}$$

$$p_e = (TP+FN)(TP+FP) + (FP+TN)(TN+FN) \tag{7}$$

where, $p_o$ and $p_e$ are observed and expected agreements, respectively (Franklin, 2001; Alatorre et al., 2011).

## 3. RESULTS

The study area was covered by 1,076 images with spatial resolution of 3.47 cm taken by Phantom-4 UAV. The images were georeferenced by 12 GCPs with RMSE of 8 cm (Fig. 5).

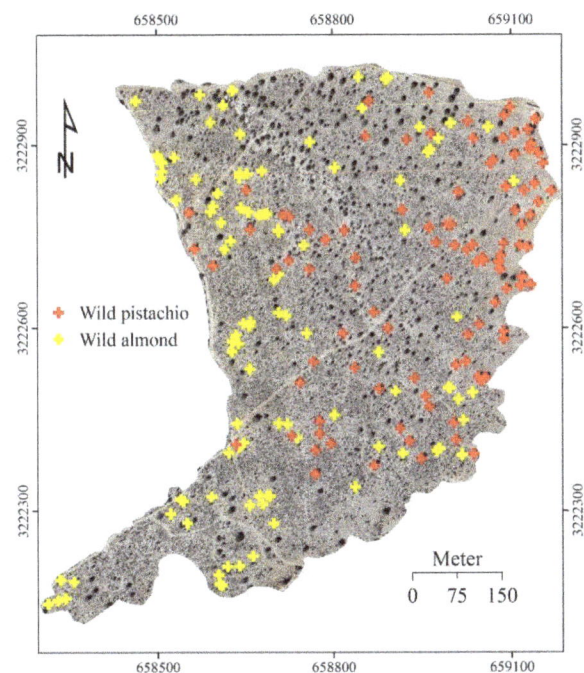

Figure 5. The UAV orthoimagery of the study area with spatial resolution of 3.5 cm (orthorectification RMSE = 8 cm). Red and

yellow crosses indicate 100 random wild pistachios and 100 wild almonds selected for assessment of the results.

The UAV orthoimagery was splitted into homogeneous segments by object-oriented classification method using scale parameter of 30, color/shape parameter of 0.5, and smoothness/compactness parameter of 0.9. These parameters were optimized for similar conditions in another investigation. Due to the great number of segments automatically generated on the UAV orthoimagery of the study area, it was not feasible to exhibit all segments in one layer. Therefore, Fig. 6a presents a small subset of the orthoimagery and the corresponding segments. Following the segmentation process and merging the homogeneous segments, the objects were converted into vector polygons and the area of each polygon was calculated for further analyses. All trees and shrubs and their type of species were well delineated during the classification process, as illustrated in Fig. 6b.

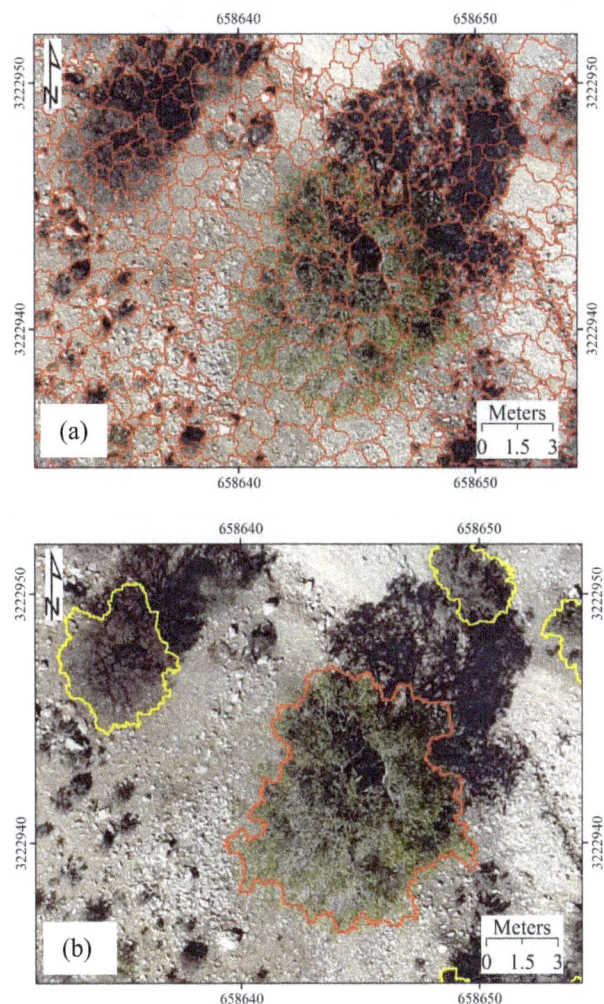

Figure 6. Segmentation used to discriminate crowns of wild pistachio trees and wild almond shrubs on the UAV orthoimagery. (a) A sample of splitting the UAV orthoimagery into segments, (b) The results of merging the segments to obtain vector map of crowns of studied species (red polygon: wild pistachio, yellow polygons: wild almond).

In the first step, 100 random wild pistachio trees and 100 random wild almond shrubs were selected within the study area to assess the accuracy of crown area delineation (Fig. 5). The crown area of individual tree estimated by segmentation of the UAV orthoimagery was compared with observed crown area in field measurements (Fig. 7). The validation of results showed that the predicted and observed crown areas of wild pistachios ($R^2$=0.86) and wild almonds ($R^2$=0.82) were highly correlated. Moreover, there was no significant difference between mean (±standard deviation) predicted and mean (±standard deviation) observed crown area of wild pistachio trees (52.09±24.67 $m^2$ and 56.74±24.49 $m^2$, respectively) (p-value=0.182). Similarly, mean (±standard deviation) predicted and mean (±standard deviation) observed crown area of wild almonds were not significantly different (3.97±1.69 $m^2$ and 3.80±1.86 $m^2$, respectively) (p-value=0.499) (α=0.05).

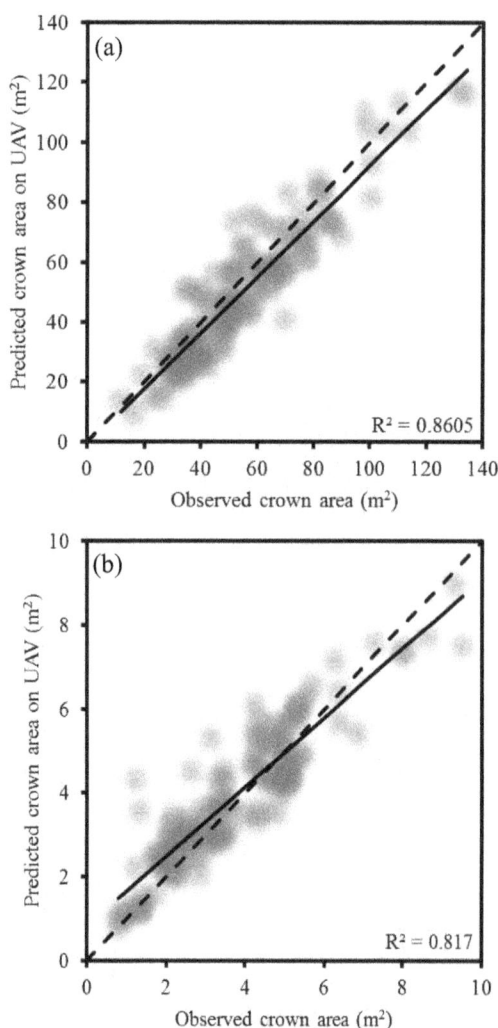

Figure 7. Predicted vs. observed crown area of 100 randomly selected wild pistachios (a) and 100 randomly selected wild almonds (b). Solid lines are the linear regression lines and dotted lines are the 1:1 lines.

In the second step, the 200 randomly selected wild pistachios and wild almonds were used to evaluate species recognition by the classification procedure. Although, the accuracy of species recognition was similar in both studied species (0.90), the precision was higher in wild pistachio (0.92), followed by wild almond (0.89). Wild pistachios had higher specificity (0.92) and less sensitivity (0.89) compared to specificity (0.89) and sensitivity of wild almonds (0.92),

indicating that wild almonds were classified more correctly than wild pistachios. In addition, Cohen's Kappa index explained a significant agreement between recognized and observed species individuals of wild pistachio and wild almond (Table 2).

| Indices | Wild pistachio | Wild almond |
|---|---|---|
| TP | 92 | 89 |
| FP | 8 | 11 |
| FN | 11 | 8 |
| TN | 89 | 92 |
| Accuracy | 0.90 | 0.90 |
| Precision | 0.92 | 0.89 |
| Sensitivity | 0.89 | 0.92 |
| Specificity | 0.92 | 0.89 |
| κ | 0.99 | 0.99 |

Table 2. The indices applied for performance assessment of object-oriented classification method on UAV orthoimagery to delineate wild pistachios and wild almonds (Total number of samples=200).

## 4. CONCLUSION

Vegetation maps exhibiting the present status of biometric characteristics of plant species (e.g., spatial distribution of canopy cover, types of species) are critical datasets essential for sustainable management and monitoring of valuable vegetation cover in Zagros arid and semi-arid woodlands (Owji and Hamzepour, 2012; SaghebTalebi et al., 2014). In order to produce these maps, laborious and intensive field measurements is necessary. However, it is not possible to collect such data with appropriate temporal and spatial resolution in vegetation cover distributed in large areas, e.g. Zagros open woodlands. In such conditions, remotely sensed datasets offer economical and practical tools for mapping vegetation cover (Franklin, 2001; Schowengerdt, 2007). This research was aimed to map single species of Zagros Woodlands applying orthoimagery taken by UAVs.

In the first step, the UAV orthimagery was obtained with spatial resolution of 3.47 cm. Previous investigations indicated that the spatial resolution of orthoimagery significantly influences on the accuracy of crown delineation of trees (Ouyang et al., 2011). Therefore, flight mission was planned to acquire images with very high spatial resolution appropriate for discrimination of single trees and shrubs in the study area. This was consistent with previous studies on application of UAVs to map vegetation cover (Dandois and Ellis, 2013; Cunliffe et al., 2016; Ivosevic et al., 2017).

The UAV orhoimagery was classified by object-oriented classification method optimized in another study. The scale parameter determines the heterogeneity allowed for delineated tree crowns (i.e., 30 in this study) which indirectly controls the size of crowns delineated by classification method. As wild pistachio trees with large crown area have been observed in the study area, it seems necessary to reduce color/shape parameter to represent variation of height distribution better. In addition, larger amount of compactness/smoothness parameter (i.e., 0.9 in the present study) makes delineation of tree crowns possible more efficiently (Kuzmin et al., 2016; Piazza et al., 2016; Trang et al., 2016). However, more detailed investigations reveal the optimized amounts of these parameters to recognize wild pistachios and wild almonds due to different canopy structure of these two species.

A comparison with mean crown areas of studied species from field observations and object-oriented classification of UAV orthoimagery revealed that there was no significant difference between estimated and observed mean crown areas of both species. This outcome was also confirmed by linear regression analyses of estimated and observed crown area values of 200 random samples ($R^2$=0.86 for wild pistachio and $R^2$=0.82 for wild almond). It also has been concluded previously that UAV orthoimagery is an appropriate means to estimate crown area of single trees (Chianucci et al., 2016). Moreover, classification of wild pistachios and wild almonds exhibited excellent discrimination accuracy (0.90 and 0.90, respectively) and precision (0.92 and 0.89, respectively), indicating that the types of species were well characterized by object-oriented classification on the UAV orthoimagery of study area. This was in accordance with the findings of previous studies on species recognition from remotely sensed datasets with very high spatial resolution (Åkerblom et al., 2017).

Thus, the results of the present study revealed that very high resolution orthoimagery obtained by UAVs are valuable datasets for detailed mapping of vegetation cover in open woodlands at single-tree levels. In addition, our findings indicated that application of object-oriented classification method on UAV orthoimagery provides a reliable tool for species recognition and canopy cover estimation of single species in open woodlands with low diversity.

## 5. REFERENCES

Åkerblom, M., Raumonen, P., Mäkipää, R. and Kaasalainen, M., 2017. Automatic tree species recognition with quantitative structure models. *Remote Sensing of Environment*, 191, pp. 1-12.

Alatorre, L.C., Andrés, R., Cirujano, S., Beguería, S. and Carrillo, S., 2011. Identification of mangrove areas by remote sensing: The ROC curve technique applied to the Northwestern Mexico coastal zone using Landsat imagery. *Remote Sensing*, 3, pp. 1568-1583.

Blaschke, T., Lang, S., Lorup, E., Strobl, S., and Zeil, P., 2000. Object-oriented image processing in an Integrated GIS/remote sensing environment and perspectives for environmental application. *Environmental Information for Planning*, 2, pp. 555–559.

Blaschke, T. and Strobl, J., 2001. What's wrong with pixels? Some recent developments interfacing remote sensing and GIS. *GeoBIT/GIS*, 6, pp. 12 17.

Carreiras, J., Pereira, J. and Pereira, J., 2006. Estimation of tree canopy cover in evergreen oak woodlands using remote sensing. *Forest Ecology and Management*, 223, pp. 45-53.

Chianucci, F., Disperati, L., Guzzi, D., Bianchini, D., Nardino, V., Lastri, C., Rindinella, A. and Corona, P., 2016. Estimation of canopy attributes in beech forests using true colour digital images from a small fixed-wing UAV. *Applied Earth Obsevation and Geoinformation*, 47, pp. 60-68.

Chopping, M., Moisen, G.G., Su, L., Laliberte, A., Rango, A., Martonchik, J.V. and Peters, D.P.T., 2008. Large area mapping of southwestern forest crown cover, canopy height, and biomass using the NASA multiangle imaging spectro-radiometer. *Remote Sensing of Environment*, 112, pp. 2051-2063.

Colomina, I. and Molina, P., 2014. Unmanned aerial systems for photogrammetry and remote sensing. *ISPRS Journal of Photogrammetry and Remote Sensing*, 92, pp. 79-97.

Cunliffe, A.M., Brazier, R.E. and Anderson, K., 2016. Ultra-fine grain landscape-scale quantification of dryland vegetation structure with drone-acquired structure-from-motion photogrammetry. *Remote Sensing of Environment*, 183, pp. 129-143.

Dandois, P. and Ellis, E.C., 2013. High spatial resolution three-dimensional mapping of vegetation spectral dynamics using computer vision. *Remote Sensing of Environment*, 136, pp. 259-276.

Díaz-Varela, R., de la Rosa, R., León, L. and Zarco-Tejada, P., 2014. High-resolution airborne UAV imagery to assess olive tree crown parameters using 3D photo reconstruction: Application in breeding trials. *Remote Sensing*, 7, pp. 4213-4232.

Féret, J. and Asner, G.P., 2012. Semi-supervised methods to identify individual crowns of lowland tropical canopy species using imaging spectroscopy and LiDAR. *Remote Sensing*, 4(8), pp. 2457–2476.

Franklin, S.E., 2001. *Remote sensing for sustainable forest manangment*. Lewis Publishers, New York, 407 p.

Gleason, C.J. and Im, J., 2011. A review of remote sensing of forest biomass and biofuel: options for small-area applications. *GIScience & Remote Sensing*, 48 (2), pp. 141-170.

Hanberry, B.B., Yang, J., Kabrick, J.M. and He, H.S., 2012. Adjusting forest density estimates for surveyor bias in historical tree surveys. *American Midland Naturalist*, 167, pp. 285-306.

Ivosevic, B., Han, Y. and Kwon, O., 2017. Calculating coniferous tree coverage using unmanned aerial vehicle photogrammetry. *Ecology and Environment*, 41, pp. 10.

Korhonen, L., Korhonen, K.T., Rautiainen, M. and Stenberg, P., 2006. Estimation of forest canopy cover: a comparison of field measurement techniques. *Silva Fennica*, 40(4), pp. 577-588.

Kuzmin, A., Korhonen, L., Manninen, T. and Maltamo, M., 2016. Automatic segment-level tree species recognition using high resolution aerial winter imagery. *European Journal of Remote Sensing*, 49, pp. 239-259.

Lu, B. and He, Y., 2017. Species classification using Unmanned Aerial Vehicle (UAV)-acquired high spatial resolution imagery in a heterogeneous grassland. *ISPRS Journal of Photogrammetry and Remote Sensing*, 128, pp. 73-85.

McNeil, B., Pisek, J., Lepisk, H. and Flamenco, E., 2016. Measuring leaf angle distribution in broadleaf canopies using UAVs. *Agricultural and Forest Meteorology*, 218, pp. 204-208.

Mlambo, R., Woodhouse, I., Gerard, F. and Anderson, K., 2017. Structure from motion (SfM) Photogrammetry with drone data: A low cost method for monitoring greenhouse gas emissions from forests in developing countries. *Forests*, 8, pp. 68.

Ouyang, Z., Zhang, M., Xie, X., Shen, Q., Guo, H. and Zhao, B., 2011. A comparison of pixel-based and object-oriented approaches to VHR imagery for mapping saltmarsh plants. *Ecological Informatics*, 6, pp. 136-146.

Owji, M. and Hamzepour, M., 2012. *Vegetation profile of wild pistachio experimental forest*. Research Institute of Forests and Rangelands Press, Tehran. 240 p. (In Persian)

Piazza, G.A., Vibrans, A.C., Liesenberg, V. and Refosco, J.C., 2016. Object-oriented and pixel-based classification approaches to classify tropical successional stages using airborne high-spatial resolution images. *GIScience & Remote Sensing*, 53, pp.

Panta, M., Kim, K. and Joshi, Ch., 2008. Temporal mapping of deforestation and forest degradation in Nepal: Applications to forest conservation. *Forest Ecology and Management*, 256, pp. 1587-1595.

Rautiainen, M., Stenberg, P. and Nilson, T., 2005. Estimating canopy cover in Scots pine stands. *Silva Fennica*, 39 (1), pp. 137-142.

Rezayan, F. and Erfanifard, Y., 2016. Estimating biophysical parameters of Persian oak coppice trees using UltraCam-D airborne imagery in Zagros semi-arid woodlands. *Arid Environments*, 133, pp. 10-18.

Rougier, S., Puissant, A., Stumpf, A. and Lachiche, N., 2016. Comparison of sampling strategies for object-based classification of urban vegetation from very high resolution satellite images. *Applied Earth Observation and Geoinformation*, 51, pp. 60-73.

Sagheb Talebi, K., Sajedi, T., and Pourhashemi, M., 2014. *Forests of Iran*. Iranian Research Institute of Forests and Rangelands, Tehran, 152 p. (In Persian)

Schowengerdt, R.A. 2007. *Remote sensing: models and methods for image processing*. Academic Press, New York, 558 p.

Torres-Sanchez, F., Lopez-Granados, F. and Pena, J.M., 2015. An automatic object-based method for optimal thresholding in UAV image: Application for vegetation detection in herbaceous crops. *Computers and Electronics in Agriculture*, 114, pp. 43-52.

Trang, N., Toan, L., Ai, T., Giang, N. and Hoa, P., 2016. Object-based vs. pixel-based classification of mangrove forest mapping in Vien An Dong Commune, Ngoc Hien District, CaMau Province using VNREDSat-1 Images. *Advances in Remote Sensing*, 5, pp. 284-295.

Vega, F., Ramírez, F., Siaz, M. and Rosua, F., 2015. Multi-temporal imaging using an unmanned aerial vehicle for monitoring a sunflower crop. *Biosystems Engineering*, 132, pp. 19-27.

Wallace, C., Webb, R. and Thomas, K.A., 2008. Estimation of perennial vegetation cover distribution in the Mojave Desert using MODIS-EVI data. *GIScience & Remote Sensing*, 45(2), pp. 167-187.

Xiao, P., Zhang, X., Wang, D., Yuan, M., Feng, X. and Kelly, M., 2016. Change detection of built-up land: A framework of combining pixel-based detection and object-based recognition. *ISPRS Journal of Photogrammetry and Remote Sensing*, 119, pp. 402-414.

Yan, G., Mas, J.F., Maathuis, B., Xiangmin, Z. and Van Dijk, P., 2006. Comparison of pixel-based and object-oriented image classification approaches - A case study in a coal fire area, Wuda, Inner Mongolia, China. *Remote Sensing*, 27(18), pp. 4039-4055.

# Permissions

# List of Contributors

**S.Abbaszadeh and H. Rastiveis**
School of Surveying and Geospatial Engineering, University of Tehran, College of Engineering,, Tehran, Iran

**Nurollah Tatar, Mohammad Saadatseresht, Hossein Arefi**
School of Surveying and Geospatial Engineering, College of Engineering, University of Tehran

**E.Tamimi**
MS.c. Student, Faculty of Geodesy & Geomatics Engineering, K.N.Toosi University of Technology, Tehran, Iran

**H. Ebadi**
Prof., Faculty of Geodesy & Geomatics Engineering, K.N.Toosi University of Technology, Tehran, Iran

**A. Kiani**
PhD student, Faculty of Geodesy & Geomatics Engineering, K.N.Toosi University of Technology, Tehran, Iran

**Shima Toori**
M.Sc. in GIS Engineering, Graduate University of Advanced Technology, Kerman, Iran

**Ali Esmaeily**
Assistant professor, Dept. of Remote Sensing Engineering, Graduate University of Advanced Technology, Kerman, Iran

**N.Zarrinpanjeh**
Department of Geomatics Engineering, Qazvin Branch, Islamic Azad University, Qazvin, Iran

**F. Dadrassjavan**
School of Surveying and Geospatial Engineering University College of Engineering University of Tehran

**P.Azari and M. Karimi**
Faculty of Geodesy and Geomatics Engineering, K.N.-Toosi University of Technology, Tehran, Iran

**Aref Ebrahimi**
MSc. Student in GIS Division, School of Surveying and Geospatial Eng., College of Eng., University of Tehran, Tehran, Iran

**Parham Pahlavani**
Assistant Professor, Center of Excellence in Geomatics Eng. in Disaster Management, School of Surveying and Geospatial Eng., College of Eng., University of Tehran, Tehran, Iran

**Zohreh Masoumi**
Assistant Professor, Department of Earth Sciences, Institute for Advanced Studies in Basic Sciences, Zanjan, Iran

**A. A. Heidari, O. Kazemizade, F. Hakimpour**
School of Surveying and Geospatial Engineering, College of Engineering, University of Tehran, Iran

**H.Jafari**
Dept. of GIS, K.N. Toosi University Of Technology, Tehran, Iran

**Ali A. Alesheikh**
Dept. of GIS, K.N. Toosi University Of Technology, Tehran, Iran

**S.Khoshahval, M. Farnaghi and M. Taleai**
Faculty of Geodesy and Geomatics Engineering, K.N. Toosi University of Technology, Tehran, Iran

**N.Mijani**
MSc. Student of Remote Sensing and GIS, Department of Remote Sensing and GIS, University of Tehran,Tehran, Iran

**N. Neysani Samani**
Assis. Prof of Remote Sensing and GIS, Department of Remote Sensing and GIS, University of Tehran,Tehran, Iran

**S.Nadi**
Assistant Professor, Department of Geomatics Engineering, Faculty of Civil and Transportation Engineering, University of Isfahan, Isfahan, Iran

**A. H. Houshyaripour**
M.Sc. Remote Sensing, Department of Geomatics Engineering, Faculty of Civil and Transportation Engineering, University of Isfahan, Iran

**M.M. Rahimi**
MSc student of GIS, Department of Surveying and Geomatics Engineering, College of Engineering, University of Tehran

**F. Hakimpour**
Assistant Professor, Department of Surveying and Geomatics Engineering, College of Engineering, University of Tehran

**Ali Shah-Heydari pour and Parham Pahlavani**
School of Surveying and Geospatial Engineering, College of Engineering, University of Tehran, Tehran, Iran

**Behnaz Bigdeli**
School of Civil Engineering, Shahrood University of Technology, Shahrood, Iran

**A.Jouybari and A. A. Ardalan**
University of Tehran, School of Surveying and Geospatial Engineering, Tehran, Iran

**M-H. Rezvani**
University of Tasmania, School of Land and Food, Hobart, Tasmania, Australia

**F.Zangeneh-Nejad and M. A. Sharifi**
School of Surveying and Geospatial Engineering, Research Institute of Geoinformation Technology (RIGT), College of Engineering, University of Tehran, Iran

**A. R. Amiri-Simkooei and J. Asgari**
Department of Geomatics Engineering, Faculty of Civil Engineering and Transformation, University of Isfahan, 81746-73441 Isfahan, Iran

**Aliyu Ja'afar Abubakar ,Mazlan Hashim and Amin Beiranvand Pour**
Geoscience and Digital Earth Centre (INSTeG) Research Institute for Sustainability and Environment (RISE) Universiti Teknologi Malaysia (UTM) 81310 UTM Skudai, Johor Bahru, Malaysia

**T.Bibi**
Department of Geoinformation, Faculty of Geoinformation and Real Estate, Universiti Teknologi Malaysia, Skudai 81310, Johor, Malaysia
UTM RAZAK School of Engineering and Advanced Technology, Universiti Teknologi Malaysia, 54100 Jalan Sultan Yahya Petra, Kuala Lumpur, Malaysia

**K. Azahari Razak A. Abdul Rahman and A. Latif**
Dept. of Education, Azad Jammu and Kashmir, Muzaffaraba

**Bienvenido G. Carcellar III**
Department of Geodetic Engineering, College of Engineering, University of the Philippines

**Gerard. A. Domingo, Mayann M. Mallillin and Anjillyn Mae C. Perez**
Department of Geodetic Engineering, University of the Philippines, Diliman, Quezon City 1101

**Ayin M. Tamondong and Alexis Richard C. Claridades**
Phil-LiDAR 2 Program Project 2: Aquatic Resources Extraction from LiDAR Surveys, University of the Philippines

**Mohamed Elsobeiey**
Department of Hydrographic Surveying, Faculty of Maritime Studies, King Abdulaziz University P. O. Box 20807 - Jeddah 21589, Kingdom of Saudi Arabia

**N.Z.A. Halim, S.A. Sulaiman, K. Talib, O.M. Yusof and M.A.M. Wazir**
Center of Studies for Surveying Science and Geomatics, Faculty of Architecture, Planning and Survey, University Technology Mara, Shah Alam, Malaysia

**M.K. Adimin**
Cadastral Division, Department of Survey and Mapping Malaysia, Kuala Lumpur, Malaysia

**N.M. Hashim, A. H. Omar, S. N. M. Ramli, K.M. Omar and N. Din**
Dept. of Geomatic Engineering, University Technology Malaysia, Skudai, Johor Bahru, MALAYSIA

**Mohd Asri Hakim Mohd Rosdi, Ainon Nisa Othman, Muhamad Arief Muhd Zubir Zulkiflee Abdul Latif and Zaharah Mohd Yusoff**
Centre of Studies for Surveying Science and Geomatics, Faculty of Architecture, Planning and Surveying, Universiti Teknologi MARA, 40450 Shah Alam, Selangor Darul Ehsan, MALAYSIA

**Shahrin Amizul Samsudin and Rozaimi Che Hasan**
UTM Razak School of Engineering and Advanced Technology, Universiti Teknologi Malaysia, Kuala Lumpur

**M.N. Uti and A. H. Omar**
Geomatic Innovation Research Group (GIG)

**A. H. M. Din**
Geomatic Innovation Research Group (GIG)
Geoscience and Digital Earth Centre (INSTEG), Faculty of Geoinformation and Real Estate, Universiti Teknologi Malaysia, 81310 Johor Bahru, Johor, Malaysia
Associate Fellow, Institute of Oceanography and Environment (INOS), Universiti Malaysia Terengganu, Kuala Terengganu, Terengganu, Malaysia

**Siamak Talebi Nahr and Mohammad Saadatseresht**
School of Surveying and Geospatial Engineering, College of Engineering, University of Tehran

**Jamshid Talebi**
Pooya-Naghsh-Omid Consulting Engineers Company

**S.Mola Ebrahimi, H. Arefi and H. Rasti Veis**
School of Surveying and Geospatial Engineering, University of Tehran, Tehran, Iran

**S.Daneshtalab and H.Rastiveis**
School of Surveying and Geospatial Engineering, College of Engineering, University of Tehran, Tehran, Iran

**A. Sh. Amini**
Dept. of Geomatic Engineering, Islamic Azad University – South Tehran Branch, Piroozi Street, Abuzar Boulevard, Dehhaghi Avenue, P. O. Box: 1777613651, Tehran, Iran

**A.Chenari , Y. Erfanifard and H.R. Pourghasemi**
Dept. of Natural Resources and Environment, College of Agriculture, Shiraz University, Shiraz, Iran

**M. Dehghani**
Dept. of Civil and Environmental Engineering, School of Engineering, Shiraz University, Shiraz

# Index